THERMAL MANAGEMENT HANDBOOK

Electronic Packaging and Interconnection Series
Charles M. Harper, Series Advisor

Alvino • PLASTICS FOR ELECTRONICS

Classon • SURFACE MOUNT TECHNOLOGY FOR CONCURRENT ENGINEERING AND MANUFACTURING

Di Giacomo • RELIABILITY OF ELECTRONIC PACKAGES

Garrow and Turlik • MULTICHIP MODULE TECHNOLOGY HANDBOOK

Ginsberg and Schnoor • MULTICHIP MODULES AND RELATED TECHNOLOGIES

Harmon • WIRE BONDING, 2/E

Harper • ELECTRONIC PACKAGING AND INTERCONNECTION HANDBOOK

Harper and Miller • ELECTRONIC PACKAGING, MICROELECTRONICS, AND INTERCONNECTION DICTIONARY

Harper and Sampson • ELECTRONIC MATERIALS AND PROCESSING HANDBOOK, 2/E

Hwang • MODERN SOLDER TECHNOLOGY FOR COMPETITIVE ELECTRONICS MANUFACTURING

Lau • BALL GRID ARRAY TECHNOLOGY

Lau • FLIP CHIP TECHNOLOGIES

Lau • SOLDER JOINT RELIABILITY FOR BGA, CSP, FLIP CHIP

Licari • MULTICHIP MODULE DESIGN, FABRICATION, AND TESTING

Smith and Whitehall • OPTIMIZING QUALITY IN ELECTRONICS ASSEMBLY

Related Books of Interest

Boswell • SUBCONTRACTING ELECTRONICS

Boswell and Wickam • SURFACE MOUNT GUIDELINES FOR PROCESSING CONTROL, QUALITY, AND RELIABILITY

Byers • PRINTED CIRCUIT BOARD DESIGN WITH MICROCOMPUTERS

Capillo • SURFACE MOUNT TECHNOLOGY

Chen • COMPUTER ENGINEERING HANDBOOK

Coombs • PRINTED CIRCUITS HANDBOOK, 4/E

Di Giacomo • DIGITAL BUS HANDBOOK

Di Giacomo • VLSI HANDBOOK

Fink and Christiansen • ELECTRONICS ENGINEERS' HANDBOOK, 3/E

Ginsberg • PRINTED CIRCUITS DESIGN

Juran and Gryna • JURAN'S QUALITY CONTROL HANDBOOK

Jurgen • AUTOMOTIVE ELECTRONICS HANDBOOK

Manko • SOLDERS AND SOLDERING, 3/E

Rao • MULTILEVEL INTERCONNECT TECHNOLOGY

Sze • VLSI TECHNOLOGY

Van Zant • MICROCHIP FABRICATION

THERMAL MANAGEMENT HANDBOOK

for Electronic Assemblies

Jerry Sergent

Al Krum

McGRAW-HILL

New York San Francisco Washington, D.C. Auckland Bogotá
Caracas Lisbon London Madrid Mexico City Milan
Montreal New Delhi San Juan Singapore
Sydney Tokyo Toronto

Library of Congress Cataloging-in-Publication Data

Sergent, Jerry E.
Thermal management handbook : for electronic assemblies / Jerry Sergent, Al Krum.
p. cm. — (Electronic packaging and interconnection series)
Includes index.
ISBN 0-07-026699-9
1. Electronic apparatus and appliances—Temperature control. 2. Electronic packaging. 3. Multichip modules (Microelectronics)—Design and construction. I. Krum, Al. II. Title. III. Series.
TK7870.25.S47 1998
621.381—dc21 98-10108
CIP

McGraw-Hill

*A Division of The **McGraw-Hill** Companies*

1 2 3 4 5 6 7 8 9 0 DOC/DOC 9 0 3 2 1 0 9 8

ISBN 0-07-026699-9

The sponsoring editor for this book was Steve Chapman, the editing supervisor was Bernard Onken, and the production supervisor was Claire Stanley. It was set in Times Roman by North Market Street Graphics.

Printed and bound by R. R. Donnelley & Sons Company.

Jerry Sergent
To my grandchildren; Brian, Ian, Channing, and Ivy. May your generation explore new horizons beyond our wildest dreams.

Al Krum
This book is dedicated to William T. Campbell who has provided insight, inspiration, and a multitude of ideas for packaging high-power electronics.

CONTENTS

PREFACE

The demands of the electronics industry have dictated that electronic assemblies operate at increased speeds in a smaller package with higher reliability. These requirements create extreme challenges in thermal management for the packaging engineer. Faster speeds generate more heat, while smaller packages increase the difficulty of removing it from sensitive devices which are adversely affected by increased temperature.

This book emphasizes microelectronic circuits, including hybrids and multichip modules, and is intended as a practical reference for designers and users alike who need to know the impact of heat on the operation and reliability of electronic circuits. It was the goal of the editors to provide a reference which provided a basic explanation of the science of thermal management along with the data and formulas to enable fundamental analysis by engineers who may not have a background in this technology. To this end, the material has been written by practicing engineers with extensive experience in both microelectronics and thermal management.

For the convenience of the reader, the authors were encouraged to make each chapter an independent, stand-alone document with all the pertinent data included for that topic. Accordingly, a small degree of redundancy from chapter to chapter occurs which prevents the reader from having to jump back and forth between chapters. The authors were also encouraged to provide practical examples from their experience to enhance the understanding of the phenomena associated with thermal management.

The book is divided into three basic parts: the generation of heat within an electronics circuit, the calculation of the circuit temperature, and the removal of heat from the circuit. Chapter 1 is an introduction to the science of thermal management and describes the structure of the book in more detail. Chapter 2 describes the effect of heat on the operation of the active and passive circuit components enabling the effect of heat on the performance and reliability of these devices to be modeled during the analysis phase. Chapter 3 considers the thermal properties of the materials used to fabricate electronic circuits and in the thermal management of these circuits. Chapter 4 looks at the generation of heat in electronic circuits by components and signal traces. Chapter 5 describes a simplified method of thermal analysis utilizing only a calculator or simple computer program which can be used in many cases to determine whether a thermal problem may exist. Chapter 6 is a more analytical study of thermal analysis. The various approaches to thermal analysis are described along with more advanced computer solutions. Chapter 7 considers thermal management approaches. Simplified methods for calculating the thermal stress on components in common configurations are presented. Also considered are various methods for drawing the heat away from the circuit, including fins, thermoelectric coolers, fans, cold plate, and heat pipes. Methods of assembling microelectronic circuits for improved thermal management are also presented. Chapter 8 is an in-depth summary of thermal management and the cooling techniques.

It is the hope of the editors that this book will provide the information which enables practitioners to understand and apply the concepts of thermal management and provide references for those who wish to explore the topic in more depth.

Jerry E. Sergent
Al Krum

THERMAL MANAGEMENT HANDBOOK

CHAPTER 1
INTRODUCTION

Jerry Sergent
Al Krum

1.1 WHY THERMAL MANAGEMENT

The term *thermal management* encompasses the technology of the generation and control of heat in electronics circuits. Heat is an unavoidable by-product of every electronic device and circuit and is usually detrimental to performance and reliability. Heat may be generated by the devices themselves or may be present from other sources, internal or external. The trend in packaging electronic systems and subsystems has been to reduce size and increase performance, both of which contribute to heat generation and concentration. Evidence of this trend can be seen in the higher levels of integration in semiconductors and the increased usage of hybrids and multichip modules (MCMs). Placing more functions in a smaller package has resulted in higher heat densities, mandating that thermal management be given a high priority in the design cycle in order to maintain system performance and reliability.

The role of the packaging engineer is to predict the temperature that an electronic device will attain during performance as a result of operational losses or ambient heat and to apply the principles of thermal management to keep the temperature below the level which will cause circuit performance to degrade or cause the circuit to fail prematurely. Unfortunately, thermal management is not well understood by most engineers except for the regular practitioner, and the task usually falls to one who does not have the necessary background. The outcome is usually one of two results: the thermal management scheme is woefully inadequate and the circuit fails soon after it begins to operate, or the thermal design is massive overkill and the circuit is too large, too heavy, and too expensive.

The packaging engineer must understand the effect of heat on the performance and reliability of an electronics circuit in order to optimize the thermal design. Each circuit will behave differently, depending on the components, the layout, and the materials, and must be analyzed on an individual basis. Virtually every parameter of every component is a function of temperature to some extent, and the cumulative effect must be known in order to establish a maximum operating temperature for the circuit. Changing the temperature of passive devices typically changes their values. For example, film resistors have temperature coefficients that range from several to several hundred ppm/°C. Ceramic capacitors, depending upon the dielectric material, have temperature variations from less than 30 ppm/°C to as much as 60 percent over the military temperature range. The gain of bipolar transistors may change by a factor of more than three over the same range, and leakage currents may double for every 10°C rise in temperature.

Heat also has a detrimental effect on the reliability of an electronics circuit. Following the same mathematical pattern as leakage currents, the rate of chemical reactions doubles with every 10°C temperature rise, accelerating such failure mechanisms as junction failure, metallization failure, corrosion, and electromigration diffusion. The rise in leakage currents may ultimately result in thermal runaway and catastrophic failure. Excess heat also accelerates the drift mechanisms in resistors and may ultimately result in circuit failure.

Clearly, thermal management is one of the more important tasks of the packaging engineer. Developing a systematic process leading to a thermal design meeting the requirements of the circuit without being excessive will result in a circuit meeting not only the performance requirements, but the cost and reliability requirements as well.

There are three fundamental questions which must be considered in thermal management.

1. *How and where is the heat generated?* There are many sources of heat in an electronics circuit, all of which must be considered. It is common to consider only the heat created by the operation of active devices, but such components as resistors, conductors, and even small wire bonds can generate significant heat. The location of the heat is just as important as the amount. Heat-generating components in proximity may cause a temperature rise greater than the heat generated by the component itself.
2. *How is the temperature at a given point in the circuit determined?* Thermal analysis is not an exact science. Heat conduction is reasonably well defined, but most thermal models relating to radiation and convection are empirically determined for all but the simplest of geometries. Approximate methods solvable on a calculator or PC exist which are reasonably accurate for single devices, but fall short when applied to multiple devices or when convection and/or radiation must be considered, and they serve only to determine where the trouble spots are. To achieve the requisite degree of accuracy needed for optimal thermal management, more sophisticated approaches requiring a workstation or mainframe computer must be employed. These methods utilize finite element analysis (FEA) or finite volume analysis (FVA) and require a degree of mathematical expertise by the user.
3. *How is the heat removed?* The generated heat must be carried away from the device and dissipated safely. This requires the design and construction of an efficient heat path from the device through the mounting surface(s) to the outside world and can be as simple as a clip-on heat sink or as complicated as a thermoelectric cooler attached to a finned heat sink with circulating liquid nitrogen. This discipline requires an extensive knowledge of material properties and thermodynamics.

The individual skilled enough to have expertise in all the disciplines required to develop thermal management solutions is a true Renaissance person. This book follows the sequence described here, also providing information on material properties and circuit functions to enable the packaging engineer to make analytical decisions, and is intended not only for thermal management experts, but also for those in related fields who have a need to model and optimize their designs.

CHAPTER 2
THERMAL EFFECTS ON ELECTRONIC CIRCUITS

Jerry Sergent

2.1 INTRODUCTION

Both the performance and reliability of electronic circuits are strongly influenced by temperature. Exposure to temperatures beyond which the circuit is designed to withstand may result in failure of the circuit to perform to specification or in failure altogether. The maximum temperature to which a circuit may be exposed can be defined in two ways: the maximum *operating* temperature, defined as the maximum temperature at which the circuit will meet the electrical specification with power applied, and the maximum *storage* temperature, defined as the maximum temperature, when the power is off, to which the circuit may be exposed for a given period of time without detrimental effects.

The detrimental effects of excessive temperatures may be divided into three categories as shown in Table 2.1. Soft failures happen as a result of the tendency of the parameters of both active and passive components to exhibit a degree of sensitivity to temperature. As the temperature increases, the cumulative effects of component parameter drift may eventually cause the circuit output variables to deviate from the specification. Hard failures in the short term may occur as a result of component overload as a result of excessive heat or as a result of the breakdown of component attach or packaging materials. Hard failures in the long term may occur for a variety of reasons, such as corrosion, chemical reactions, and intermetallic compound formation, all of which are accelerated by elevated temperature. Hard failures may also occur as a result of mechanical stresses due to differences in the temperature coefficient of expansion (TCE) between two materials joined together, such as a component mounted to a circuit board.

In order to maximize circuit performance and reliability, it is important to be able to predict the maximum operating and storage temperatures and to design and package electronic circuits to minimize the heat generated during operation and to minimize circuit temperature by employing adequate thermal management methods.

2.2 TEMPERATURE EFFECTS ON PASSIVE COMPONENTS

The fundamental material properties, such as conductivity, permittivity, or permeability, which determine the parameters of passive components, exhibit a dependence on temperature to a greater or lesser degree. These changes may be positive or negative, linear or nonlinear, permanent or temporary, and may be due to a variety of factors, including physical changes in the material or variations in the lattice structure.

TABLE 2.1 Classification of Temperature-Related Failures

Failure mode	Characteristics
Soft failures	Circuit continues to operate, but does not meet specification when the temperature is elevated beyond the maximum operating temperature. Circuit returns to normal operation when the temperature is lowered. Failure is due to changes in component parameters with temperature.
Hard failures (short term)	Circuit does not operate. Circuit may or may not return to normal operation when temperature is lowered. Failure is likely due to component or interconnection breakdown, but may also be due to changes in component parameters with temperature.
Hard failures (long term)	Circuit does not operate at any temperature. Failures are irreversible. Failures may be caused by corrosion, intermetallic formation, or similar phenomena. Failures may also be caused by mechanical stresses due to differences in temperature coefficient of expansion (TCE) between a component and the substrate.

2.2.1 Resistors

The electrical resistance of a rectangular solid is defined as

$$R = \frac{L}{\sigma A} \tag{2.1}$$

where R = electrical resistance in ohms
L = length of the solid
A = area of the solid
σ = conductivity in siemens/m

While the dimensions may vary with temperature, the factor which primarily determines the temperature dependence of the resistor is the conductivity.

In basic terms, the conductivity of a material is given by

$$\sigma = q\mu \tag{2.2}$$

where q = carrier concentration in coulombs/m^3
μ = mobility of the carriers in m^2/volt-second

For conductors and n-type semiconductors, the carriers are electrons only, for p-type semiconductors, the carriers are holes only, and Eq. 2.2 applies. For intrinsic semiconductors, the carriers may be both electrons and holes and Eq. 2.2 must be modified accordingly.

A qualitative view of the temperature dependence of a resistor may be obtained by examining the effects of temperature on carrier concentration and mobility. These parameters have substantial variation among materials.

Carriers are created by the ionization of atoms within the lattice comprising the solid. Conductors are easily ionized by nature and have an abundance of free electrons. At the temperatures of interest, essentially all the atoms in a conductor are ionized and the supply of electrons is virtually constant with temperature. In a semiconductor, the carrier concentration is an exponential function of temperature and increases rapidly as temperature increases.

The mobility is a measurement of the ease with which carriers move through the lattice. Carriers do not move in a straight line as they traverse through a material. They are influenced by such factors as defects in the lattice, impurities, grain boundaries, and fixed ions. A carrier deflected from a straight path is said to have undergone a *collision*. As temperature increases, the carriers are more active and suffer more collisions, thereby decreasing the mobility.

For conductors, the mobility is due almost entirely to ionic scattering and depends on the characteristics of the particular material. The temperature dependence of resistance is expressed as

$$R(T_2) = R(T_1)[1 + \alpha(T)(T_2 - T_1)] \tag{2.3}$$

where $R(T_2)$ = resistance at a temperature, T_2
$R(T_1)$ = resistance at a temperature, T_1
$\alpha(T)$ = temperature coefficient of resistance (TCR)
T_2 = temperature in K
T_1 = base temperature in K

Mathematically, the TCR is the slope of the $R(T)$ versus T curve normalized to $R(T)$ as defined by Eq. 2.4.

$$\alpha(T) = \frac{1}{R(T)} \frac{dR(T)}{dT} \tag{2.4}$$

For cases where the TCR is nonlinear, it is often presented as a linearized average over a range of temperature and may be derived from Eq. 2.3.

$$\alpha = \text{TCR} = \frac{R(T_2) - R(T_1)}{R(T_1)(T_2 - T_1)} \times 10^6 \frac{\text{ppm}}{°\text{C}} \tag{2.5}$$

Typically, there are two values given: a *hot TCR* and a *cold TCR*. The hot TCR is usually defined between 25 and 125°C, and the cold TCR is defined between –40 and 25°C.

The TCR of conductors (metals) is typically a positive number in the range of several thousand ppm and, for the temperatures of general interest, is a constant. The resistivities and TCRs of common conductors used in electronics applications are shown in Table 2.2.

TABLE 2.2 Resistance Properties of Selected Metals

Material	Resistivity, μohm-cm	TCR, ppm/°C
Alloy 42	66.5	1,000
Alloy 52	43.0	2,900
Aluminum	2.83	3,400
Copper	1.72	3,900
Gold	2.44	3,400
Kovar	48.9	3,700
Nickel	7.80	6,000
Silver	1.63	3,800

For semiconductors, the mobility is a function of both ionic and phonon scattering and can be approximated as

$$\mu(T) = \frac{1}{\frac{1}{a}T^{3/2} + \frac{1}{b}T^{-(3/2)}} \tag{2.6}$$

where a = ionic constant
b = phonon constant
T = absolute temperature

At the temperatures of interest the $T^{3/2}$ term dominates, so that the mobility varies by $T^{-3/2}$. As a function of temperature, the conductivity of a semiconductor can be expressed as

$$\sigma(T) = ke^{cT}T^{-(3/2)} \tag{2.7}$$

where k, c = material constants

A plot of the normalized conductivity is shown in Fig. 2.1. The slope of the curve indicates that the resistance is nonlinear and decreases with temperature, which indicates a negative TCR.

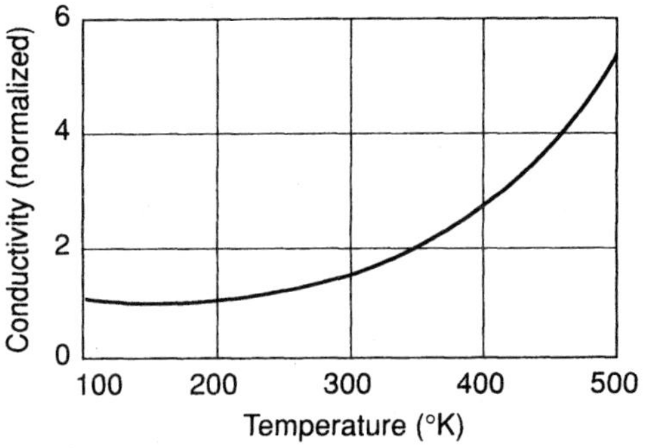

FIGURE 2.1 Normalized conductivity for a semiconductor.

Thick film resistors have a complex structure as shown in Fig. 2.2 [1], with a combination of metallic and semiconductor contacts. Consequently, the TCR is a nonlinear function of temperature and may change polarity across the temperature range of interest. The hot and cold TCRs of the same resistor may vary in magnitude as well as in polarity. Typical values of the TCR may range from ±50 ppm and up. It is common for the TCR to become significantly more negative as the temperature approaches –40°C as the semiconductor mechanisms become more dominant. At the extreme, the TCR at –40°C may be in excess of the stated value for the cold TCR and may result in soft failures. The properties of a particular family of thick film resistors are shown in Table 2.3.

The TCR of thin film resistors is a function of the material and the deposition parameters. For a given resistor, the TCR is a constant and the hot TCR and the cold TCR will be the same. The characteristics of thin film resistors are summarized in Table 2.4.

The TCR of carbon composition resistors used in through-hole applications is quite high in comparison to thick and thin film resistors, approaching ±600 ppm in

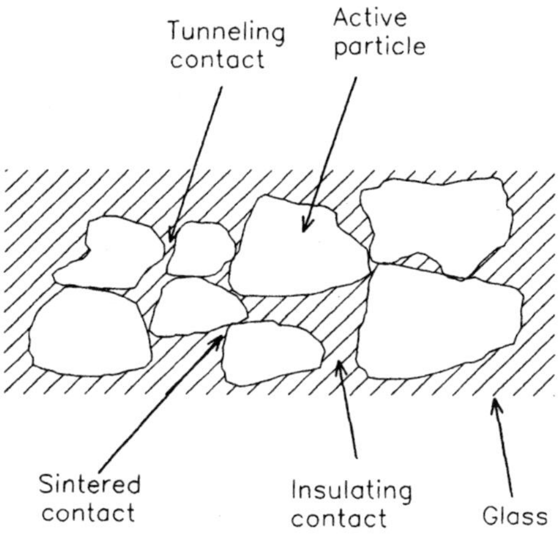

FIGURE 2.2 Contacts in thick film resistors.

some families. Leaded film resistors are available which have a considerably smaller TCR than carbon composition.

The other resistor parameter of interest is the *stability,* which is usually defined as the permanent change in resistor value, as shown in Eq. 2.8, after exposure to elevated temperature for a period of time, usually 150°C for 1000 hours, which represents end-of-life conditions.

$$\text{Stability} = \frac{\Delta R}{R} \times 100 \tag{2.8}$$

Film resistors are extremely stable. A thin film resistor will typically drift less than 0.1 percent over life. The stability of thick film resistors depends on the value and on whether the resistor has been trimmed. Higher-value resistors tend to drift more than low value, and laser-trimmed resistors tend to drift more than untrimmed. As a rule, a thick film resistor will drift less than 0.5 percent, with lower-value and untrimmed resistors drifting less than 0.1 percent. As a further note, thick film resistors printed

TABLE 2.3 Typical Properties of Thick Film Resistors

Property	10Ω	100Ω	1K	10K	100K	1 Meg	10 Meg
Sheet resistance	10	100	1K	10K	100K	1M	10M
Tolerance	±10%	±10%	±10%	±10%	±10%	±10%	±20%
TCR hot/cold(ppm/°C)	±100	±100	±100	±100	±100	±100	±200
Drift on ESD (2000 V)	<0.1%	<0.1%	<0.1%	<0.1%	<0.1%	<0.1%	<0.1%
Drift on thermal aging	<0.1%	<0.1%	<0.1%	<0.1%	<0.1%	<0.5%	<0.5%
Drift on thermal cycling	<0.5%	<0.5%	<0.5%	<0.5%	<0.5%	<0.5%	<0.5%
Drift on thermal shock	<0.5%	<0.5%	<0.5%	<0.5%	<0.5%	<0.5%	<0.5%
Drift on humidity aging	<0.1%	<0.1%	<0.1%	<0.1%	<0.1%	<0.5%	<0.5%
STOL voltage (V/mm)	8	25	75	250	380	370	—
Power dissipation (mW/mm²)	900	600	800	1000	230	22	—
Tech noise (dB)	<–30	<–30	<–20	<–15	<–10	—	—

TABLE 2.4 Characteristics of Thin Film Resistor Materials

Film material	Deposition technique	Sheet resistance R_S (Ω/sq)	TCR (ppm/°C)
Cr-Ni (20-80)	Evaporation	10–400	100–200
Cr-Si (24-76)	Flash evaporation	100–4000	±200
Cr-Ti (35-65)	Flash evaporation	250–600	±150
CrSiO (70-30)	Flash evaporation	Up to 600	−15 to −200
SnO_2	Spray pyrolysis/CVD	Up to 10^4	−50 to −200
w, Mo, Ru	Sputtering	10–500	−20 to −100
Ta	Sputtering	Up to 100	±100
Ta_2N, TaN	Reactive sputtering	10–100	−85

and fired as part of a hybrid circuit will generally be trimmed, while those in chip form designated for use in surface-mount applications will be untrimmed.

2.2.2 Capacitors

There are three types of capacitors commonly used in hybrid microcircuits, multichip modules, and surface-mount applications: ceramic capacitors, tantalum electrolytic capacitors, and MOS (metal-oxide-silicon) capacitors.[1] Aluminum electrolytic capacitors are not generally used in hybrid microelectronics circuits since they are not widely available in the surface-mount configuration and may emit hydrogen gas during the charge and discharge cycles.

Ceramic Chip Capacitors

Manufacturing Process. Ceramic chip capacitors are manufactured in much the same manner as multilayer thick film or cofired ceramic substrates in that alternating layers of conductor (electrodes) and dielectrics are formed. There are two primary processes used to fabricate ceramic chip capacitors: the dry process and the wet process.

In the dry process, the dielectric material is milled into a fine powder, mixed with a binding ingredient to form a viscous liquid, and formed into a thin film several inches wide on a Mylar sheet using a doctor-blade technique to control the thickness. The ceramic films are dried, cut into lengths approximately two feet long, and peeled away from the Mylar. The sheets are cut into shorter lengths and placed in a template with locating pins to allow the electrodes to be screen-printed onto the film with a high degree of precision. The electrodes are typically a PdAg thick film composition similar to those used to fabricate thick film interconnections. The electrodes are screen-printed, forming a number of capacitor electrodes simultaneously, and dried. A second ceramic film is placed over the electrodes and the screen-and-dry process with the conductor is performed again. The cycle is repeated until the required number of layers are formed. The resulting structure is dried and laminated at an elevated temperature, pressed to force out any air bubbles, and separated into individual capacitors by cutting between the electrodes. The capacitors are fired in a two-step process: first at 800°C to remove the organic binders and then at 1200°C to sinter the PdAg electrodes and dielectric films together to form a monolithic structure. After firing, the capacitors are milled to remove any burrs or irregularities and to ensure that the edges of the electrodes are exposed at the ends. End terminations are applied by dipping the capacitors in a PdAg paste and firing at about 850°C. The structure of a multilayer ceramic chip capacitor is shown in Fig. 2.3.

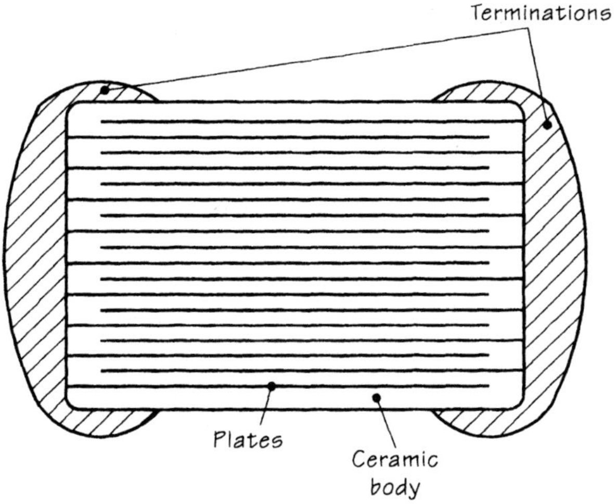

FIGURE 2.3 Cross section of a typical ceramic chip capacitor.

In the wet process, the dielectric film is formed by spraying a slurry of the dielectric material onto a surface and drying. The electrodes are then screened onto the dielectric film and dried. The sequence is repeated until the requisite number of layers are formed. The drying, pressing, firing, and electroding operations are identical to those used in the dry process.

The dielectric thickness of capacitors made by the wet process can be made thinner than those made by the dry process, and they will consequently have a higher capacitance/unit volume. However, they are more susceptible to dielectric breakdown and leakage for the same reason.

Dielectric materials used to fabricate ceramic capacitors may be divided into two types: paraelectric and ferroelectric. When an electric field is applied to a paraelectric material and removed, the internal polarization immediately drops to zero. By contrast, when an electric field is removed from a ferroelectric material, a residual polarization remains. To reduce this remnant polarization to zero, an electric field of the opposite polarity must be applied. If the opposing field is increased beyond the point where the polarization is zero, a polarization in the opposite direction is created. Paraelectric materials tend to have a lower dielectric constant, be less sensitive to temperature, be more stable with time, and have better performance at high frequencies.

The ceramic materials based on the titanates (TiO_2, $BaTiO_3$, or the rare earth titanates) exhibit the highest usable dielectric constant with good electrical characteristics. The temperature dependence of $BaTiO_3$ is given in Fig. 2.4. The transformation between the paraelectric and ferroelectric phases is called the *Curie point* (T_c). Pure barium titanate exhibits a high K at the Curie point and a large negative temperature coefficient on both sides of the peak.

To produce materials with a range of dielectric constants and useful temperature characteristics, additives are introduced that either shift the Curie temperature or depress the peak value of K at T_c. Depressors are not substitutionally soluble in the perovskite structure of the $BaTiO_3$ family of oxides and tend to act as grain boundary impurities. These additives (such as bismuth and magnesium) react with $BaTiO_3$ to increase sintering rates. Shifters are usually materials that can form solid solution

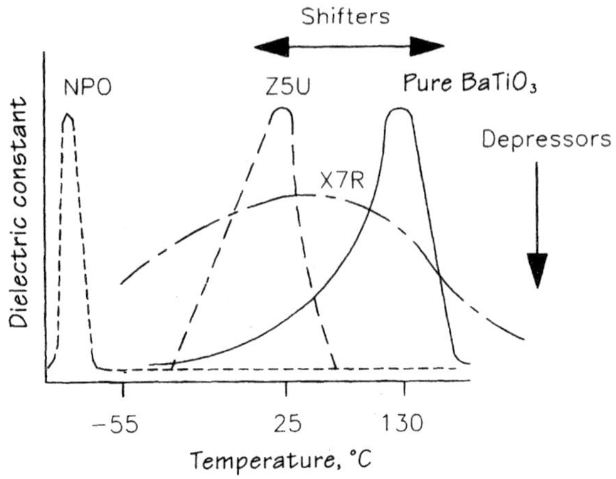

FIGURE 2.4 Temperature dependence of barium titanate.

with $BaTiO_3$ (such as titanates and zirconates of Ba, Ca, Sr, and the rare earth elements) and are used to shift the Curie point of the dielectric. A listing of shifters and depressors is given in Table 2.5.

Certain of the parameters of capacitors are dependent on temperature.

Temperature Coefficient of Capacitance (TCC). The TCC is a measure of the change in capacitance as a function of temperature and is defined by the following equation:

$$\mathrm{TCC} = \frac{C_2 - C_1}{C_1(T_2 - T_1)} \tag{2.9}$$

where C_2 = capacitance at temperature T_2
C_1 = capacitance at temperature T_1 (reference temperature)
T_2 = temperature at which C_2 is measured
T_1 = reference temperature (usually 25°C)

When the TCC is high, it may be expressed in percent (10,000 ppm = 1 percent).

The TCC is a function of temperature, but is usually expressed as a *hot* or *cold* TCC, both of which use room temperature as the reference temperature. The hot TCC is the average TCC between room temperature and an arbitrary elevated temperature, usually +125°C, and the cold TCC is the average TCC between room temperature and an arbitrary lowered temperature, usually −55°C.

Dielectric Class. Dielectric classes are categorized by the performance of the TCC of the materials. In ceramic chip capacitors, two types (Class I and Class II) are

TABLE 2.5 Additives for $BaTiO_3$ Formulation

Shifters		Depressors	
$BaSnO_3$	$SrTiO_3$	$CaTiO_3$	$MgSnO_3$
$BaZrO_3$	$PbTiO_3$	$MgTiO_3$	$Bi_2(SnO_3)$
$CaZrO_3$	La_2O_3	$MgZrO_3$	$Bi_2(TiO_3)_3$
$CaSnO_3$	$NaTiO_3$		

used. Class I dielectrics have low TCC, low dielectric constants, and are fabricated from paraelectric materials, while Class II materials have high dielectric constants, higher TCC, and are ferroelectric in nature. Code classifications have been developed by the Electronic Industries Association (EIA), defined in Standard RS-198-B, the DOD, and in MIL Specification C-55681, for both Class I and Class II dielectrics for capacitance change over a range of temperature performances. EIA designations for Class I and Class II dielectrics are shown in Tables 2.6 and 2.7.

Class I materials are made with low-K ($K < 150$) dielectric materials and are paraelectric in nature. These materials have these characteristics in common:

- Linear TCC but tending to be more negative
- High Q (DF < 1 percent)
- High stability of both capacitance and DF with respect to time, voltage, and frequency
- No aging effects since they are paraelectric
- VCC near zero

Class I dielectrics can be made to have a TCC as low as ±30 ppm (NPO) with variations ranging to several thousand ppm with both positive and negative values.

Class II dielectrics are ferroelectric in nature and have these characteristics:

- Nonlinear TCC expressed in percent (Fig. 2.5)
- Low Q (DF > 2 percent)

TABLE 2.6 Capacitor Construction Variables

Type	Dielectric material	Dielectric constant	Area/ volume	Dielectric thickness	Capacitance value
MOS	Silicon dioxide	Fixed, small	Variable	Fixed	<50 pF
Ceramic	Various	Variable, 80–10,000	Variable	Fixed	1 pF–10 μF
Tantalum	Tantalum pentoxide	Fixed	Fixed	Variable	1–100 μF

TABLE 2.7 EIA Designations for Class I Dielectrics (–55 to +125°C)

Significant figure of temperature coefficient of capacitance, (ppm/°C) (a)	Letter code for (a) (b)	Multiplier applicable to column (a) (c)	Numerical codes for (c) (d)	Tolerance of temperature coefficient (ppm/°C) (e)	Letter code for (e) (f)
0.0	C	–1.0	0	±30	G
1.0	M	–10	1	±60	H
1.5	P	–100	2	±120	J
2.2	R	–1,000	3	±250	K
3.3	S	–10,000	4	±500	L
4.7	T	+1.0	5	±1,000	M
7.5	U	+10	6	±2,500	N
		+100	7		
		+1,000	8		
		+10,000	9		

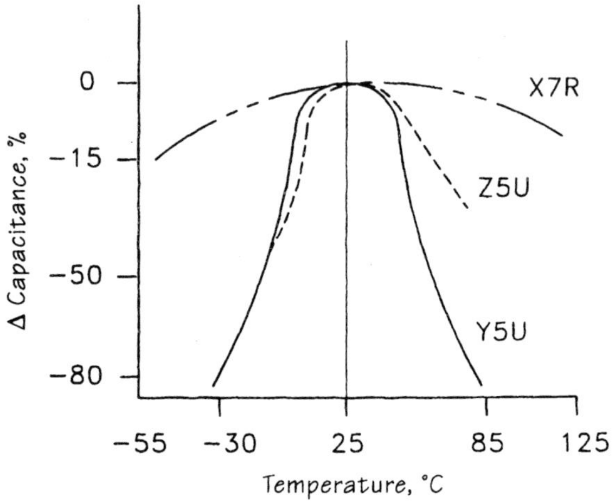

FIGURE 2.5 Temperature coefficient of Class II dielectrics.

- VCCs that decrease (DC) and increase (AC) with increase in effective dielectric constant
- Aging effects, increasing from 1 to 2 percent per decade hour for BX/X7R to 5 percent per decade hour for high-K dielectrics

Class II dielectrics are divided into two subcategories based on the dielectric constant, the stability, and the temperature characteristics. The so-called mid-K materials ($500 < K < 4000$) include the BX class, which meets military standards, and the X7R class, which does not. The mid-K materials are more stable than the high-K materials ($K > 4000$), which include the Z5U class. Class II materials have a higher capacitance per unit volume than Class I materials and are used primarily in coupling and decoupling applications. Class I materials are used in applications where stability and TCC are at a premium, such as tuned circuits and timing circuits. The electrical specifications for Class I and Class II dielectrics are summarized in Table 2.8. Graphs of the TCC versus temperature for X7R, Z5U, and various Class I dielectric materials are shown in Figs. 2.6, 2.7, and 2.8, respectively.

Dissipation Factor (DF). Changing the polarity of the dipoles requires a finite amount of energy. This energy is dissipated as internal heat, which is quantified by a parameter called the *dissipation factor,* or *DF.* In addition, dielectric materials are not perfect insulators. These phenomena may be modeled as a resistor in parallel with the capacitor. The DF, as expected, is a strong function of the applied frequency, increasing as the frequency increases.

In alternating current applications, the current and voltage across an ideal capacitor are 90° out of phase. In actuality, the resistive component of the capacitor causes the current to lead the voltage by an angle less than 90° as illustrated in Fig. 2.9. The DF is a measure of the real or resistive component of the capacitor and is given by

$$\text{DF} = \tan(\delta) = \tan(90° - \theta) \tag{2.10}$$

where θ = the angle by which the current leads the voltage

The dissipation factor is often given as a percentage as calculated by Eq. 2.11.

$$\text{DF}(\%) = \text{DF} \times 100 \tag{2.11}$$

TABLE 2.8 Electrical Specifications for Class I and Class II Dielectrics—EIA RS 198 and MIL-C-55681

	Dielectric characteristic			
Properties	NPO or COG Mil-C-55681/ EIA RS 198	BX MIL-C-55681	X7R EIA RS 198	Z5U EIA RS 198
Operating temperature range, °C	–55 to +125°C	–55 to +125°C	–55 to +125°C	+10 to +85°C
Temperature coefficient, % ΔC max, 0 bias	0 ± 30 ppm/°C			
Temperature-voltage coefficient, % ΔC max at V_{DCW}	0 ± 30 ppm/°C			
Dissipation factor at 25°C	0.10% max			
Insulation resistance at V_{DCW}, 25°C	Lesser of 100 GΩ or 1,000 MΩ·μF	Lesser of 100 GΩ or 1,000 MΩ·μF	Lesser of 100 GΩ or 1000 MΩ·μF	Lesser of 100 GΩ or 500 MΩ·μF
Insulation resistance at V_{DCW}, 125°C	Lesser of 1 GΩ or 10 MΩ·μF	Lesser of 10 GΩ or 100 MΩ·μF	Lesser of 10 GΩ or 100 MΩ·μF	Lesser of 10 GΩ or 50 MΩ·μF
Dielectric withstanding test voltage, 25°C*	$2.5 \times V_{DCW}$	$2.5 \times V_{DCW}$	$2.5 \times V_{DCW}$	$2.5 \times V_{DCW}$
Aging rate, % ΔC/decade hour, max %	0	–2.5	–2.5	–5.0
Test frequency, 25°C	<100 pf, 1 MHz >100 pf, 1 kHz	1 kHz	1 kHz	1 kHz
Test voltage, 25°C	$1.0 \pm 0.2\ V_{rms}$	$1.0 \pm 0.2\ V_{rms}$	$1.2\ V_{rms}$ max $1.0 \pm 0.2\ V_{rms}$, typical	$1.0 \pm 0.2\ V_{rms}$

* $1.2 \times V_{DCW}$ for capacitors rated at or above 1,000 V_{DC}.

The DF is related to the power factor (PF) and the quality factor (Q).

$$PF = \cos(\theta) \tag{2.12}$$

For very small values of θ, PF = DF.

$$Q = \frac{1}{\tan(\delta)} \tag{2.13}$$

These factors are of interest only when the capacitor is excited by alternating current.

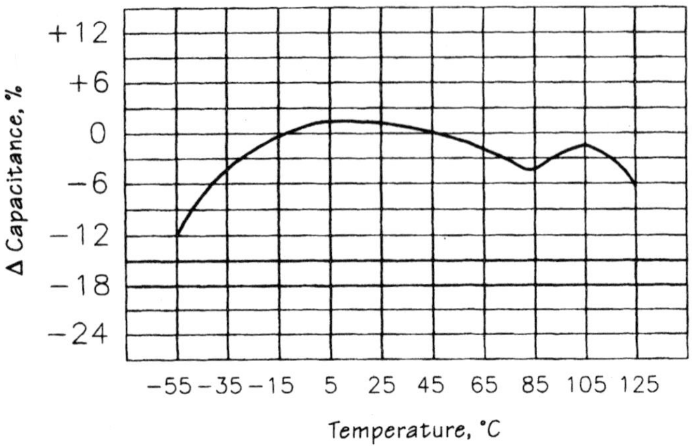

FIGURE 2.6 Temperature coefficient of capacitance for an X7R dielectric. *(Courtesy of AVX Corporation.)*

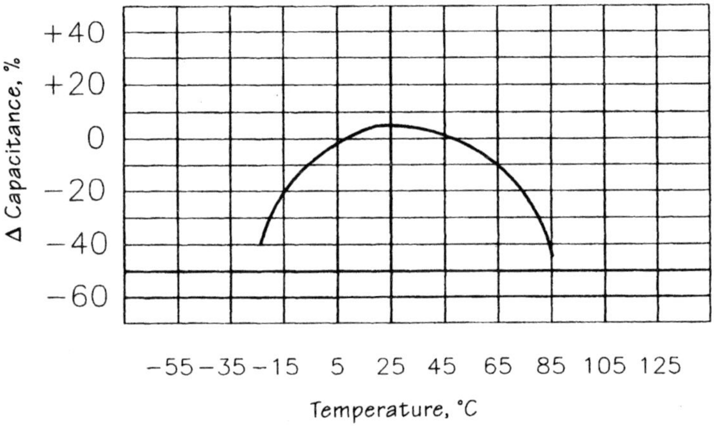

FIGURE 2.7 Temperature coefficient of capacitance for a Z5U dielectric. *(Courtesy of AVX Corporation.)*

Referring to Fig. 2.10, the DF is also strongly dependent on temperature. At low temperatures, the frequency component dominates and the DF is nearly constant with temperature, and, at frequencies of 10^{10} Hz and above, the ionic component of the polarization vector dominates, and the effect of temperature is negligible. At high temperatures and low frequencies, the DF is exponentially related to the temperature. For alumina, the slope is approximately 0.00633 as determined from Fig. 2.10, and the relationship between DF and temperature for $f = 1$ kHz is given by

$$\mathrm{DF}(T) = 0.000141 \times 10^{0.00633} \tag{2.14}$$

For $f = 100$ Hz and $f = 10$ kHz at elevated temperatures, the slope is the same and only the multiplication coefficient is different.

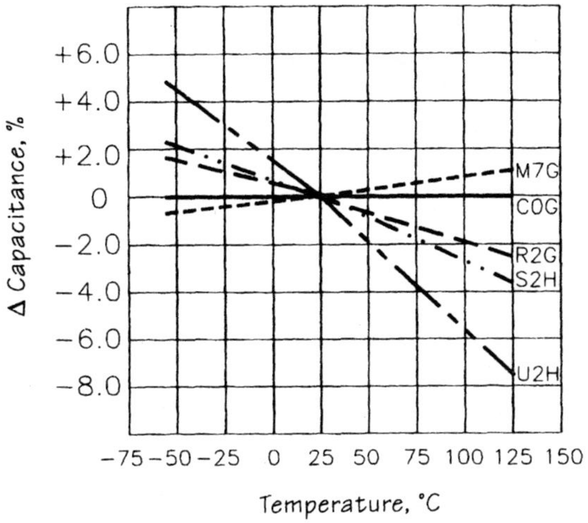

FIGURE 2.8 Temperature coefficient of Class I dielectrics. *(Courtesy of Novacap.)*

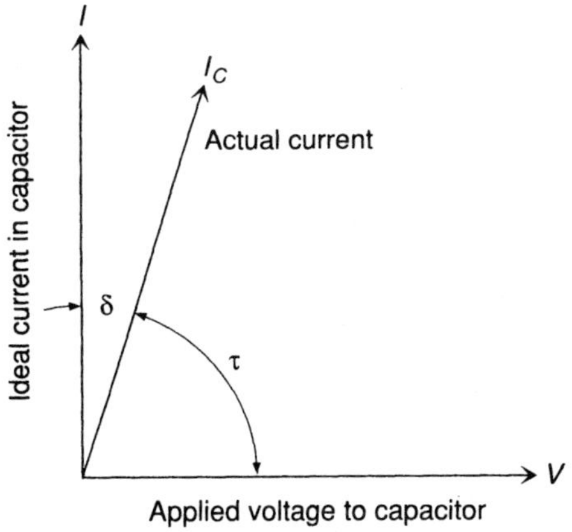

FIGURE 2.9 Current and voltage relationships in a capacitor.

Class I dielectric materials exhibit negligible change in DF with respect to temperature. The dissipation factor versus temperature relationships for various Class II dielectrics is presented in Fig. 2.11. Note that the DF for ferroelectric materials decreases with temperature, as opposed to alumina, which is a paraelectric material.

Insulation Resistance (IR). The IR is the direct current analog of the DF. The IR is largely a bulk effect and is inversely proportional to the plate area. The resistance is usually expressed as the product of the IR and the capacitance with units of megohm-microfarads (Mohm-mF), a term that is constant for a given dielectric material. The IR has an inverse logarithmic relationship with the temperature due to increased ionization and may vary by several orders of magnitude between +25 and +125°C. The IR is also inversely proportional to the leakage current. Typical values

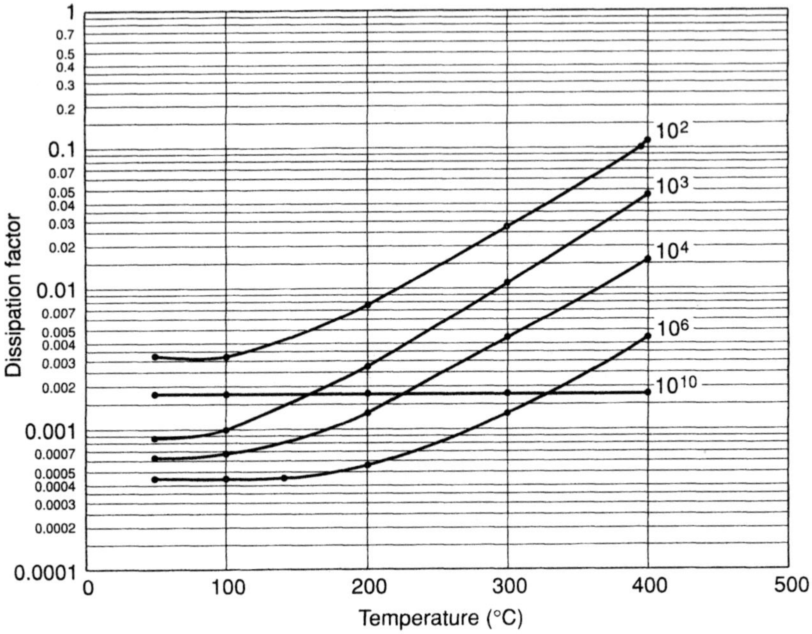

FIGURE 2.10 Dissipation factor versus temperature and frequency for aluminum oxide.

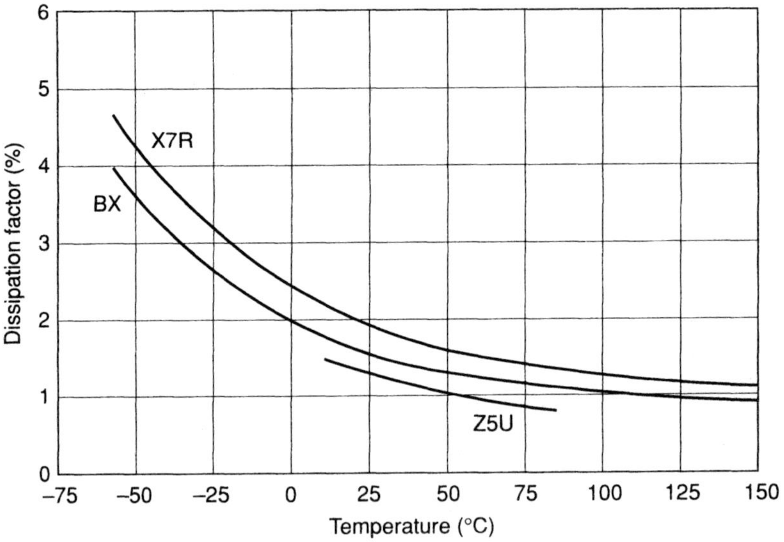

FIGURE 2.11 Dissipation factor versus temperature for Class II dielectrics.

of IR for NPO, X7R, and Z5U dielectrics are presented in Figs. 2.12, 2.13, and 2.14, respectively.

Dielectric Strength. The *dielectric strength* is a measure of the voltage required to cause breakdown for a given configuration. The dielectric strength is actually a field effect and the term V/unit length is constant for a given dielectric material. The dielectric strength is also a function of the temperature, but to a lesser degree than the insulation resistance.

There are two mechanisms which contribute to dielectric breakdown: [2]

1. *Thermal failure.* The application of a high electric field will generate a current in a weaker, more conductive area of the dielectric which generates heat due to I^2R losses. The heat further lowers the resistivity of the surrounding material, thereby increasing both the current and the heat, which continues until the dielectric fails. This phenomenon is obviously exacerbated by elevated temperature, but is difficult to quantify except in a specific case.
2. *Electronic or intrinsic failure.* The electric field concentrates on localized free electrons at various points within the dielectric structure. These energized electrons collide with other electrons, creating an avalanche effect and ultimately causing breakdown. The electronic dielectric strength increases with temperature due to the decrease in mobility of the electrons which are inhibited by increased lattice vibrations.

Quality Factor. The *quality factor,* or Q, of a component may also be defined as the ratio of the amount of energy stored to the amount of energy dissipated per cycle. For a capacitor, this equates to

$$Q = \frac{1}{\omega RC} \tag{2.15}$$

where R is the equivalent series resistance of the capacitor.

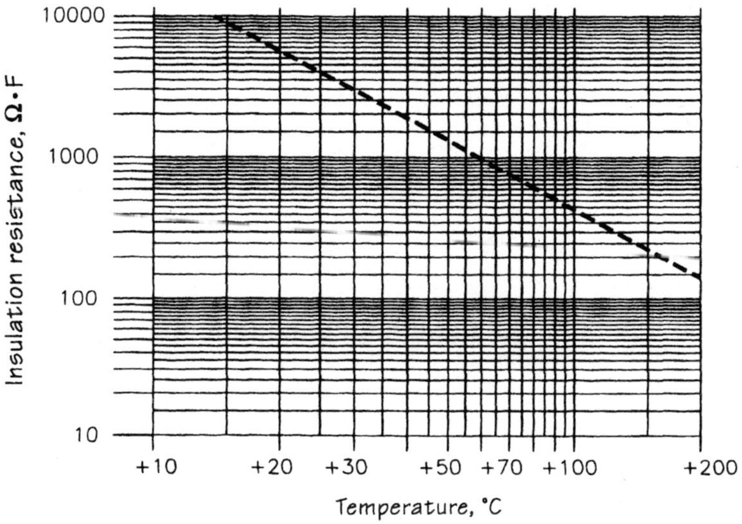

FIGURE 2.12 Insulation resistance versus temperature for NPO dielectric. *(Courtesy of AVX Corporation.)*

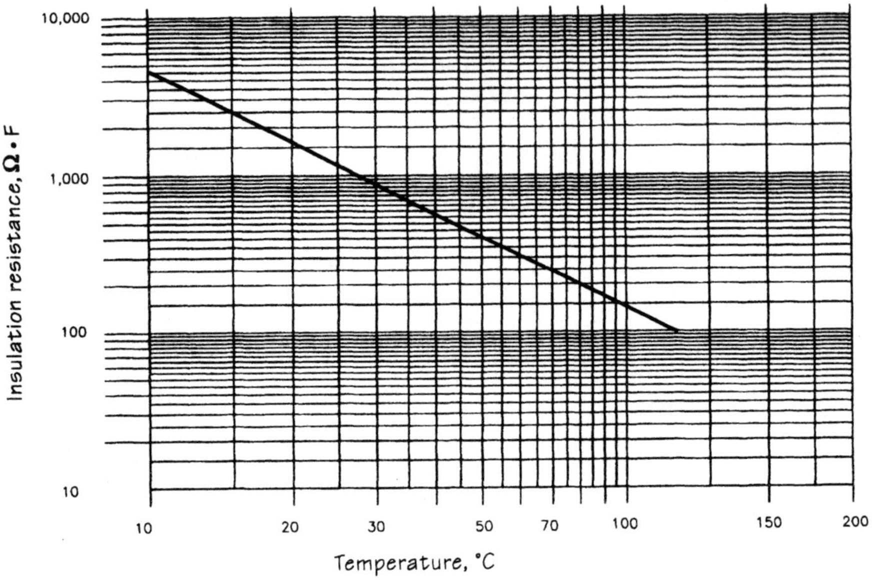

FIGURE 2.13 Insulation resistance of a typical X7R dielectric. *(Courtesy of AVX Corporation.)*

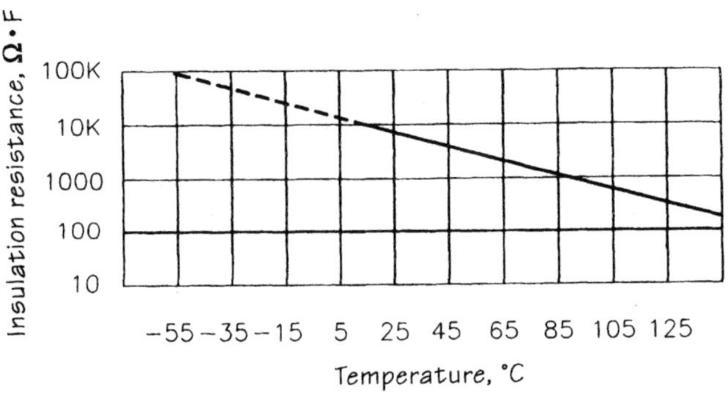

FIGURE 2.14 Insulation resistance versus temperature for a Z5U dielectric. *(Courtesy of AVX Corporation.)*

The extent to which these parameters vary with temperature depends largely on the type of dielectric material used to fabricate the component.

MOS Capacitors. MOS capacitors are fabricated by thermally growing a thick layer of SiO_2 on a wafer of low-resistivity silicon. The oxide is etched away over the capacitor areas and carefully regrown to obtain the proper thickness. To improve the stability, the breakdown voltage, and the capacitance/unit area, a thin layer of Si_3N_4 may be grown by chemical vapor deposition. Aluminum is deposited over the wafer by evaporation, and bonding pads are formed over the capacitor areas using a photoetching process. The bottom electrode is formed by evaporating gold on the bottom of the wafer and alloying. The device may be mounted onto the circuit by eutectic bonding or with conductive epoxy. The top electrode is wire-bonded to the circuit metallization at the appropriate point, making the MOS capacitor compatible with active chip mounting.

Capacitance values are limited by processing considerations to a few pF with a DF as low as 0.015 percent. Insulation resistance and stability are extremely high with respect to time, frequency, temperature, and voltage. The TCC is typically positive, varying from about +30 to +70 ppm/°C. The parameters are not generally strong functions of temperature. MOS capacitors may be actively trimmed in two ways, both involving the disconnection of discrete sections of the capacitor. One method utilizes a laser trimmer to remove connections between capacitors of successively increasing values until the proper tolerance is achieved. A second method uses a capacitor with top electrodes of different areas that are wire-bonded to a common point. Trimming is achieved by removing wire bonds from the different pads. The structure of an MOS capacitor with a range from 1 to 15 pF is depicted in Fig. 2.15.

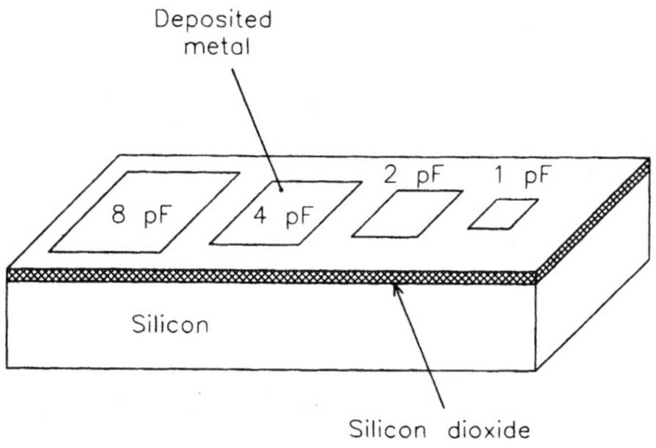

FIGURE 2.15 MOS capacitor with a range of 1 to 15 pF.

Tantalum Solid Electrolytic Capacitors. Tantalum capacitors have the highest capacitance × voltage (CV) rating of any type of capacitor for hybrid or surface-mount applications. Whereas the ceramic capacitor obtains a high capacitance value from the use of a high-*K* dielectric material, tantalum capacitors use large effective surface area and a very thin dielectric to obtain high capacitance values.

The basic structure of a tantalum capacitor is a porous sintered tantalum anode body with an electrochemically formed tantalum oxide dielectric. The counter electrode is formed by utilizing a conductive electrolyte and a conductor. Although liquid electrolytes are used in many commercial applications because of their high CV ratio, this discussion is limited to solid electrolyte devices, since this type is primarily found in hybrid/surface-mount applications. For a solid electrolyte device, a semiconducting film of manganese dioxide serves as the electrolyte and a graphite/silver coating is used to fabricate the counter-electrode (see Fig. 2.16).

The effective area of a tantalum slug is large due to the open structure of the sintered tantalum anode. For any given volume, the effective area is constant and area can only be considered a variable by change in component size. Since the effective area for any given volume of tantalum is the same, the capacitance is changed by

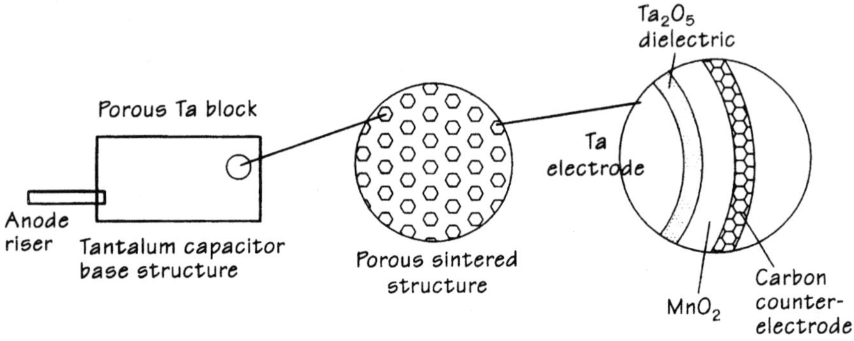

FIGURE 2.16 Solid tantalum electrolytic capacitor.

varying the thickness of the dielectric. Dielectric films have typical thicknesses that range from 200 to 1000 Å (two orders of magnitude smaller than a ceramic capacitor). Since the thickness of dielectric determines the dielectric breakdown voltage and, therefore, the effective working voltage, tantalum capacitors are selected for both capacitance and voltage because the CV product is a constant for any given size (volume efficiency of 100,000 uFV/in^3).

Tantalum capacitors are found in various configurations. The high-reliability solid tantalum devices are covered in MIL-C-55365. Table 2.9 represents the sizes and the minimum/maximum range of working voltage available. In this range, at least eight voltage ranges exist.

Tantalum capacitors are polar and catastrophic failure will occur if a reverse bias is applied to the device which is over 15 percent of the working voltage at room temperature. The polar nature of the device occurs because of the operation of the solid electrolyte, MnO_2. After manufacture, defect sites occur in the dielectric. Upon application of voltage, current flows through the defects and into the MnO_2. The increased localized temperature causes the MnO_2 to decompose into an insulating oxide phase, thus healing the defect site. Because the electron flow is highly localized at the MnO_2 layer, sufficient heating occurs. On a reverse bias, the current flow is not localized at the MnO_2 layer but in the tantalum layer, leading to heating and ultimately ignition of the tantalum without the clearing mechanism for protection.

TABLE 2.9 Sizes of Tantalum Capacitors per MIL-C-55365

		Voltage rating	
Case	Size (L × W × T, ± 0.015 in)	4 V_{DC}	50 V_{DC}
A	0.100 × 0.050 × 0.050	2.2	0.1
B	0.150 × 0.050 × 0.050	4.7	0.33
C	0.200 × 0.050 × 0.050	6.8	0.47
D	0.150 × 0.100 × 0.050	10.0	0.68
E	0.200 × 0.100 × 0.050	15.0	1.0
F	0.220 × 0.135 × 0.070	33.0	2.2
G	0.265 × 0.150 × 0.110	68.0	3.3
H	0.285 × 0.150 × 0.110	100.0	4.7

Tantalum capacitors exhibit a high series inductance and are prone to leakage current failures if subjected to severe mechanical and environmental forces. The distributed inductance limits their frequency application to about 100 kHz.

Typical leakage current and capacitance change versus temperature curves are shown in Figs. 2.17 and 2.18, respectively.

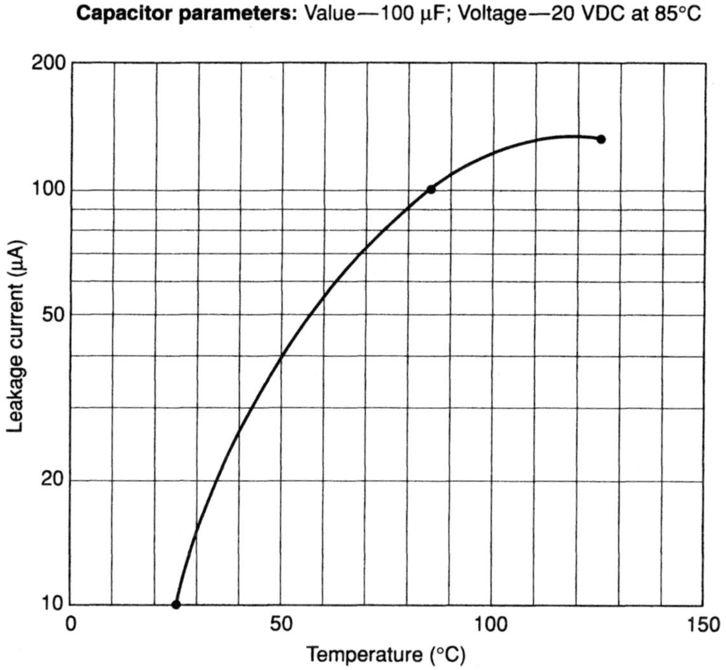

FIGURE 2.17 Typical leakage current versus temperature curve for solid tantalum capacitor.

2.2.3 Inductors and Transformers

An electric current flowing in a conductor creates a magnetic field in the space surrounding the current. The fundamental relationship which describes this phenomenon in a vacuum is

$$B = \mu_0 H \tag{2.16}$$

where B = magnetic flux density in webers/m²
μ_0 = permeability of free space 4p × 10^{-7} henrys/meter
H = magnetic field strength in ampere-turns/meter

The total magnetic flux in webers in a given area is calculated by integrating the expression for the flux density as a function of position over the area of interest.

$$\Phi = \oiint \overline{B} \cdot \overline{dS} \tag{2.17}$$

where Φ = magnetic flux in webers
dS = differential unit of area expressed as a vector

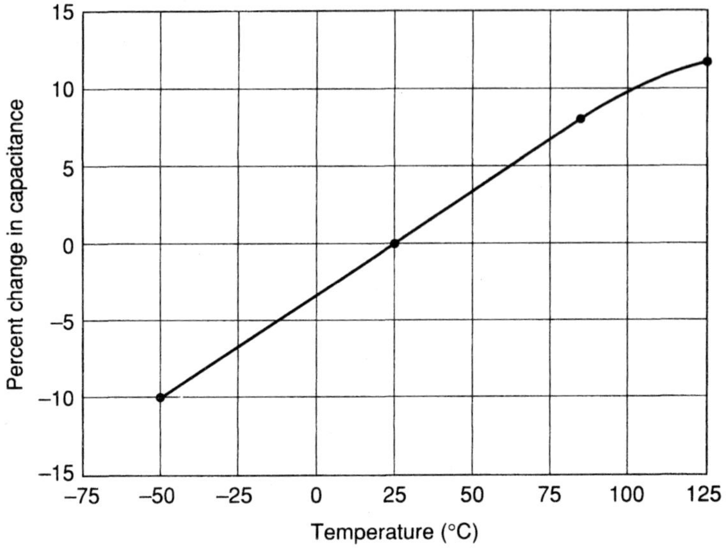

FIGURE 2.18 Percentage change in capacitance versus temperature for a solid tantalum capacitor.

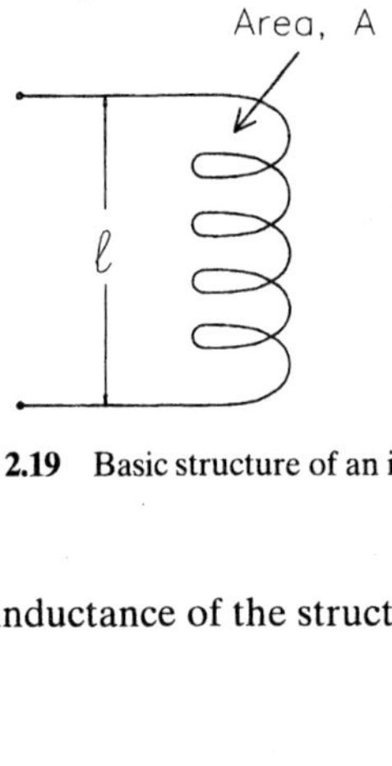

FIGURE 2.19 Basic structure of an inductor.

The amount of flux generated by a current in a conductor may be increased by winding the conductor in the form of a spiral coil as shown in Fig. 2.19. The efficiency of the coil in generating magnetic flux is called the *inductance* and is defined as

$$L = \frac{\Phi}{I} \tag{2.19}$$

where L = inductance in henrys
Φ = flux in webers
I = current in amperes

The inductance of the structure in Fig. 2.19 is given by Eq. 2.20.

$$L = \frac{\mu_0 N^2 A}{l} \tag{2.20}$$

where L = inductance of coil in henrys
A = area of core in m^2
l = length of coil in meters
N = number of turns

When a material which has the capability of forming magnetic dipoles is placed in the magnetic path, an additional magnetic field proportional to the field strength, as given in Eq. 2.21, is created in the material.

$$M = \chi_m H \tag{2.21}$$

where M = magnetization of the material in webers/m^2
χ_m = magnetic susceptibility

The magnetization is added to the field already present, making the total field equal to

$$B = \mu_0 H + \mu_0 M \tag{2.22}$$

The overall relationship between B and H is therefore

$$B = \mu H \tag{2.23}$$

where $\mu = \mu_0 \mu_r$
μ_r = permeability of the material relative to vacuum = $1 + \chi_m$

There are four basic types of magnetic materials: diamagnetic, paramagnetic, ferromagnetic, and ferrimagnetic, which are classified as to the nature of the B-H curve and the degree of permanent magnetism.

Diamagnetism is a weak, nonpermanent form of magnetism induced by changes in the orbits of the outer electrons of atoms of certain materials. The induced magnetic field in this instance is opposite that of the applied magnetic field, and the resultant magnetic field is actually lowered. Diamagnetic and paramagnetic materials have a linear relationship between B and H (Φ and I), as shown in Fig. 2.20, and do not exhibit residual magnetism when the applied field is removed. The permeability of certain diamagnetic and paramagnetic materials is listed in Table 2.10.

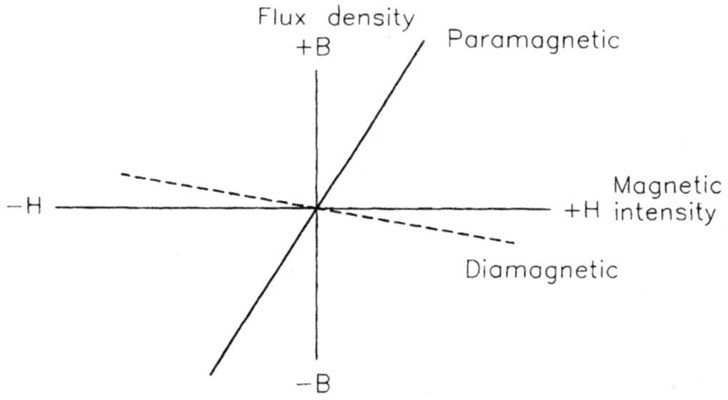

FIGURE 2.20 Magnetic curves for paramagnetic and diamagnetic materials.

Ferromagnetic materials are metallic in nature and are characterized by a nonlinear relationship between B and H, a degree of hysteresis, and a residual permanent magnetism component, as shown in Fig. 2.21. As the field is increased, substantially all the magnetic dipoles eventually become aligned, resulting in no further increase in the magnetic field as the current is increased. This phenomenon is referred to as *saturation.* Since the magnetic field is nonlinear, the permeability is a function of the magnetic field strength and no single number can be assigned. At the maximum value, the relative permeability of certain ferromagnetic materials may range as high as 10^6 henrys/meter. Typical ferromagnetic materials are iron, cobalt, and nickel, and certain of the rare earth metals.

TABLE 2.10 Relative Permeability of Diamagnetic and Paramagnetic Materials

Material	Type	Relative permeability
Al_2O_3	Diamagnetic	0.999982
Silicon	Diamagnetic	0.999996
Silver	Diamagnetic	0.999976
Gold	Diamagnetic	0.999966
Copper	Diamagnetic	0.999990
Aluminum	Paramagnetic	1.000002
Molybdenum	Paramagnetic	1.000012
Chromium	Paramagnetic	1.000031

Ferrimagnetism is a form of permanent magnetism exhibited by some ceramic materials. The *B-H* curve of ferrimagnetic materials (*ferrites*) is similar to that exhibited by the ferromagnetic materials shown in Fig. 2.21, except that the saturation levels are somewhat lower. Another significant difference is that ferrites, being generally metallic oxides, are good electrical insulators, a desirable property for inductors or transformers used in high-frequency applications. One of the most common ferrite materials is magnetite, which has the chemical formula Fe_3O_4, often referred to as *lodestone.*

As a result of the hysteresis loop exhibited in ferromagnetic and ferrimagnetic materials, energy in the form of heat is absorbed by the core. This is an undesirable property, and considerable effort has been directed toward the development of so-called *soft* magnetic materials, or materials which have a relatively small hysteresis loop, as shown in Fig. 2.22. By contrast, *hard* magnetic materials have a larger hysteresis loop and are utilized in permanent magnets.

An increase in the temperature of a ferromagnetic or ferrimagnetic material causes the random motion of the magnetic dipoles to increase accordingly. This has the effect of randomizing the internal magnetic field, lowering the effective permeability and the saturation level. As the temperature is further increased, the saturation magnetism will gradually drop until the temperature reaches a critical temperature, called the Curie temperature, T_c, at which point it abruptly drops to zero and the material

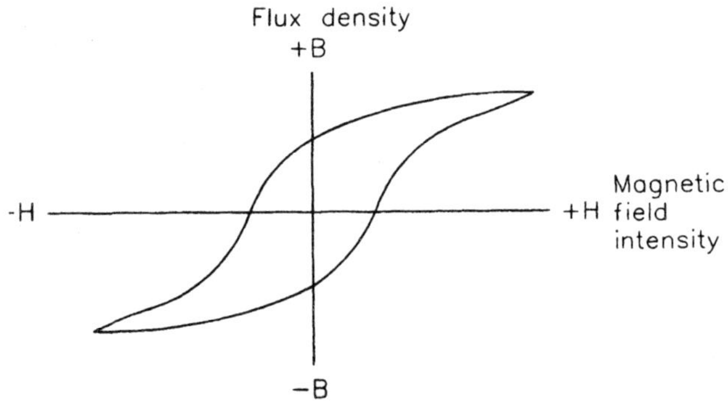

FIGURE 2.21 Magnetization curve for ferromagnetic and ferrimagnetic materials.

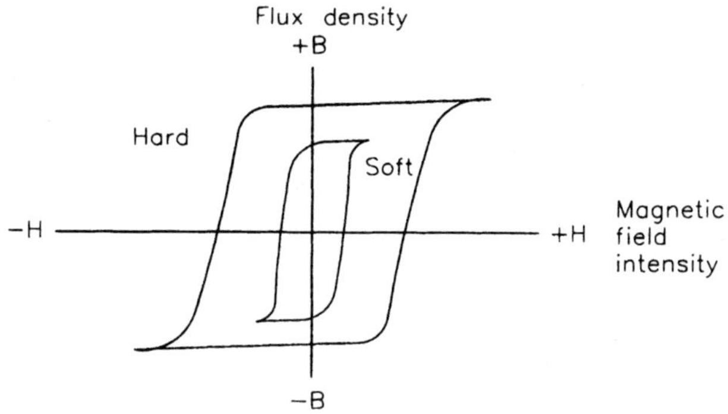

FIGURE 2.22 Comparison of soft and hard magnetization curves.

becomes paramagnetic in nature. For most applications, the Curie temperature is sufficiently high that it is not a factor. For example, the T_c for Fe_3O_4 is 585°C, and the permeability drops to approximately 90 percent of the room-temperature level at 150°C.

Soft Ferrite Materials. *Soft ferrites* are the most common magnetic materials used to manufacture coils and transformers for hybrid microelectronic applications. The basic formula for a soft ferrite is $MO{\cdot}Fe_2O_3$, where MO represents a divalent metal oxide, such as manganese, manganese-zinc, or nickel-zinc. These are blended with 48 to 60 mole percent of iron oxide. [3] Ferrites typically have a low TCE (8 to 10 ppm/°C) and a high bulk resistivity.

Manganese ferrites are very stable materials with a very square hysteresis loop and are used in such applications as switched-mode power supplies and high-frequency power supplies. Manganese-zinc ferrites are highly vitrified and have the highest permeability. The volume resistivity may range as high as several thousand Ω-cm. They are useful up to the low MHz range. Nickel-zinc ferrites may contain oxides of other materials, such as magnesium, manganese, copper, or cobalt. The volume resistivity may be >10 megohm-cm, and they may be used in higher-frequency applications.

The electrical properties of soft ferrites are a function of temperature and frequency, although the temperature variations are generally small over the range of interest. The equivalent circuit of an inductor is shown for reference in Fig. 2.23. Due

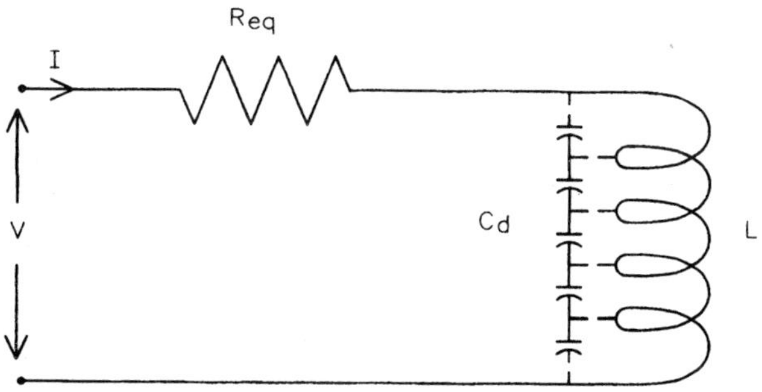

FIGURE 2.23 Equivalent circuit of an inductor.

to the distributed capacitance and resistance, the phase angle between the voltage and current will vary from the normal 90° exhibited in a perfect inductor. The actual relationship will be

$$\theta = 90° \pm \lambda \tag{2.24}$$

where θ = phase angle between the voltage and current
λ = deviation from 90°

Note that λ may be positive or negative, depending on the frequency. The effective resistance of the inductor is dependent on the frequency due to a phenomenon known as the *skin effect.* As the frequency increases, the current density becomes nonuniform within the conductor, increasing toward the surface. This has the net effect of decreasing the effective cross section of the conductor, thereby increasing the resistance. The expression for the skin depth is

$$\delta = \frac{1}{\sqrt{\pi f \mu \sigma}} \tag{2.25}$$

where δ = skin depth
f = frequency in Hz
μ = permeability in henrys/meter
σ = conductivity in siemens/m

A table of the skin depth for copper at various frequencies is shown in Table 2.11 for 25 and 150°C. Note that the skin depth increases with temperature by a significant amount.

A resistive loss component exists in the core due to the hysteresis element and, to a lesser degree, eddy currents. *Eddy currents* are circulating currents which exist in the core of an inductor or transformer when alternating current is applied. Due to the high resistivity of ferrites, eddy-current losses are substantially less than those in ferromagnetic materials. As a result of these losses, a resistive component in parallel with the inductance exists, which causes the magnetic flux to be out of phase with the current by an angle φ.

Loss Factor. The *loss factor* is defined as

$$\text{Loss Factor} = \frac{\tan \varphi}{\mu_i} \tag{2.26}$$

where μ_i is the initial permeability, measured at $\Phi < 10$ gauss.

Quality Factor. The *quality factor Q* is defined as the ratio of the energy stored per cycle to the energy dissipated and is given by

TABLE 2.11 Skin Depth of Copper versus Frequency and Temperature

Frequency	Skin depth at 25°C, m	Skin depth at 150°C, m
60 Hz	0.008	0.010
1 kHz	0.002	0.0025
10 kHz	6.51×10^{-4}	7.97×10^{-4}
1 MHz	6.51×10^{-5}	7.97×10^{-5}
10 MHz	6.51×10^{-6}	7.97×10^{-6}

$$Q = \frac{\omega L}{R_{\text{eff}}} \tag{2.27}$$

where $\omega = 2\pi f$, the frequency in radians
R_{eff} = the effective resistance at the frequency of interest

Referring to Eq. 2.27, R_{eff} varies inversely with the square root of the frequency and, therefore, the Q varies directly with the square root of frequency. Consequently, the frequency must be specified when the Q is given. The method of attach can strongly affect the Q due to the contact resistance. Solder attach or wire bonding is preferable to epoxy attach.

Note that the R_{eff} term in Eq. 2.27 will increase with temperature, thereby reducing the Q. The extent to which the Q is lowered is a function of the frequency, the temperature, and the size of the wire.

Self-Resonant Frequency. The distributed capacitance in combination with the inductance will be resonant at a frequency f_0 which is defined by

$$f_0 = \frac{1}{2\pi\sqrt{LC_D}} \tag{2.28}$$

where C_D is the distributed capacitance of the windings, referred to in Fig. 2.23. Above the self-resonant frequency, the inductor actually becomes capacitive in nature and f_0 may be considered to be the maximum usable frequency of the inductor regardless of the core characteristics. The distributed capacitance is also dependent on the surrounding environment and may be increased by potting or conformally coating the inductor or by proximity to a ground plane.

The characteristics of typical soft ferrites are summarized in Table 2.12. There are many variations of these materials, including certain rare earth additives, which may be formulated for specific applications.

Inductors for Hybrid and Surface-Mount Applications. Inductors for surface-mount applications (chip inductors) are more troublesome to terminate than capacitors due to the difficulty in attaching the wire leads to the circuit. Most frequently,

TABLE 2.12 Selected Properties of Typical Soft Ferrite Materials

Property	Unit	Symbol	NiZn[a]	Mn[b]	MnZn[c]	MnZn[d]
Initial permeability		μ_i	20	300	800	10,000
Flux density	mT	B	200	390	250	400
at field strength	A/m	H	3,200	800	800	160
Residual flux density	mT	B_r	100	345	135	125
Loss factor at	10^{-6}	$\tan \delta/\mu_i$	400	50	35	15
frequency	MHz		100	0.1	0.2	0.025
Curie temperature	°C	T_c	>500	>300	>150	>120
Resistivity	Ω·cm	ρ	10^7	1.5×10^3	10^2	50
Recommended frequency range	MHz		<400[e]	<0.15[f]	<3[e]	<0.5[e]

[a] Fair-Rite Material 68.
[b] Fair-Rite Material 83.
[c] Fair-Rite Material 33.
[d] Fair-Rite Material 76.
[e] For low flux density inductors and transformers.
[f] Special square-loop ferrite.
Source: Courtesy of Fair-Rite Products Corp.

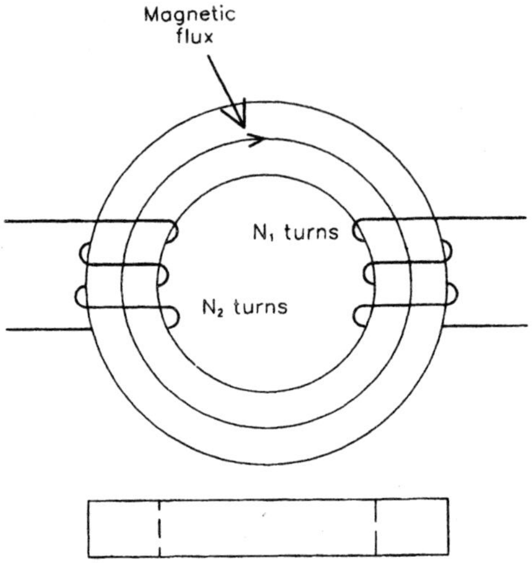

FIGURE 2.24 Transformer wound on a toroidal core.

these are wound as toroids or on square cores and attached to a lead frame which accommodates bonding to a substrate. A *toroid* is a doughnut-shaped structure which effectively contains the magnetic flux within the structure as shown in Fig. 2.24. The toroid is especially convenient for transformers, as it couples the flux from winding to winding with a very high efficiency. The inductor may be potted or left open to reduce distributed capacitance. Properties of typical chip inductors are shown in Tables 2.13 and 2.14.

The following guidelines are beneficial for using chip inductors in hybrid circuits.

1. Attach with solder or wire bonding where possible. The higher ohmic resistance of conductive epoxy may be detrimental to the quality factor.
2. When breadboarding the circuit in the design phase, use the exact inductor or transformer intended for use in the actual hybrid, as characteristics of these components may vary widely between manufacturers.

TABLE 2.13 Representative Parameters of TDK Multilayer Chip Inductors

Part number	Thickness, in	L, μH	Q, nominal	Test frequency, MHz	f_o, nominal	DC resistance, Ω	Rated current, mA
MLF3225C390	0.043	39	45	0.4	14.0	0.70	5
MLF3225C470	0.043	47	45	0.4	11.5	0.80	5
MLF3225C560	0.043	56	45	0.4	10.5	0.90	5
MLF3225C680	0.098	68	70	2.0	10.5	1.75	10
MLF3225C820	0.098	82	70	2.0	10.0	1.95	10
MLF3225C101	0.098	100	70	1.0	9.0	2.20	10
MLF3225C121	0.098	120	45	0.2	7.0	1.30	5
MLF3225C151	0.098	150	45	0.2	6.5	1.50	5
MLF3225C181	0.098	180	45	0.2	6.0	1.65	5
MLF3225C221	0.098	220	45	0.2	5.5	1.75	5

Source: Courtesy of TDK Corporation.

TABLE 2.14 Representative Parameters of TDK Ferrite Chip Inductors

Part number	Thickness, in	L, μH	Q, nominal	Test frequency, MHz	f_o, nominal	DC resistance, Ω	Rated current, mA
ACL3225S390	0.043	39	50	2.52	11.5	2.90	70
ACL3225S390	0.043	47	50	2.52	10.5	3.15	65
ACL3225S390	0.043	56	50	2.52	10.0	3.40	45
ACL3225S390	0.098	68	50	2.52	9.0	3.80	55
ACL3225S390	0.098	82	50	2.52	8.5	4.30	40
ACL3225S390	0.098	100	50	2.52	7.8	4.70	40
ACL3225S390	0.098	120	25	0.796	7.6	3.05	14
ACL3225S390	0.098	150	25	0.796	6.8	3.50	13
ACL3225S390	0.098	180	25	0.796	6.2	3.95	10
ACL3225S390	0.098	220	25	0.796	5.6	4.20	10

Source: Courtesy of TDK Corporation.

2.3 THERMAL EFFECTS ON ACTIVE SEMICONDUCTOR DEVICES

The fundamental difference between active and passive devices is that passive devices may affect the magnitude and phase of a waveform, do not affect the shape, and cannot produce a transfer function of the input power signal with a magnitude greater than one. Active devices, however, can alter the signal shape by rectification or other means and can produce a power gain.

2.3.1 Diodes

The most fundamental of the active devices is the diode. Intrinsic (pure) semiconductors have a high electrical resistivity with relatively few free carriers. The conduction characteristics can be dramatically modified by adding an impurity with a valence of one more or one less to the semiconductor. For example, silicon has a valence of four. Adding phosphorous, with a valence of five, leaves an electron which is weakly bound and which can act as a free carrier, creating n-type material. Adding boron, with a valence of three, leaves a hole which has the capability of transporting charge, thereby creating p-type material. Impurities which produce n-type semiconductors are referred to as *donor* impurities, while impurities which produce p-type semiconductors are referred to as *acceptor* impurities.

A diode consists of p-type material in contact with n-type material as shown in Fig. 2.25. Diodes may be physically realized in two ways:[4]

1. By diffusing acceptor impurities into n-type material in sufficient quantities that the acceptor impurities are the majority. This is the most common.
2. By diffusing donor impurities into p-type material in sufficient quantities that the donor impurities are the majority.

As the diode is formed, free electrons combine with free holes in the vicinity of the junction, resulting in a region with few free carriers, but with fixed positive donor ions in the n-region and fixed negative acceptor ions in the p-region. These fixed ions, spread over a region approximately 0.5 μ in length, create a potential barrier to current. When the diode is forward-biased, as shown in Fig. 2.26, the potential bar-

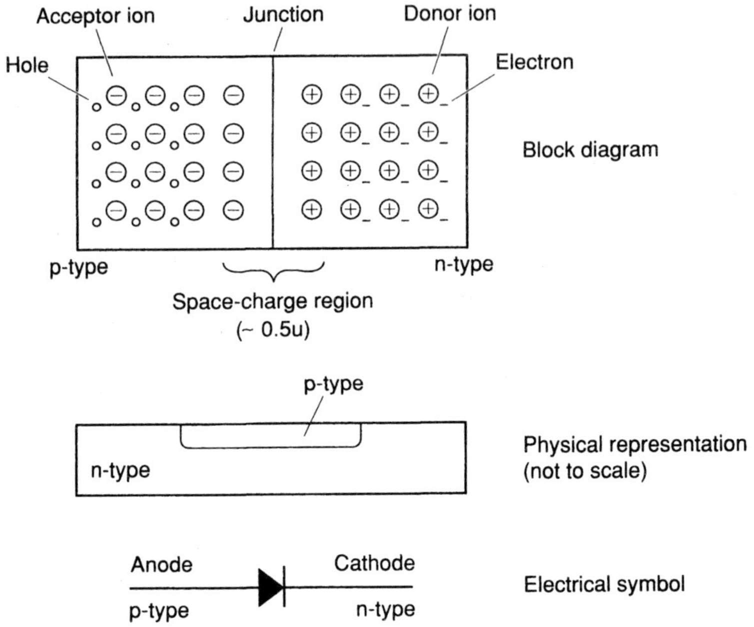

FIGURE 2.25 Block diagram and physical representation of a PN junction.

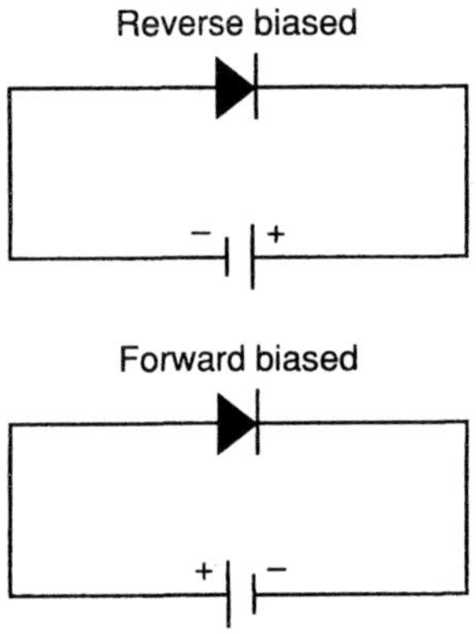

FIGURE 2.26 Reverse- and forward-biased diodes.

rier is partially overcome, allowing current to flow. When the diode is reverse-biased, the potential barrier is increased, and the current is limited to that created by random hole-electron recombination, referred to as the *reverse saturation current,* I_s. At a given temperature, I_s is independent of the applied voltage until the diode reaches the breakdown, or avalanche voltage, at which time the current increases rapidly. The reverse saturation current is a complex function of the carrier concentration, the carrier mobility, the diffusion constant, and the diffusion length, all of which are functions of temperature.

The current in an ideal diode, as a result of diffusion alone, is given by

$$I = I_s(e^{qV/kT} - 1) \tag{2.29}$$

where I = diode current
I_s = reverse saturation current
V = diode voltage
q = electron charge

$= 1.6 \times 10^{-19}$ coulomb
k = Boltzmann's constant
$= 1.38 \times 10^{-23}$ joules/K
$= 8.625 \times 10^{-5}$ electron volts/K

Other parameters which affect the current-voltage relationship in a diode follow.

- The voltage drop across the bulk p-type and n-type regions outside the junction
- Carrier generation and recombination in the junction area
- Leakage current across the surface

These factors result in deviations from the ideal case expressed in Eq. 2.29.

The behavior of the ideal diode as a function of temperature can be derived from an approximation of the reverse saturation current.

$$I_s \cong Kn_i^2 = KT^3e^{-(E_g/2kT)} \tag{2.30}$$

where K = constant
n_i = intrinsic carrier concentration in carriers/m^3
E_g = energy gap in electron volts (1.1 eV for silicon)
k = Boltzmann's constant in electron volts/K

Differentiating Eq. 2.30 and normalizing to obtain the relative change results in

$$\frac{1}{n_i^2}\frac{d}{dT}(n_i^2) = \frac{3}{T} + \frac{E_g}{2kT^2} \tag{2.31}$$

A plot of Eq. 2.31 is shown in Fig. 2.27. Note that, in the vicinity of room temperature (300 K), the percentage change is about 8 percent. Since

$$(1.08)^9 = 2 \tag{2.32}$$

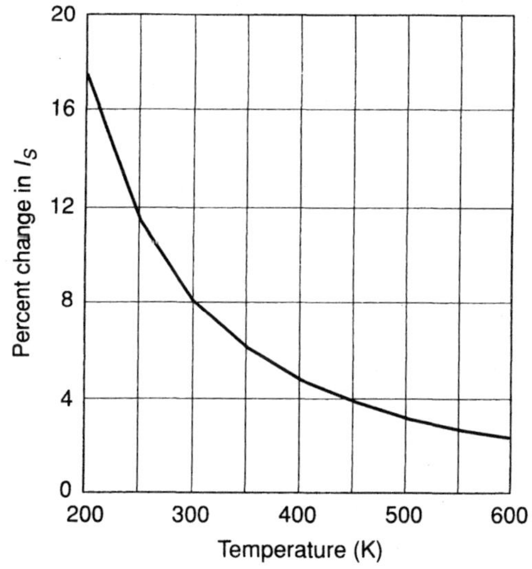

FIGURE 2.27 Percent change in I_s versus temperature.

it can be concluded that the reverse saturation current doubles with a change in temperature of about 9°C at room temperature. The other factors affecting the *V-I* relationship increase this figure to about 10°C in practice, and the expression for I_s as a function of temperature becomes

$$I_s(T) = I_{s1}2^{(T - T_1)/10} \tag{2.33}$$

This steep change has a profound effect on the diode current as a function of temperature. In the forward bias condition,

$$e^{qV/kT} >> 1 \tag{2.34}$$

If Eq. 2.29 is differentiated with respect to temperature with the voltage held constant under the assumption stated in Eq. 2.34, the result is

$$\frac{1}{I}\frac{\partial I}{\partial T}\bigg|_V \cong \frac{E_g - qV}{kT^2} \tag{2.35}$$

which describes an approximate exponential increase in current with temperature. Figure 2.28 shows the V-I relationship (Eq. 2.29) of an ideal diode with temperature as a parameter, while Fig. 2.29 shows the reverse saturation current as a function of temperature. Note the rapid increase in current as the temperature increases. In practice, the current is somewhat less due to the bulk resistance of the semiconductor material.

Differentiation of Eq. 2.29 with the current held constant yields an interesting result.

$$\frac{\partial V}{\partial T}\bigg|_I = -\frac{(E_g/q) - V}{T} \tag{2.36}$$

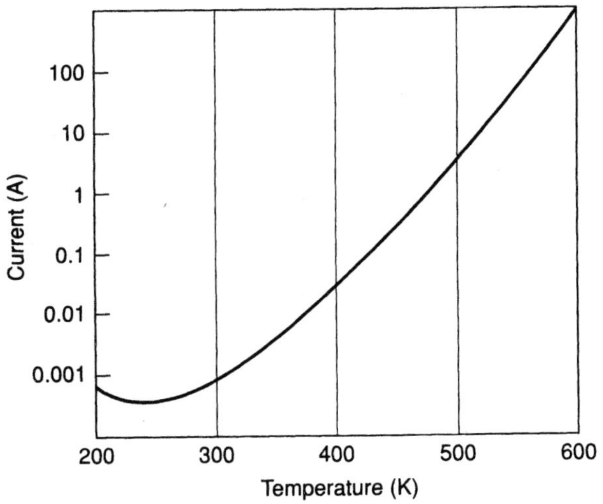

FIGURE 2.28 Current versus temperature for silicon diode.

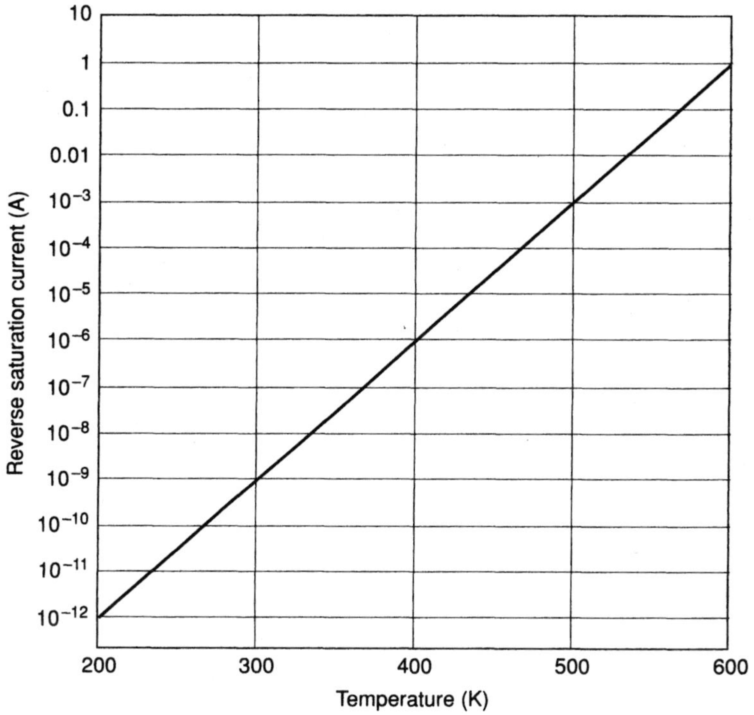

FIGURE 2.29 Diode reverse saturation current versus temperature.

Eq. 2.36 indicates that, under constant current conditions, the voltage decreases linearly with temperature, with a coefficient in the range –1 to –3 mV/K. This feature is often used in temperature-sensing applications.

In summary, the current in a diode increases approximately exponentially with temperature, while the voltage decreases in an approximately linear fashion.

The temperature dependence of so-called Zener diodes is also of significance. Actually, there are two mechanisms which can produce breakdown in reverse-biased diodes. At higher voltages (> ~6 volts), sufficient energy is imparted to a carrier such that it collides with ions to produce other carriers. The result is a rapid increase in current with very little change in voltage and is referred to as the *avalanche effect*. Added heat increases the kinetic energy of the carriers, increasing the number of collisions, and the avalanche effect has a positive temperature coefficient. A second mechanism, the Zener effect, occurs at lower voltages, where the construction of the diode is such that an applied field is sufficient to remove carriers from the covalent bond. Once the voltage is increased above the threshold level, the result is a large current at a virtually constant voltage. The Zener effect has a negative temperature coefficient. In diodes constructed to have a breakdown voltage of approximately 6 volts, both the Zener effect and the avalanche effect are in play and the temperature coefficient is nearly zero. The temperature coefficient of avalanche diodes can be partly overcome by placing a forward-biased diode in series. From Eq. 2.36, the negative temperature coefficient compensates in large part for the positive temperature coefficient of the avalanche effect, and reference diodes can be manufactured which have an extremely low temperature coefficient.

2.3.2 Bipolar Transistors

Bipolar transistors (BPT) consist of two junctions and three terminals as shown in Fig. 2.30. The emitter (E) is the most heavily doped region, followed by the base (B) and the collector (C). There are two basic types of BPT: npn and pnp, depending on the structure. The characteristics of the BPT depend on the biasing conditions of the E-B junction and the C-B junction. There are four possible operating states as shown in Table 2.15.

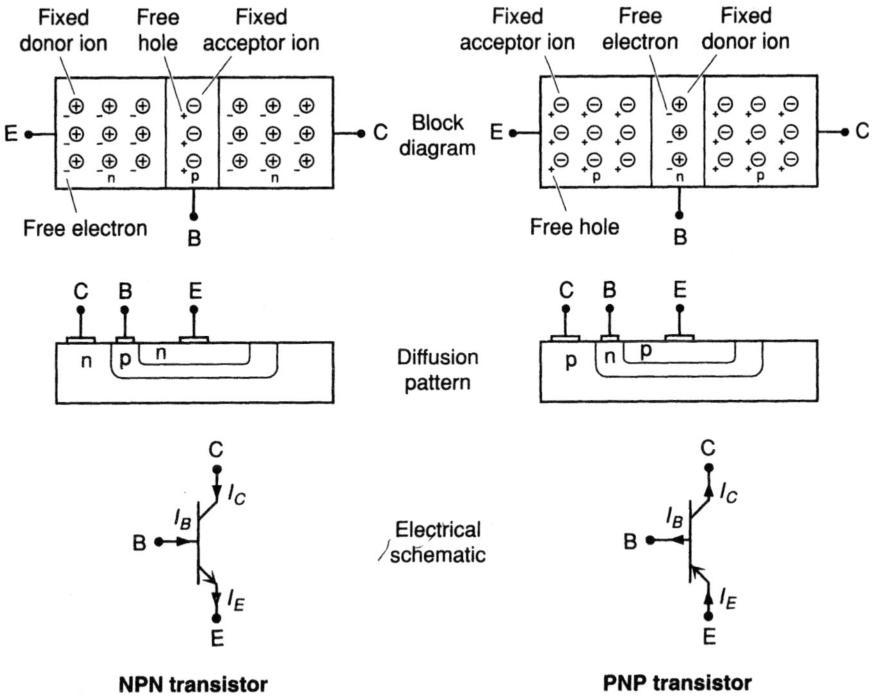

FIGURE 2.30 NPN and PNP transistors.

In the active region, the E-B junction is forward-biased, resulting in an excess of minority carriers in the base region, most of which diffuse across to the collector region where they are attracted by the relatively high collector potential. A few of these carriers recombine with the opposite carrier in the base region, resulting in a much smaller base current and an effective current gain. Most amplifiers are biased in the linear region.

When the BPT is operating in the cutoff mode, both junctions are reverse-biased, allowing only the respective reverse saturation currents to flow. This mode is often used in digital circuits and represents a *high* or a *1*.

TABLE 2.15 Operating States of a Bipolar Transistor

Operating state	Emitter-base junction	Collector-base junction
Active	Forward biased	Reverse biased
Cutoff	Reverse biased	Reverse biased
Saturation	Forward biased	Forward biased
Inverted	Reverse biased	Forward biased

By contrast, the saturation mode represents a *low* or *0*. With both junctions forward-biased, the collector voltage is near zero, while the current is relatively high.

In the inverted mode, with the E-B junction reverse-biased and the C-B junction forward-biased, the collector acts as the emitter and the emitter acts as the collector. The current gain in this mode is very low compared to the active mode, as the relative doping levels make the structure less efficient in generating and transporting carriers.

Applying Kirchhoff's current law to the BPT in Fig. 2.30,

$$I_E = I_C + I_B \tag{2.37}$$

A generalized equation describing the collector current as a function of collector voltage can be derived by applying the principles of conduction and recombination across a junction to Eq. 2.37.

$$I_C = -\alpha I_E + I_{CO}(1 - e^{qV_{CB}/kT}) \tag{2.38}$$

where I_C = collector current
α = current gain, emitter to collector (Eq. 2.39)
I_E = emitter current
I_{CO} = reverse saturation current, base to collector
V_{CB} = collector-base voltage

The term α, as defined in Eq. 2.39, is the so-called common base current gain, [5] as depicted by the configuration in Fig. 2.31, and is always less than one. Typical figures are in the 0.9 to 0.99 range.

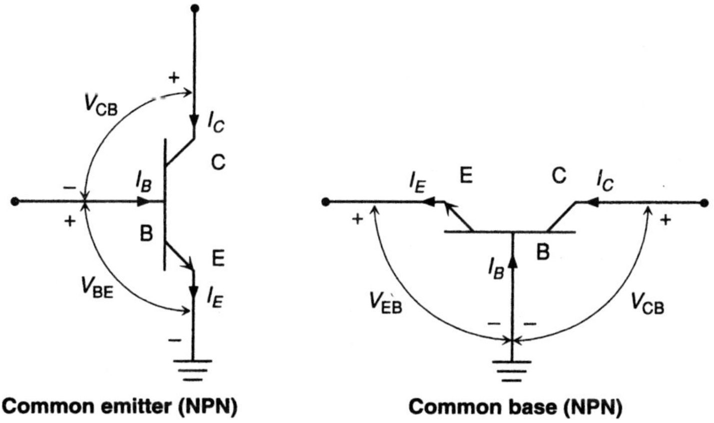

FIGURE 2.31 Common emitter and common base configurations.

$$\alpha \equiv \frac{I_C - I_{CO}}{I_E} \tag{2.39}$$

In the common emitter configuration, the collector current may be defined as

$$I_C = \frac{I_{CO}}{1-\alpha} + \frac{\alpha I_B}{1-\alpha} \tag{2.40}$$

The term β, the common emitter current gain, is defined by

$$\beta \equiv \frac{\alpha}{1-\alpha} \tag{2.41}$$

and Eq. 2.40 may be written as

$$I_C = (1+\beta)I_{CO} + \beta I_B \tag{2.42}$$

It is difficult to obtain an analytical expression for the current as a function of temperature from Eqs. 2.38, 2.40, or 2.42, as I_{CO} and α are both functions of temperature, except to note that α (and therefore β) increases with temperature. Figure 2.32 illustrates the typical change in β with temperature for a small-signal transistor.

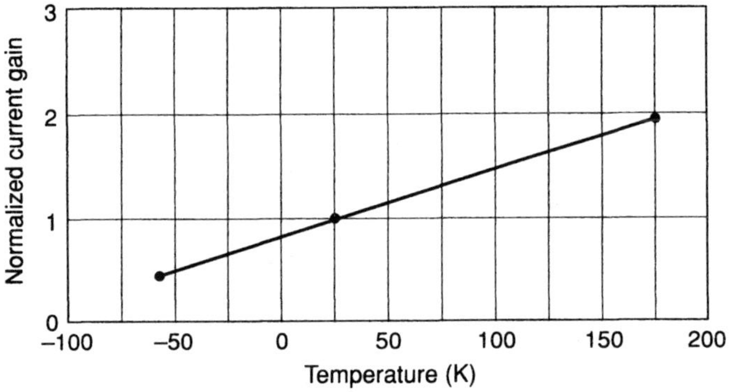

FIGURE 2.32 Normalized current gain versus temperature for a small-signal transistor.

Equation 2.40 has considerable ramifications at elevated temperature. Under normal circumstances, $I_B >> I_{CO}$ and the first term in Eq. 2.40 is negligible. As the temperature increases, both I_{CO} and α increase and this term becomes more significant, and several features which are detrimental to circuit performance can occur.

One of these is an instability in the operating point in the circuit. As the temperature increases, the collector current increases, which increases the voltage drop across the load resistor, changing the collector potential. The accompanying increase in I_{CO} may also affect the potential across the resistors which determine the base current.

In the extreme, elevated temperatures may produce a phenomenon called *thermal runaway,* in which the increased heat causes the collector current to increase which, in turn, causes the temperature to further increase, raising the current still higher. The end result is failure of the transistor as a result of excessive heat, excessive current, or both.

Bias shifting and thermal runaway may both be minimized by placing a resistor in the emitter circuit. As the current increases, the voltage across the emitter resistor increases, lowering the effective base-emitter voltage, which tends to lower the current. The emitter resistor thus provides negative feedback to limit the current.

In power applications, it is common to operate transistors in parallel to increase the available current. If the two transistors are not perfectly matched, one will tend to carry more current than the other, producing a rise in temperature, which further increases the imbalance. Ultimately, one transistor will be carrying the bulk of the current and failure will result. To operate bipolar transistors in parallel, negative feedback is required, usually in the form of fractional ohm resistors in the emitter circuitry. The disadvantage of this approach is that the resistors limit the available gain and dissipate a significant amount of power, lowering the overall efficiency of the amplifier.

In summary, elevated temperature has a profound effect on the operation of bipolar transistors. For maximum circuit performance and reliability, improved thermal management and design procedures must be implemented.

2.3.3 Field Effect Transistors

Bipolar transistors may be said to be current-driven devices while field effect transistors (FETs) are considered voltage-driven devices. An applied voltage is used to either repel or attract majority carriers, which alters the conductivity of the channel which carries the current. Like BPTs, FETs are three-terminal devices. The gate (G), controls the field between the source (S) and the drain (D), as shown in Figs. 2.33 and 2.35. Although the FET is symmetrical to a degree, it is designed to be more efficient when the source is at a negative potential compared to the drain, much as a BPT is more efficient in the active mode than in the inverted mode. The analogous terminals between the FET and the BPT are shown in Table 2.16.

There are two basic types of FETs: junction FETs (JFETs), which rely on the field created by a reverse-biased junction, and metal-oxide-silicon FETs (MOSFETs), which are constructed by depositing a metal gate on a layer of silicon dioxide grown between the drain and source.

JFETs may also be divided into two categories: n-channel and p-channel. Figure 2.33 represents an n-channel JFET in which a p-type diffusion is made into n-type

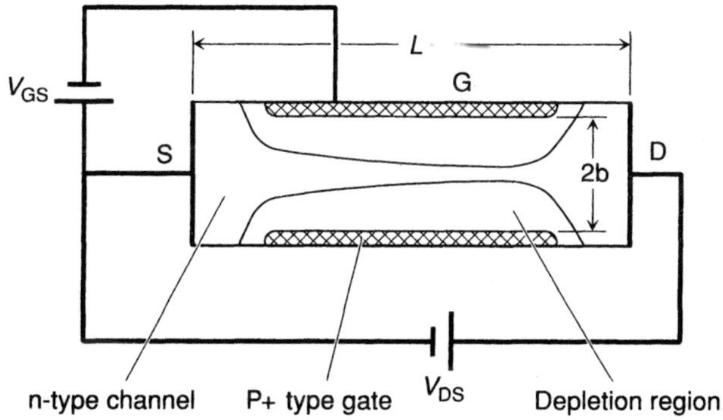

FIGURE 2.33 Cross section of n-channel FET.

TABLE 2.16 Analogous Terminals Between a Bipolar Transistor and a Field Effect Transistor

Device	
BPT	FET
B	G
C	D
E	S

silicon such that it can influence the electric field throughout the device. By applying a voltage to the gate which is negative with respect to the source, the depletion region can be made to extend deeper into the channel, limiting the current. Typical characteristics of a small-signal JFET are depicted in Fig. 2.34. The characteristics in Fig. 2.34 may be divided into four regions: the ohmic region, the saturation region, the breakdown region, and the cutoff region.

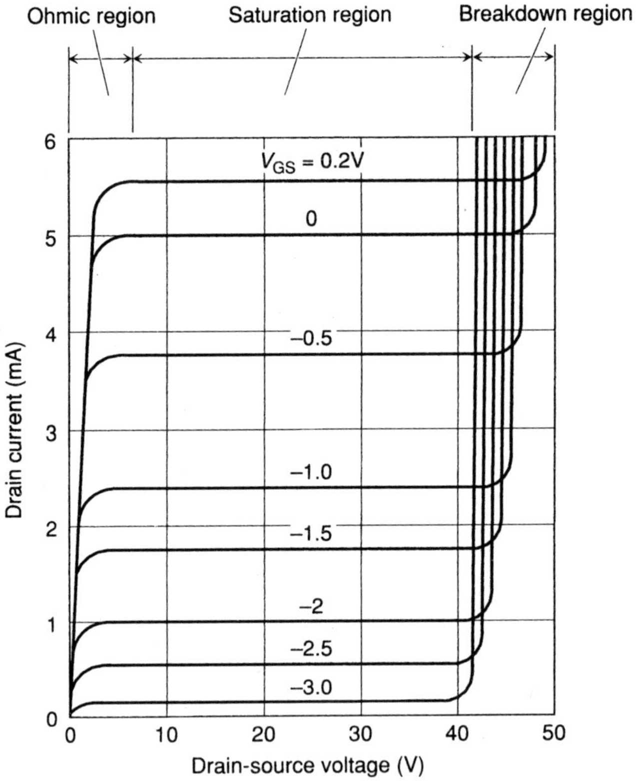

FIGURE 2.34 Typical output characteristics of a small-signal n-channel JFET.

The ohmic region is prevalent at small current levels where the induced field has no significant effect on the channel width. In this region, the current may be simply calculated by Ohm's law as shown in Eq. 2.43. [5]

$$I_D = 2bwqN_D\mu_n \frac{V_{DS}}{L} \tag{2.43}$$

where b = channel height
w = channel width
q = electron charge
N_D = donor concentration
μ_n = electron mobility
V_{DS} = drain-source voltage
L = channel length

The only temperature-sensitive term in Eq. 2.43 is the mobility μ_n, which, as shown in Eq. 2.6, varies with temperature by $T^{-3/2}$. The drain current, therefore, varies by approximately a like amount. Unlike the BPT, the JFET current decreases with temperature, providing a natural source of negative feedback. A parameter $r_{DS(ON)}$ can be defined which can be used to characterize both JFETs and MOSFETs.

$$r_{DS(ON)} = \frac{V_{DS}}{I_D} = \frac{L}{2bwqN_D\mu_n} \tag{2.44}$$

Values of $r_{DS(ON)}$ may vary from a few ohms for small signal devices down to a few milliohms for power MOSFETs. This parameter increases with temperature due to the decrease in mobility.

As the gate voltage increases, the channel width decreases to the point where it becomes a constant, which defines the saturation region. Here, the current varies little over a wide range of V_{DS}.

In the breakdown region, the voltage exceeds the avalanche breakdown voltage of the gate junction, and the current increases dramatically. In cutoff, the gate voltage is increased to the point where the drain current is at minimum, referred to as $I_{D(OFF)}$, usually in the range of a few ma.

There are four types of MOSFETs, depending on the structure, as shown in Table 2.17. The basic construction of an n-channel enhancement MOSFET is shown in Fig. 2.35. In normal operation, the substrate electrode B is connected to the source. As the gate voltage is made more positive with respect to the source, the majority carriers, holes, in this case, are repelled, leaving an excess of electrons. As the gate voltage is increased above a threshold voltage V_T the channel becomes capable of maintaining a drain current. The characteristics of a MOSFET are similar to those shown in Fig. 2.34, except that the transition from the ohmic to the saturation region is more gradual, following a square-law relationship.

TABLE 2.17 Types of MOSFETs

Channel	Mode
n-type	Enhancement
n-type	Depletion
p-type	Enhancement
p-type	Depletion

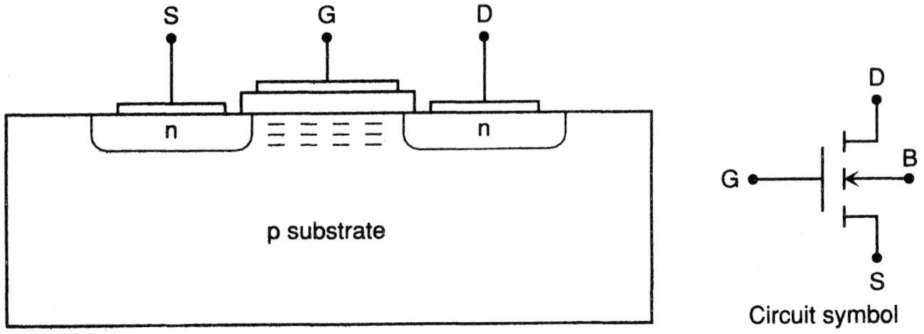

FIGURE 2.35 Basic construction of n-channel MOSFET.

In the ohmic region, the drain current is given by

$$I_D = \frac{\mu C_0 w}{2L}\,[2(V_{GS} - V_T)V_{DS} - V_{DS}^2] \tag{2.45}$$

where C_0 = gate capacitance/unit area

The transition region is defined at the point where $V_{GS} - V_{DS} = V_T$ and Eq. 2.45 becomes

$$I_D = \frac{\mu C_0 w}{2L}\,V_{DS}^2 \tag{2.46}$$

In the saturation region,

$$I_D = \frac{\mu C_0 w}{2L}\,(V_{GS} - V_T)^2 \tag{2.47}$$

which is independent of V_{DS}.

Equations 2.45, 2.46, and 2.47 share one common trait: the primary temperature dependence is the mobility μ, which decreases with temperature. MOSFETs, therefore, like JFETs, exhibit a decrease in drain current as the temperature increases, making them inherently more stable than BPTs with no danger of thermal runaway. Correspondingly, power MOSFETs may be operated in parallel in any number without current-sharing issues.

2.4 RELIABILITY CONSIDERATIONS

2.4.1 Reliability and Failure Mechanisms

The *reliability* of a system is the ability of that system to meet the required specifications for a given period of time. The reliability is measured in a number of ways, including the failure rate, expressed as a probability distribution function $f(t)$, and the mean time to failure (MTTF), which represents the length of time the system is expected to operate until the first failure. The MTTF is related to the failure rate by the expression in Eq. 2.48.

$$\text{MTTF} = \bar{t} \int_0^\infty t f(t)\,dt \tag{2.48}$$

The failure rate of an electronics circuit typically varies with time as shown in Fig. 2.36. During the initial phase, referred to as the *infant mortality* phase, the failure rate is relatively high and is decreasing, as inherent defects in the devices and the manufacturing processes surface. This is followed by a phase in which the failure rate is somewhat lower and is relatively constant, referred to as the *operating life* of the product, which in turn is followed by a phase characterized by an increasing failure rate, referred to as the *end-of-life* phase.

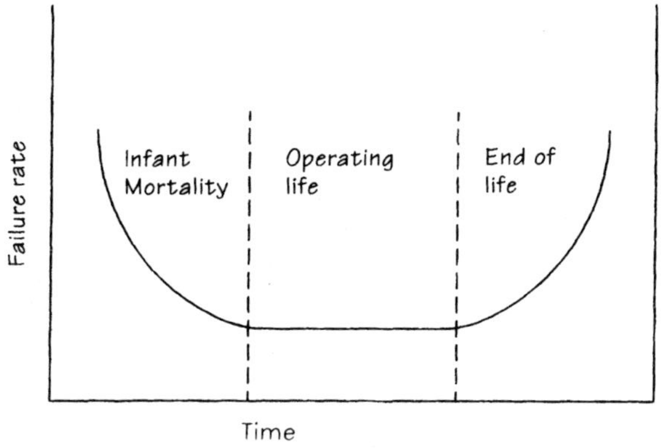

FIGURE 2.36 Failure rate versus time.

When $f(t)$ in Eq. 2.48 is a constant t_f as in the operating life phase, it can be shown that the MTTF is simply the reciprocal of the failure rate.

$$\text{MTTF} = \frac{1}{t_f} \tag{2.49}$$

This is not the case, however, where the failure varies with time, as in the infant mortality and end-of-life phases.

There are a number of mechanisms which can cause failure in a hybrid circuit. The majority of these can be divided into four categories.

1. *Failures due to mechanical interconnections.* This includes failures due to wire bonds, die bonds, lead frame bonds, TAB bonds, and other bonding schemes. These may be due to inadequate process controls, leading to bonds which are not optimally formed, or to selection of materials which are not compatible with each other or with the process.
2. *Failures due to chemical reactions.* This includes failures such as corrosion or the formation of intermetallic compounds, which can be manifested as a mechanical failure.
3. *Failures due to inherent manufacturing defects in active devices.* This includes defects due to pinholes in the insulating oxide, defects or impurities in the body of the semiconductor, or mask defects.
4. *Failures due to electrical overstress.* Failures in this category can be created either by overstress during operation or test or by exposure to electrostatic discharge (ESD).

This section describes the methods by which reliability can be predicted, methods of accelerating the failure rate (screening), and designing for reliability.

2.4.2 Failure-Rate Prediction

Many of the chemical reactions involved in the processes used to fabricate hybrid microcircuits continue to occur throughout the life of the circuit and will eventually lead to failure if the circuit operates long enough. One example of this is the gold-aluminum intermetallic compounds that form when a gold wire is thermosonically bonded to an aluminum pad on an active device. The aluminum continues to diffuse into the gold throughout the life of the device and will eventually cause failure. However, under normal operating conditions, the diffusion rate is so low that the circuit will probably become obsolete before failure occurs! The thermodynamic theory of chemical reactions states that many reactions that are dependent on time, current, and temperature follow the relationship given in Eq. 2.50, the so-called Arrhenius equation.

$$R(i, t, T) = R(i)R(t)e^{(-E_a)/kT} \tag{2.50}$$

where $R(i, t, T)$ = reaction rate of the chemical reaction
i = electrical current in amperes
t = time in seconds
T = temperature in K
k = Boltzmann's constant = 8.625×10^{-5} eV/K
E_a = activation energy in eV

Some reactions, particularly those involving corrosion and/or migration, are also accelerated by the presence of electric current. The electric current can transport active ions to the vicinity of other oppositely charged ions, causing a reaction to occur.

The rate of the reaction with respect to time depends on the amount of material present and its ease of movement. In a liquid, which can be kept constantly agitated and where there is a vast amount of material present, the reaction rate with respect to time is essentially constant. In a solid where a limited amount of material is present and where agitation is impossible, the rate will decrease with respect to time. A rather simplistic explanation for these phenomena is that the probability of a reaction occurring between molecules is proportional to the probability that the molecules will be in close enough proximity for the reaction to occur. For a fixed amount of material, as a reaction progresses, there is less material left to react and, therefore, the probability of the molecules being in proximity decreases. Analogously, at an increased temperature, the molecules are more active and are more likely to come in proximity to other molecules and react with them.

For most cases, the dependence on temperature is substantially greater than either the dependence on current or time, and Eq. 2.51 may be reduced to the following relationship:

$$R(T) = R_0e^{(-E_a)/kT} \tag{2.51}$$

where R_0 is a constant, which depends on a number of variables, including the amount of material available and the physical configuration. R_0 is a difficult parameter to define analytically for a number of practical cases, and Eq. 2.51 is generally used to extrapolate data points obtained for a specific situation at a specific temperature to another temperature.

The temperature in Eq. 2.51 is the temperature at the point of reaction. In an electronics circuit, the temperature may vary widely from one point to another during normal operation, and a thermal analysis must be performed to determine the temperature at a given point.

If the cumulative percentage of failures is plotted against time on a log-log scale, most failure mechanisms common to hybrid circuits plot as a straight line as shown in Fig. 2.37. This is indicative of the so-called normal or bell-shaped distribution of failures. An S-shaped curve indicates that the failure rate is initially higher and that the sample is prone to infant mortality. If a plot is made at a second temperature, and if the slope is the same, the Arrhenius equation, Eq. 2.50, applies and the following relationships also apply.

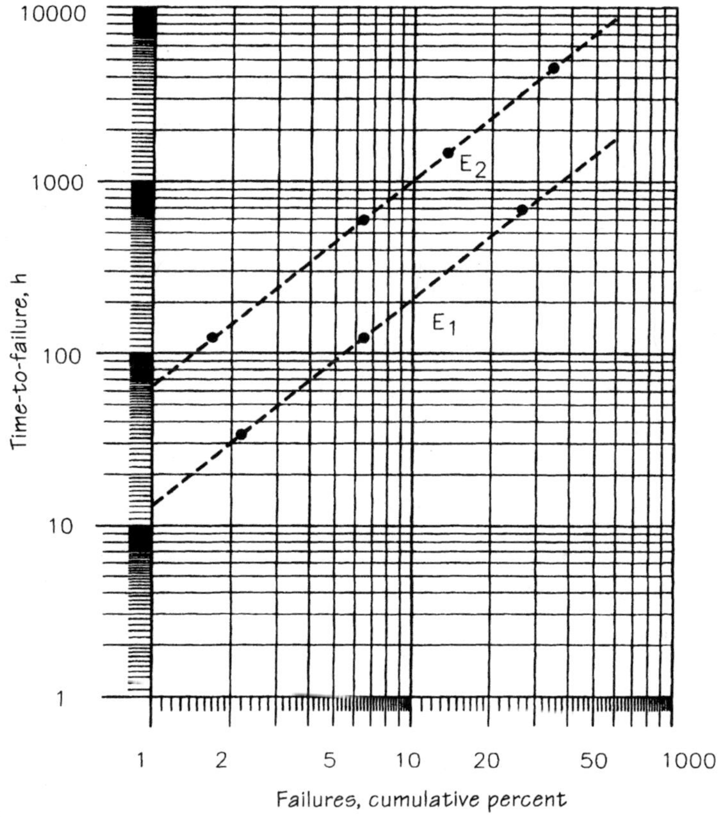

FIGURE 2.37 Cumulative percent failures versus time for two different activation energies.

The acceleration equation compares the rate of reaction at different temperatures.

$$\frac{r_1}{r_2} = e^{(E_a/k)[(1/T_2) - (1/T_1)]} \tag{2.52}$$

Once the acceleration rate is known, the activation energy may be calculated.

$$E_a = k \ln \frac{r_1/r_2}{1/T_2 - 1/T_1} \tag{2.53}$$

If the time to failure t_1 is known for a temperature T_1, the time to failure t_2 for a temperature T_2 is given by

$$t_2 = t_1 e^{[(-E_a)/k][(1/T_2) - (1/T_1)]} \tag{2.54}$$

The same basic form applies for the failure rate.

$$F_2 = F_1 e^{[(-E_a)/k][(1/T_2) - (1/T_1)]} \tag{2.55}$$

where F is the failure rate at a given temperature.

Typical values of the activation energy for common failure modes of semiconductor devices in hybrid circuits are shown in Table 2.18. [6]

Failure modes with activation energies in the range of 0.30 eV are characteristic for failures of the infant mortality type and are only slightly affected by temperature, while those in the 1.50-eV range are characteristic of end-of-life phenomena and are greatly accelerated by temperature. As a rule of thumb, the lower the activation

TABLE 2.18 Activation Energies for Hybrid Circuit Failure Modes

Device type or test condition	Activation energy, eV
Transistors	
Germanium	0.88–1.24
Silicon	
Surface inversion failures	1.02
Au-Al bond failures	1.02–1.04
Metal penetration into silicon	1.65
p n p n	1.65
Diodes	
Four-layer	1.41
Varactors	2.31–2.38
Others	1.13–2.77
Integrated circuits	
Oxide defects	0.30
Silicon defects	0.30
Mask defects	0.50
Ball bond lifts*	0.35–0.44
Stitch bond lifts*	0.80–1.10
Electromigration	1.00–1.10
Contamination	1.00–1.40
Surface charge	0.50–1.00
Charge injection	1.30
Electrolytic corrosion	0.30–0.70
MIL-STD-883	
Method 1015.3 burn-in	0.44
Method 1008.1 high-temperature storage	1.00
Method 1005.3 steady-state life	1.00

* Dependent on cause of failure mode.

energy, the higher will be the failure rate. If the Arrhenius equation applies, a plot of time to failure on a log scale against temperature on a reciprocal scale is a straight line as shown in Fig. 2.37. From observation of Table 2.18, it is apparent that the range of activation energies is quite broad. This factor makes it difficult to extrapolate Eq. 2.55 accurately across a range of temperatures. Referring to Fig. 2.38, each of the three different failure mechanisms dominates at a given temperature. To accurately predict the failure rate at a different temperature, the nature of the failures must be known. Each failure mode can then be individually predicted and the net failure rate is the sum of the individual failure rates.

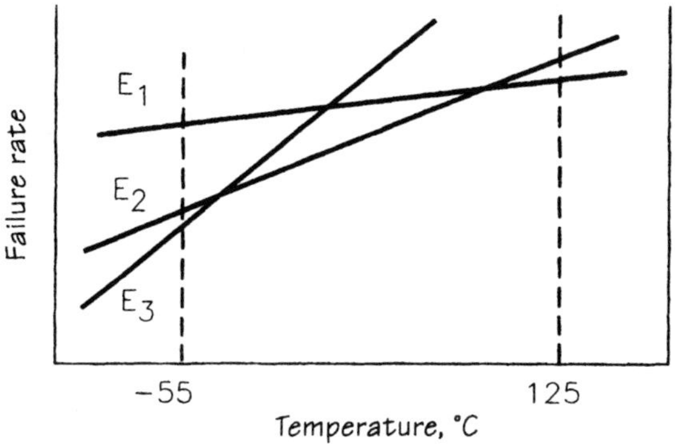

FIGURE 2.38 Failure rate versus temperature for three different activation energies.

For example, assume that a sample of 500 active devices is subjected to a powered life test for 1000 hours at 125°C and that 32 fail. The net failure rate is, therefore, 6.4 percent/1000 hr.

Referring to Table 2.18, the activation energy assumed in Method 1005.3 of MIL-STD-883 is 1.0 eV. This value may be used in Eq. 2.55 to estimate the failure rate at 25°C.

$$F_2 = 6.4\%\ e^{(-1.0/8.625 \times 10^{-3})(1/298 - 1/398)} = 0.000364\%/1000 \text{ hr at } 25°\text{C}$$

Further assume that 20 of the failures were caused by oxide defects, 8 by contamination (assume $E_a = 1.00$ eV for this failure), and 4 by wire-bond lifts (assume $E_a = 0.80$ eV for this failure). In this case, the failure rates for each of the failure modes are

Oxide defects	= 20/500	= 4.0 percent
Contamination	= 8/500	= 1.6 percent
Wire bonds	= 4/500	= 0.8 percent

The failure rate for oxide defects at 25°C is

$$F_2 = 4.0\%\ e^{(-0.30/8.625 \times 10^{-3})(1/298 - 1/398)} = 0.21\%/1000 \text{ hr}$$

The failure rate for contamination at 25°C is

$$F_2 = 1.6\%\ e^{(-0.30/8.625 \times 10^{-3})(1/298 - 1/398)} = 0.00009\%/1000 \text{ hr}$$

The failure rate for wire bonds at 25°C is

$$F_2 = 0.8\%\ e^{(-0.30/8.625 \times 10^{-3})(1/298 - 1/398)} = 0.000321\%/1000 \text{ hr}$$

The net failure rate at 25°C is

$$0.21 + 0.00009 + 0.000321 = 0.210411\%/1000 \text{ hr}$$

This failure rate is substantially higher than that predicted by MIL-STD-883! This is due to the preponderance of failures due to oxide defects that have a low activation energy and that dominate at lower temperatures.

A further consideration is the position of the devices on the failure rate versus time curve shown in Fig. 2.37. A larger sample run for a shorter time may accumulate the same number of device hours as a smaller sample run for a longer time, but will show a substantially higher failure rate since it appears on the infant mortality portion of the curve.

In summary, to obtain accurate failure rates of active devices, it is necessary to obtain actual failure mode data at the actual time of occurrence. Given the extremely high reliability of most semiconductor devices made today, this can be a lengthy and expensive proposition. To accomplish this task, it is necessary either to operate a small number of devices for a very long time or a large number of devices for a shorter time. Translation of failure rate data from one temperature to another must take the activation energy of the actual failure mode into account to avoid serious errors. Failure rates, to be usable, must be expressed in terms of infant mortality, normal operation, and wearout.

2.4.3 Accelerated Testing

In certain applications, such as medical or other industries which require electronic circuits with very high reliability, it is very desirable to remove those units which are prone to infant mortality prior to releasing the remainder of the units to the field. This is usually accomplished by subjecting the circuits to some form of accelerated testing. Accelerated testing is designed to increase the failure rate of circuits by subjecting them to one or more extremes of temperature, humidity, bias, or mechanical stress. In this manner, units with inherent defects will fail earlier and may be removed from further processing.

Before accelerated testing can be effective, the acceleration factor must be known in order to insure that the units are just far enough into the useful life region to insure that all units prone to infant mortality are removed without stressing them so hard that a considerable part of the useful life is also removed. It must also be assured that the stresses placed on the units are not so severe that failure mechanisms are introduced which would not be observed in normal use.

High-Temperature Storage. High-temperature storage is effective in removing those failures which are caused by chemical reaction. These failures include intermetallic formation, diffusion, and corrosion. It can be shown from Eq. 2.52 that the rate of a chemical reaction approximately doubles every time the temperature is increased by 10°C. By this rule of thumb, one hour at 125°C is equivalent to 2^{10} hours, or 1024 hours, at 25°C. Many failures are further accelerated by the presence of bias or moisture. Elevated temperature in combination with an applied potential will accelerate inversion phenomena, particularly in the vicinity of device contacts.

Temperature storage in the presence of moisture will accelerate corrosion, and, with the further addition of electric potential, electromigration.

Temperature Cycling. Temperature cycling, or alternating the unit between hot and cold extremes at a predetermined rate, is an effective method for stressing wire bonds, solder joints, die bonds, and hermetic seals. [7] The alternating hot and cold temperatures act to flex the junctions, promoting the propagation of microcracks or voids which occur as a result of intermetallic compound formation, mismatched TCEs, improper wire-bonding parameters, and similar phenomena. Temperature cycling under temperature is becoming more prevalent, particularly in the automotive industry. At least two failure mechanisms are exhibited in this procedure which are reflective of actual performance conditions. As a circuit is cooled, any moisture present tends to condense on the surface of the hybrid. In ordinary temperature cycling with no bias, this would likely have no effect. In the presence of bias, however, the conditions exist for metal migration. The most stressful condition which a die bond, particularly a power die bond, can be subjected to is to have power applied at cold temperature. The resulting heat generated creates larger temperature gradients and higher stresses between the die and the substrate than exhibited under any other conditions. From observation, a significant percentage of circuit failures occur at the point when power is applied. Typical temperature-cycling schedules are 10 to 50 cycles from –55 to +150°C for use as a screen, and up to 1000 cycles at the same extremes to simulate end of life.

Temperature Shock. Temperature shock is similar to temperature cycling, except that the circuit is transferred abruptly from one temperature extreme to another. This may be performed air to air or liquid to liquid, which is a more severe shock, as heat is transferred much more efficiently in a liquid than in air.

Centrifuge. The centrifuge test subjects the circuits to high-g forces by spinning them rapidly about an axis. This process applies additional stress to bonds which have been subjected to previous temperature cycling.

Burn-In. Burn-in simulates operational conditions by applying potential to a circuit under conditions of elevated temperature. This process may extend from several hours up to 168 hours (1 week). End of life is considered to be 1000 hours.

2.4.4 Derating Factors

A common practice to improve the reliability of electronic circuits is to use devices at lower than the rated limits. MIL-HDBK-217 uses an acceleration factor defined by

$$A = e^{m(p_1 - p_0)} \tag{2.56}$$

where p_1 = percent of maximum rated electrical stress
p_0 = reference percent of rated stress (usually 25 percent)
m = component factor (see expressions in MIL-HDBK-217)

The term p refers to the stress variable of primary interest to the component in question. For capacitors, p is the applied voltage, and for thick film resistors, p may refer to either the applied voltage or dissipated power, depending on which parameter is of most significance in a given application.

TABLE 2.19 Considerations in Designing for Reliability

Design parameter	Comments
TCE matching	Matching of TCEs at die bond interfaces, substrate to package interfaces, and other attachment points will minimize the number of interconnection failures by lowering the mechanical stress on the joint.
Use of derating curves	Using components at lower than the maximum ratings decreases the number of failures due to component breakdown.
Compatible metals	Certain metals in combination form intermetallic compounds which increase the electrical resistance and lower the mechanical strength of the bond. A significant difference in electrochemical potential between two metals can accelerate the corrosion rate and lower the time to failure dramatically.
Conformal coatings	Water in the liquid form is one of the greatest detriments to the reliability of an electronic circuit. Even pure distilled water can cause corrosion by leaching phosphorus from the protective oxide of semiconductor devices, forming phosphoric acid which attacks the aluminum device metallization. Hybrid circuits designed for use in a moist environment should be protected by applying a compatible conformal coat which has a low mobile ion (LMI) content, by applying a suitable encapsulation, or by hermetically sealing the circuit in a metal or ceramic package.

2.4.5 Designing for Reliability

While the acceleration tests described earlier are effective in preventing most potentially bad units from being placed in service, they are no substitute for good design practices, which prevent them from being manufactured in the first place. The hybrid design engineer must design high reliability into the circuits by using design practices which promote high performance. Some of these are described in Table 2.19.

REFERENCES

1. Jerry E. Sergent, and Charles A. Harper, *Handbook of Hybrid Microelectronics,* McGraw-Hill, New York, 1995.
2. "Understanding Chip Capacitors," a publication of Johanson Dielectrics, Inc.
3. Fair-Rite Product Catalog, Fair-Rite Products Corporation, PO Box J, One Commercial Row, Wallkill, NY 12589.
4. P. E. Gray, D. DeWitt, A. R. Boothroyd, and J. F. Gibbons, *Physical Electronics and Circuit Models of Transistors,* John Wiley & Sons, New York, 1964.
5. Jacob Millman, *Microelectronics—Digital and Analog Circuits and Systems,* McGraw-Hill, New York, 1979.
6. Dan Epstein, "Application and Use of Acceleration Factors in Microelectronics Testing," *Solid State Technology,* November 1982.
7. Erwin A. Herr and Alfred Poe, "Transistor High-Reliability Program," *Proc. Industry Application Soc., IEEE,* pp. 833–839, 1977.

CHAPTER 3
THERMAL PROPERTIES OF ELECTRONIC MATERIALS

Thomas C. Evans

3.1 INTRODUCTION

Packaging engineers have more material choices and thermal constraints to consider with each new product they design. Since higher performance drives integrated circuits, active, and passive components to generate more heat, selecting the right materials can be a key element of practicing good thermal management. Along with the thermal evaluation, there are four additional factors that must be considered as part of the final decision on any given material set.

1. Performance, including thermal, must increase.
2. Reliability must be the same, or even increased.
3. Available space for most new products is shrinking.
4. Cost must always be reduced.

Therefore, the design cycle should include a comparison of materials per the preceding criteria *and* a consideration of their overall effect on heat flow through the system.

To start the design of products that have good thermal performance, this chapter serves as a practical guide to material properties. Thermophysical values of thermal conductivity, specific heat, and thermal coefficient of expansion are presented along with density, modulus of elasticity, and tensile strength to provide a quick and easy-to-use reference. Specific data sheets from material suppliers should always be reviewed for comparison.

Many times there is confusion on how to classify a material, with the difference between trade names and generic classifications becoming blurred through misuse or indiscriminate substitution. Figure 3.1 summarizes most of the material types found in electronic products and serves as an overview for the tables of data. [1–3]

3.2 UNITS

Before presenting the actual data, the tabulation of thermal properties should start with an explanation of the most important terms and units. Tables 3.1 and 3.2 help the user interpret some of the typical units and the many different factors associated with them.

These are the standard base units of the SI (Système International d'Unités) system, set up in 1960 at the Eleventh General Conference on Weights and Measures. These units comprise a coherent form of the metric system, since all other units that may be used can be derived from these base units. [2,3]

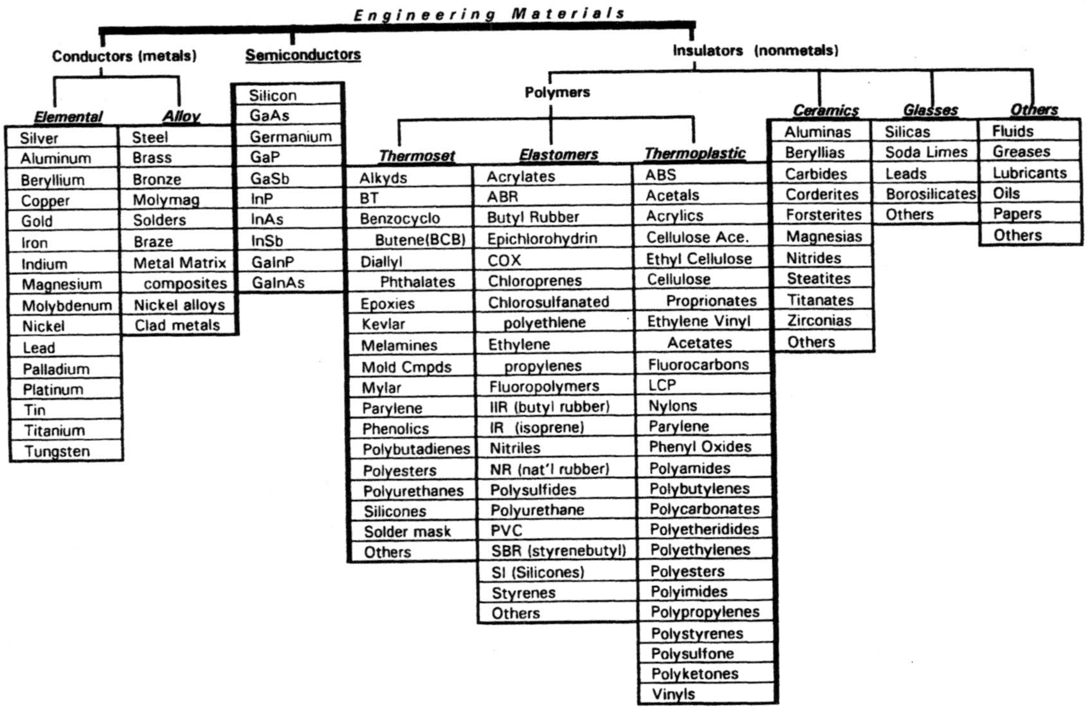

FIGURE 3.1 Electronic packaging materials classification.

TABLE 3.1 Common Units of Measurement

Quantity	Units	Symbol
Length	Meter	m
Mass	Gram	g
Time	Second	s
Temperature	Kelvin	K

TABLE 3.2 Multiplication Factors

Prefix	Symbol	Multiplying factor
Atto	a	10^{-18}
Femto	f	10^{-15}
Pico	p	10^{-12}
Nano	n	10^{-9}
Micro	u	10^{-6}
Milli	m	10^{-3}
Centi	c	10^{-2}
Hecto	h	10^{2}
Kilo	k	10^{3}
Mega	M	10^{6}
Giga	G	10^{9}
Tera	T	10^{12}

For reference, the application of energy through area, distance, and time determines the units listed in Table 3.3.

TABLE 3.3 Energy, Force, and Power Units

Force in newtons	(N)	=	m-kg/s^2
Force over area	(Pa)	= N/m^2	= kg/m-s^2
Energy in joules	(J)	= N-m	= m^2-kg/s^2
Power in watts	(W)	= J/s	= m^2-kg/s^3

One additional note on standard units of temperature measurement is applicable. The thermodynamic scale of temperature is called the Kelvin scale. Absolute zero is 0 K, defined as the absence of all molecular movement. More familiar is the centigrade scale, which has 0°C equal to +237.15 K. Temperature differences are equivalent in either scale.

3.3 THERMAL AND PHYSICAL VALUES

Heat exists as energy in a solid material, with four methods of transfer or conduction. For insulators, phonon generation from the random vibration of atoms in a lattice structure is the largest contributor, but conductors transfer most of their thermal energy from the translational motion of free electrons. Magnetic excitations and, in some cases, electromagnetic radiation can also contribute to heat transfer in solids. [6,7] From the different transfer methods and chemical structure, each material has a unique set of thermal and physical properties. The following are brief descriptions of thermophysical values used to describe common electronic materials.

3.3.1 Thermal Conductivity

Thermal conductivity is defined by the time rate of thermal energy (heat) transfer as it conducts between opposite faces of a cubic volume of material, and the temperature varies by 1 kelvin. Thermal conductivity indicates the efficiency of a material when heat flows from cooler to warmer regions and is defined as

$$q = -k\frac{dT}{dx} \tag{3.1}$$

where q = heat flux in W/m^2
k = thermal conductivity in W/(m-K)
dT/dx = temperature gradient, K/m in steady state

The negative sign denotes that heat flows from areas of higher temperature to areas of lower temperature, with lower thermal conductivity values indicating poor thermal performance. [6] It should be noted that thermal resistance is the inverse of thermal conductivity, and lower values of thermal resistance indicate a better thermal path.

Experimental study of the thermal conductivity of solids was started by Benjamin Franklin in the mid-eighteenth century. In 1753, he pointed out the different abilities of different materials "to receive and convey away the heat." Exactly 100 years later, Weidemann and Franz used a steady-state test method originally suggested by Franklin to determine the relative thermal conductivity of several metals. [7]

Today, measurement of thermal conductivity can be accomplished by direct or indirect methods. ASTM has a direct method C408-88 which uses a copper rod in vacuum, placing the sample in the middle and measuring temperatures along the length as heat is pumped down the rod. E1461-92 is the indirect method using a thermal pulse or flash of heat. [6] By combining the thermal diffusivity values with the density and specific heat capacity, the thermal conductivity can be calculated as

$$\lambda = \alpha C_p \rho \tag{3.2}$$

where λ = thermal conductivity in W/(m-K)
α = thermal diffusivity in m^2/s
C_p = specific heat capacity in (W – s)/(g – K)
ρ = density g/m^3

3.3.2 Thermal Capacity and Specific Heat

Materials can absorb energy in the form of heat and release it. *Thermal capacity* is defined as the amount of thermal energy required to raise the temperature of one mole of material by 1 kelvin. One mole of material is equal to 6.023×10^{23} atoms or molecules.

$$C = \frac{dQ}{dt} \tag{3.3}$$

where C = heat capacity in W-s/mol-K
Q = energy in W-s
T = temperature in K

A more common expression is the *specific heat,* which is the thermal capacity divided by the mass and has units of W-s/g-K. Specific heat can be measured at constant volume or at constant pressure, but for the temperatures and solid materials of interest, these values are practically equivalent.

For metals, semiconductors, and ceramics, specific heat values over temperature change gradually, but the heat capacities of plastic polymers exhibit an incremental increase at the glass transition temperature T_g. The T_g of plastic polymers represents a threshold below which the plastic is hard and rigid, and above this temperature, the plastic becomes flexible and rubbery.

Differential calorimetry in accordance with ASTM standard test methods E1296-90 and E968-63 determines the specific heat by comparing thermal energy content of an empty sample container, a sapphire crystal (standard), and the actual sample. [6]

3.3.3 Thermal Coefficient of Expansion (TCE)

TCE is a result of asymmetrical changes in the interatomic spacing of atoms due to increased energy content in the form of heat. Most metals and ceramics exhibit a linear, isotropic relationship in the temperature range of interest, while plastics are generally anisotropic in nature due to behaviors above and below T_g. TCE values for plastic polymers are usually several times larger above the T_g than below it. The temperature coefficient of expansion is defined as

$$\alpha = \frac{L(T_2) - L(T_2)}{L(T_1)(T_2 - T_1)} \tag{3.4}$$

where α = temperature coefficient of expansion, 1/°C
T_1 = initial temperature
T_2 = final temperature
$L(T_1)$ = length at initial temperature
$L(T_2)$ = length at final temperature

The term α is often multiplied by 10^6 with corresponding units of parts per million length change per kelvin, ppm/K.

Direct measurement at room temperature and elevated temperature (per ASTM standard test method C372-88) has the sample accurately measured on a thermally conductive plate at room temperature, then measured again at a higher temperature after thermal equilibrium has been reached. [6]

From a historical perspective, material expansion and contractions have been observed and even feared for quite some time. "From ghoulies and ghosties, and long leggety beasties and things that go bump in the night, Good Lord deliver us" is an old Scottish prayer. One can imagine hearing strange noises in the middle of the night, as cool night air in Scotland caused homes and furnishings to contract.

The first serious attempt to measure dimensional change over temperature was performed in 1730, to obtain more accurate clocks. As a professor of astronomy, Petrus van Musschenbrock of Utrecht determined that pendulums made of iron gave the best results. Clocks using brass, copper, or lead pendulums had inaccurate timekeeping over the course of even a single day. Investigating further, he found changes in the length of the pendulum from daily temperature variations. [7]

3.3.4 Modulus of Elasticity (*E*)

The modulus of elasticity is the measure of rigidity or stiffness of a material, which can be described as the plastic deformation under load. In more technical terms, E is the ratio of stress below the proportional limit, compared to the corresponding strain, and can be represented by the slope of the initial stress-strain curve. If a tensile stress of 13.8 MPa results in an elongation of 1.0 percent, the modulus of elasticity is 13.8 divided by 0.01, or 1380 MPa. [5,8] E can be represented by

$$\sigma = Ee \tag{3.5}$$

where σ = stress in psi or N/m^2
E = modulus of elasticity in psi or N/m^2
e = strain, in m/m, the net elongation (contraction/unit length)

The units are generally listed as GPa.

3.3.5 Tensile Strength

Tensile strength is the amount of stress applied to stretch a material to its breaking point. In tensile testing, the ratio of maximum load to original cross-sectional area is also called the *ultimate strength.* [5,8] Units are listed as MPa. Both the tensile strength and E depend upon the purity and processing of seemingly equivalent materials.

It is beyond the scope of this book to detail all the important factors having an effect on thermal and physical properties. For instance, two metal samples can have identical constituents but different tempers which affect mechanical strength. Temper is determined by the annealing (heat and cooling rates) of the original material. Looking at polymers, samples may have equivalent molecular weights, but slightly different types or amounts of crosslinking. Crosslinking generally has a large effect on values of mechanical strength and T_g.

3.4 CONVERSION FACTORS

Comparisons between numbers are often complicated by the fact that values have different units to describe them. Conversion factors for various English and mgs (meter/gram/second) units are listed in Table 3.4. [5,9]

3.5 MATERIAL THERMAL AND PHYSICAL VALUES

Many times references differ in the actual values of the physical constants. It might be hoped that test standards and handbooks would better address this issue, but for whatever reason, that does not seem to be the case. Material purity can be another

TABLE 3.4 Units and Conversion Factors

Length, area, and volume
1 inch = 25.4 mm = 0.0254 m
12 inches = 1 foot
1 foot = 305 mm = 0.305 m
1 inch^2 = 645 mm^2 = 0.6452 × 10^{-3} m^2
1 foot^2 = 92,900 mm^2 = 0.0929 mm^2
1 inch^3 = 16,400 mm^3 = 16.4 × 10^{-6} m^3
1 foot^3 = 0.0283 m^3

Mass and density
1 lb (mass) = 0.45359 kg
1 lb/in^3 = 2,768 kg/m^3
1 lb/ft^3 = 16.0185 kg/m^3

Force and stress
1 lb (force) = 4.448 N
1 lb*in^2 (psi) = 6,894.76 N/m^2
1 lb*ft^2 (psf) = 47.9 N/m^2
1 MPa = 10^6 N/m^2 = 1 N/mm^2
1 dyne = 10^{-5} N
1 dyne = 2.248 × 10^{-6} lb

Work, energy, and power
1 inch-lb (force) = 0.113 J = 0.113 Nm
1 ft-lb = 1.365 J
1 BTU = 778 ft-lb = 252 calories = 1,055 J
1 BTU/hr = 0.293 W
1 calorie (cal) = 4.186 J
1 erg = 10^{-7} J
1 ft-lb/s = 1.356 W
1 ft-lb/min = 0.0226 W
1 W-s = 1 J

Thermal conductivity
1 BTU/hr-ft-°F = 1.7303 W/m-K
1 BTU/hr-in-°F = 0.1442 W/m-K
1 W/in-K = 39.3 W/m-K
1 cal/s-cm-K = 418.4 W/m-K

Specific heat
1 cal/g-°C = 4.17 W-s/g-°C
1 BTU/lb-°F = 4.1867 W-s/g-°C)

reason for differences between sources, and it is not always presented with the thermal data. Values reported in this chapter are for 300 K and should be expected to vary between 5 and 10 percent from the true value. [10–12] The properties of metals, ceramics and glasses, polymers, and semiconductors are presented in Tables 3.5, 3.6, 3.7, and 3.8, respectively.

3.6 LAMINATE SUBSTRATES AND PRINTED CIRCUIT BOARDS (PCBS)

It can be fairly difficult to determine the thermal conductivity for a specific design of an organic laminate. Several factors contribute to differences between even fairly similar layouts and must be accounted for when performing a thermal analysis. Some of the same factors can also affect ceramic substrates, but not as much as an organic-based one.

For typical substrates or PCBs, the area covered by copper traces or planes, the thickness of the copper, and the number of vias greatly influence thermal performance. If one area of a board has more copper traces, or more vias, the local thermal conductivity may be much greater. Therefore, in critical applications, the thermal analysis should not treat the substrate/PCB as a bulk material. [13,14] Important material and geometric factors affecting thermal conductivity of a given laminate substrate or PCB follow:

1. Number of metallization layers
2. Area coverage of metallization versus substrate/board size
3. Thickness of metallization

TABLE 3.5 Properties of Metals and Alloys

Metal Lead frames Clad metal Substrates Metallization Packages Joining	Thermal conductivity, W/mK	Specific heat, Ws/gK	Thermal coefficient of expansion, ppm/K	Density, g/cc	Modulus of elasticity, GPa	Tensile strength, MPa
Ag (silver) 99.99%	427	0.23	19.2	10.5	72	125
Ag filled epoxy (varies with filler and metal)	2–2.5		45 (<88 C) 20 (>88 C)		3.5	
Ag/Pd thick film	150		14			
Alloy 42	14.6	0.477	4.5	8.12	148	569
Al (aluminum) 99.99+	237	0.91	23	2.702	69	50–110
Al 1100H18	218					
Al 6063 T6	201					
Al 6061 T0	173		23.4	2.71	70	
Al 6061 T6	156		23.6	2.72		
Al 5052	139					
Al 2024 and 7075	121–130					
Au (gold) 99.99+%	318	0.13	14.3	18.9	80	130
88% Au/12% Ge (hard solder)	88		11	15		233
80% Au/20% Sn (hard solder)	57	0.015	13.9	15	59	198
97% Au/3% Si (hard solder)	94	0.147	12	15.7	83	253
Au/Pt thick film	130		10			
Ba (barium)	18.4					
Be (beryllium)	200	1.88	11.6	1.85	119	
Bi (bismuth)	5.3					
B (boron)	27					
Brass, red (85 Cu/15 Zn)	151	0.38		8.8		
Brass, yellow (65/35)	119	0.38		8.3		

Cd (cadmium)	83–104					
Cr (chromium)	66		6.3			
Co (cobalt)	100					
Cu (copper) 99.99	386	0.38	16.5	8.96	117	115–130
Cu 102 OFHC	391	0.385	17.3	8.94	117	
Cu OLIN 194	263	0.38	16.3	8.8	121	480–519
Cu OLIN 7025	188		16.3	8.82	130	344
Cu/Invar/Cu (5-90-5)	52 (x,y) 16 (z)		2.2	8.25	138	65 44
Cu/Invar/Cu (12.5-75-12.5)	110 (x,y) 17.7 (z)		3.69	8.31	133	
Cu/Invar/Cu (20-60-20)	169 (x,y) 24 (z)	1.76	6.5	8.44	131	37
Cu/Molybdenum/Cu (13-74-13)	198 (x,y) 164 (z)	1.96	5.8	9.83	228	68
Fe (iron) 99.94%	80	0.45	12	7.87	196	180–220
Ga (gallium)	159					
Ge (germanium)	60					
Inconel X-750 (73% Ni/15% Cr/7% Fe)	11.7	0.5		8.5		
Inconel 600	15		11.5			
In (indium)	50	0.23	33	7.31		
Invar (64% Fe/35% Ni)	13.8	0.48	1.33	8.13	150	450–650
Kovar (54% Fe/29% Ni/17% Co)	16.3	0.432	5.3	8.36	138	540
Li (lithium)	85					
Mg (magnesium)	157	1		1.74		
Mn (manganese)	7.8					
Mo (molybdenum) 99.90%	138	0.25	5.2	10.22	325	690–2140
Monel	26					
Monel 500	17		13.7			
Nichrome (80% Ni/20% Cr)	12	0.42		8.4		

TABLE 3.5 Properties of Metals and Alloys (Continued)

Metal Lead frames Clad metal Substrates Metallization Packages Joining	Thermal conductivity, W/mK	Specific heat, Ws/gK	Thermal coefficient of expansion, ppm/K	Density, g/cc	Modulus of elasticity, GPa	Tensile strength, MPa
Ni (nickel) 99.8%	91	0.44	13	8.9	214	345–760
Pb (lead)	35	0.129	29	11.3	14	
95% Pb/5% Sn (soft solder)	32.3	0.134	28	11	7.4	55
92.5% Pb/2.5% Ag/ 5% In (soft solder)	39		25	12		
90% Pb/10% Sn (soft solder)	36		28		19	30
88% Pb/10% Sn/ 2% Ag (soft solder)	37		29			
Pd (palladium)	71	0.24	11.7	12	57.4	
Pt (platinum) 99.99+%	72	0.13	8.9	21.4	147	125
Sb (antimony)	24					
Sn (tin)	65	0.23	21	7.32	41	
60% Sn/40% Pb (soft solder)	50	0.18	24.2	9.29	18–25	43
63% Sn/37% Pb (soft solder)	51	0.18	24.1	9.25		45
Stainless steel						
SS AISI-347	13.8	0.423			190–200	500–800
Carbon steel AISI-1010	64	0.434	11.3	7.861	192	750
Ta (tantalum)	58					
Ti (titanium)	22	0.52	8.6	4.4	110	
W (tungsten)	201	0.13	4.5	19.4	344	1825–4050
80% W/20% Cu	248	0.184	7	15.7	330	5.5
90% W/10% Cu	170	0.16–0.21	6.5	17	330	4.9
Zn (zinc)	116					

TABLE 3.6 Properties of Ceramics, Glasses, and Metal Matrix Composites

Ceramic, glass, and metal matrix composites (MMC)	Thermal conductivity, W/mK	Specific heat, Ws/gK	Thermal coefficient of expansion, ppm/K	Density, g/cc	Modulus of elasticity, GPa	Tensile strength, MPa
AIN	140–220	0.74	4.4	3.26	300	310
Alumina 96%	21	0.88	6.5	3.8	344	172
Alumina 99.5%	37	0.88	6.8	3.9	370	500
Barium Titanate	1.25		8.1	5.5		
BeO 99.5% (very temperature dependent)	250	1.04	7.5	2.9	345	138
BN (hex)	60 12–100 (hot press) 1300 (single crystal)		0.4–3.8	1.9–2.2	43	
Borosilicate glass	1.02		3.2	2.1	63	
Cofired multilayer ceramic	15		6–6.5			380
Cordierite ($2MgO{\cdot}2Al_2O_3{\cdot}5SiO_2$)	2.5		2.5	2.7		300
Diamond; type IIA	2000					
Diamond; type IIB	1300					
Diamond film	100–1200	2		3.5	1000	
Ferrite	6.2					
Forsterite ($2MgO{\cdot}SiO_2$)	3.3		9.8	2.9	96	55
Glass, pyrex	1.02	0.837		2.23		
LTCC	3		5–7	2.9		152
Magnesium titanate	41.8		7.5	3.6		
Mullite ($3Al2O_3{\cdot}2SiO_2$)	5		4	3.1	175	186
SiC aluminum MMC (70% SiC loading)	170		6.3	3	250	300
Graphite aluminum MMC	4.08		5.8	2.5	145	
Boron aluminum MMC (20% loading)	4.65 (x,y) 0.43 (z)		12.7	2.6	186	

TABLE 3.6 Properties of Ceramics, Glasses, and Metal Matrix Composites (Continued)

Ceramic, glass, and metal matrix composites (MMC)	Thermal conductivity, W/mK	Specific heat, Ws/gK	Thermal coefficient of expansion, ppm/K	Density, g/cc	Modulus of elasticity, GPa	Tensile strength, MPa
Quartz	2		1	2.6		
Quartz fiber	1.1	0.23	0.54	2.2	7.4	200
SiC	270	0.675	3.7	3.2	407	28
SiO_2 (vitreous)	0.5–2		0.5	2.2	73	
SiN	27	0.691	2.3			
Alpha Si3N4	80 (single crystal) 10–33 (hot press)		3	3.2	310	
Spinel ($MgO.Al_2O_3$)	15		7.6			
Steatite (MgO-SiO_2)	2.5		8.2	2.7	96	55
Titania (TiO_2)	5		8.3	4		

TABLE 3.7 Properties of Polymer Materials

Polymers Plastics Laminates Glass Weave Coating and Films	Thermal conductivity, W/mK	Specific heat, Ws/gK	Thermal coefficient of expansion, ppm/K	Density, g/cc	Modulus of elasticity, GPa	Tensile strength, MPa
ABS (acrylonitrile-butadiene-styrene)	0.17–0.38	1.26	32–60	1.01–1.07	2.1	21–48
Acetals	0.3	1.5	42–53	1.42		61–68
Acrylics	0.06	1.5	30–60	1.19		34–75
Alkyds	0.7–1.0	1.3	20–50	2		20–65
BCB (benzocyclobutene)			45–70		2.3	82
BT (bismaleimide-triazine) laminate	0.17–0.21		13–15 (x,y) 40–50 (z)	1.3		80–84
Chlorotrifluoroethylene	0.17–0.25		36			41
CR (chloroprene-neoprene)	0.19		34		0.002	25–38
CSM (chlorosulfonated polyethylene)	0.11		27			4–24
DAP (diallyl pthalates)	0.36		27			68
EPDM and EPM (ethylenepropylene polymers)	0.25		83–105			4–24
Epoxies	0.17–1.0	1.7	10–35			27–186
Epoxy fiberglass laminate (FR-4 types)	0.8 (x,y) 0.3 (z)	1.0–1.3	16–20 (x,y) 50–70 (z)	1.8–2.0	17.4	250
Fluoroelastomers	0.1–2.2			1.4–2.2		13
G10 epoxy fiberglass laminate	0.294				15.2	241
E-glass fiber	0.89	0.82	5	2.54	7.4	350
S-glass fiber	0.9		2.8	2.49	8.6	475

TABLE 3.7 Properties of Polymer Materials (Continued)

Polymers Plastics Laminates Glass Weave Coating and Films	Thermal conductivity, W/mK	Specific heat, Ws/gK	Thermal coefficient of expansion, ppm/K	Density, g/cc	Modulus of elasticity, GPa	Tensile strength, MPa
Aramid fiber	0.5	0.26	–5 (along axis) 60 (radial)	1.4	13	400
IIR (butyl rubber)	0.1			0.92		13
IR (isoprene)	0.11			0.92–1.04		17–32
Kapton	0.156	1.086				172
LCP (liquid crystal polymer)			20–43	1.4–1.9		140–300
Mylar	0.19					
Plastic package mold compounds	0.07–0.10	0.7–0.9	17 (<140° C) 68 (>140° C)	1.9	0.017	20–100
NBR (nitriles elast.)	0.24		39			8–24
NR (natural rubber, polyisoprene)	0.14		37		0.001–0.008	16–28
Nylon	0.25	1.7	11–80	1.1	2.76	0.6–2.8
Parylene (vapor deposited)	0.13		35			68
Phenolics	0.1–1.0	1.2–2.0	8–22			21–117
Phenyl oxides	0.19–2.0		10–45			41–124
Polyamides	0.069–0.15	0.067–0.14	12–84			34–179
Polyamide-imides	0.26		30			152–221
Polybutadienes			37			13–16
Polybutylenes	0.18–0.29		60–95			
Polycarbonate	0.19	1.3	67–70	1.2		62–138
Polyesters; thermoplastic	0.19–0.38	0.8–1.9	10–50			55–172
Polyesters; thermoset	0.21–0.87	0.8–1.9	10–60			0.6–60

most metals and by 5 percent or less for insulators. Most materials lose thermal conductivity and gain thermal capacity as the temperature increases. To be safe, it is advisable to check specific manufacturers' data sheets for more accurate thermal values over temperature. There are some noted exceptions to be aware of, such as BeO ceramic.

3.8 MATERIALS FOR CONVECTION COOLING OF ELECTRONIC PRODUCTS

Convection cooling takes place by transferring the heat from a surface to a fluid, which may be air or a liquid such as water. The heat increases the kinetic energy of the fluid and is then carried away as fluid movement increases over the surface. There are two basic types of convection: free convection, in which the motion is generated by the heat source, and forced convection, in which the fluid is forced across the heated surface by an external source. As with thermal resistance, a closed-form solution for the convection coefficient is very difficult to obtain except for a few special geometries.

To understand the mechanics of convection, it is necessary to understand the basics of the relationship between fluid flow and heat transport. There are two fundamental types of fluid flow: laminar flow, which is due primarily to density differences, and turbulent flow, which is created primarily by external means, such as a pump or fan. The distinction between and the nature of these types of fluid flow are determined empirically by the values of a series of dimensionless numbers. The properties of air and water relating to heat transport are tabulated in Tables 3.9 and 3.10, respectively.

The Reynolds number Re is the ratio of the inertial forces to the viscous forces in a coolant flow and is defined as

$$\mathrm{Re} = \frac{\rho v d}{\mu} \tag{3.7}$$

where ρ = density of the fluid in g/cm^3
v = velocity of the fluid in cm/s
d = a critical dimension, such as the edge of the circuit, in inches
μ = viscosity of the fluid in g/(s-cm)

The Reynolds number is a function of temperature since both ρ and μ are functions of temperature. For external flow, as opposed to confined flow as in a pipe, the value of Re classifies turbulent and laminar flow.

Laminar flow	$\mathrm{Re} < 5 \times 10^5$
Turbulent flow	$\mathrm{Re} > 5 \times 10^5$

The Prandtl number Pr is the ratio of fluid and thermal boundary layer thicknesses and is equal to 0.7 for air over a flat plane. The value of the Prandtl number for air is relatively insensitive to temperature. The Prandtl number for water as a function of temperature is given in Table 3.9.

The Nusselt number Nu is defined as

$$\mathrm{Nu} = \frac{hd}{k} \tag{3.8}$$

TABLE 3.9 Thermal Transport Properties of Air

Parameter	Temperature, K	Temperature, °C	Value	
			W/cm-°C	W/in-°C
k	200	−73	0.000181	0.0000712
	300	27	0.000262	0.0001030
	400	127	0.000336	0.0001321
	500	227	0.000404	0.0001590
			W s/g-°C	
c_p	200	−73	1.006	
	300	27	1.006	
	400	127	1.014	
	500	227	1.030	
			kg/m-s	g/in-s
μ	200	−73	13.28×10^{-6}	0.000337
	300	27	18.46×10^{-6}	0.000469
	400	127	22.86×10^{-6}	0.000581
	500	227	26.70×10^{-6}	0.000678
			kg/m³	g/in³
ρ	200	−73	1.769	0.02899
	300	27	1.177	0.01929
	400	127	0.882	0.01445
	500	227	0.706	0.01157
			1/°C	
α			0.00367	

TABLE 3.10 Thermal Transport Properties of Water

Temperature, °F	Temperature, °C	c_p, Btu/lb-ft	k, Btu/h-ft·°F	μ, lb/ft-h	ρ, lb/ft³	α, 1/°F	Pr
32	0	1.029	0.337	4.320	62.54	1.10×10^{-4}	13.20
100	37.8	0.999	0.365	1.620	62.20		4.43
200	93.3	1.004	0.393	0.738	60.20		1.88
		W-s/(g-°C)	W/(m-°C)	kg/(m-s)	kg/m³	1/°C	
32	0	4.305	0.583	0.00866	1001.9	1.98×10^{-4}	13.20
100	37.8	4.180	0.631	0.00325	996.4		4.43
200	93.3	4.201	0.680	0.00148	964.4		1.88

where h = convection constant in W/m²-°C
d = critical dimension in m
k = thermal conductivity in W/m-°C

The Grashof number Gr is the ratio of buoyancy forces to viscous forces and is defined as

$$\mathrm{Gr} = \frac{\rho^2 g \alpha d^3 (T_S - T_F)}{\mu^2} \tag{3.9}$$

where g = acceleration of gravity = 9.78 m²/sec
α = coefficient of expansion of the fluid
T_S = temperature of surface
T_F = temperature of fluid

The Rayleigh number Ra is the product of the Grashof and Prandtl numbers. The dimensionless numbers defined here are related by a series of empirical equations. For free convection,

$$\mathrm{Nu} = c(\mathrm{Gr\ Pr})^n = c(\mathrm{Ra})^n \tag{3.10}$$

where c and n are constants which depend on the value of the Rayleigh number as shown in Table 3.11.

TABLE 3.11 Rayleigh and Nusselt Numbers for Free and Forced Convection

	Free convection		
Orientation	Rayleigh number	c	n
Vertical plate	10^4–10^9	0.59	0.25
	10^9–10^{12}	0.13	0.33
Horizontal plate (heated surface up)	10^4–10^7	0.54	0.25
	10^7–10^{11}	0.15	0.33
	Forced convection		
Flow type	Orientation		Nusselt number
Laminar	Flat plane		$0.664\ \mathrm{Re}^{0.5}\ \mathrm{Pr}^{0.33}$
Turbulent	Flat plane		$0.037\ \mathrm{Re}^{0.5}\ \mathrm{Pr}^{0.33}$

For forced convection,

$$\mathrm{Nu} = c\ \mathrm{Re}^n\ \mathrm{Pr}^m \tag{3.11}$$

where c, n, and m are given in Table 3.11.

For zero to moderate external airflow, heat is transported away from the source by a combination of free and forced convection. The contribution due to free convection is by far the most difficult to calculate analytically, since a trial-and-error solution is required. The Grashof number is a function of the difference between the fluid temperature and the plate temperature. In most cases, the fluid, or ambient, temperature is known and the plate temperature is unknown. To obtain a solution for equations of this type, a value must be estimated for the plate temperature and the system of equations solved for the actual plate temperature based on the estimate. If the estimate is different from the solved value, another estimate is made and the process is repeated until the two converge. The complexity of this process requires a computer routine to obtain an acceptable degree of accuracy. By contrast, the effect of forced convection can be calculated from Eq. 3.11 by determining whether or not the flow is laminar or turbulent from the Reynolds number, and using the appropriate form from Table 3.11.

3.9 INDUSTRY AND MILITARY SPECIFICATIONS

Several professional organizations have material characterization and standardization of testing methods as part of their charter. Some groups deal with the starting materials described in this chapter, and others focus on the finished semiconductor package formed by the integration of many different material types. The following is a list of well-known organizations and some of their specifications addressing thermal performance of electronic materials.

ANSI (American National Standards Institute)

11 W. 42nd St.

New York, NY 10036-8002

Phone: (212) 642-4900

Fax: (212) 398-0023

Internet: www.ansi.org

ANSI has become a focal point for coordinating and ensuring that standards are both needed and will meet with international approval.

ASTM (American Society for Testing and Materials)

100 Barr Harbor Drive

West Conshohocken, PA 19428-2959

Phone: (610) 832-9585

Fax: (610) 832-9555

Internet: www.astm.org

E-mail: webmastr@astm.org

ASTM is a scientific and technical organization for the development of consensus standards on characterization and performance of materials, products, systems, and services, and promotion of related knowledge. Organized in 1898, ASTM has grown into one of the largest voluntary standards development systems in the world.

A370-96	Mechanical Testing of Steel Products
C20-46	Specific Gravity
C108-46	Heat Transmission
C177-85	Steady State Heat Flux Measurements and Thermal Transmission Properties by Means of the Guarded Hot Plate Apparatus
C329-81	Specific Gravity of Fired Ceramic Whiteware Materials
C372-94	Linear Thermal Expansion of Porcelain Enamel and Glaze Frits and Fired Ceramic Whiteware Products by the Dilatometer Method
C373-82	Porosity, Apparent, Water Absorption, Bulk Density and Apparent Specific Gravity of Fired Whiteware Products
C408-88	Thermal Conductivity of Whiteware Ceramics
C539-84	Linear Thermal Expansion of Porcelain Enamel and Glaze Frits and Ceramic Whiteware Materials by the Interferometric Method
C623-92	Young's Modulus, Shear Modulus, and Poisson's Ration for Glass and Glass-Ceramics by Resonance

C674-81	Flexural Strength Properties of Ceramic Whiteware Materials
C1045-90	Calculating Thermal Transmission Properties from Steady-State Heat Flux Measurements
C1198-96	Dynamic Young's Modulus, Shear Modulus, and Poisson's Ratio for Advanced Ceramics by Sonic Resonance
C1259-96	Dynamic Young's Modulus, Shear Modulus, and Poisson's Ratio for Advanced Ceramics by Impulse Excitation of Vibration
C1300-95	Linear Thermal Expansion of Glaze Frits and Ceramic Whiteware Materials by the Interferometric Method
C1341-96	Flexural Properties of Continuous Fiber-Reinforced Advanced Ceramic Composites
D116-86	Vitrified Ceramic Materials for Electrical Applications
D638-96	Tensile Properties of Plastics
D696-91	Coefficient of Linear Thermal Expansion of Plastics Between –30°C and 30°C
D790-90	Test Method for Flexural Properties of Unreinforced and Reinforced Plastics and Electrical Insulating Materials
D792-91	Density and Specific Gravity (Relative Density) of Plastics by Displacement
D1708-96	Tensile Properties of Plastics By Use of Microtensile Specimens
D2343-95	Tensile Properties of Glass Fiber Strands, Yarns, and Rovings used in Reinforced Plastics
D2442-75	Alumina Ceramics for Electrical and Electronic Applications
D2766-95	Specific Heat of Liquids and Solids
D2858-70	Thermal Conductivity of Electrical Grade Magnesium Oxide
D3039-95	Tensile Properties of Polymer Matrix Composite Materials
D3386-94	Coefficient of Linear Thermal Expansion of Electrical Insulating Materials
D3552-77	Tensile Properties of Fiber Reinforced Metal Matrix Composites
D3878-95	High Modulus Reinforcing Fibers and their Composites
D4065-95	Determining and Reporting Dynamic Mechanical Properties of Plastics
D5418-95	Measuring the Dynamic Mechanical Properties of Plastics Using a Dual Cantilever Beam
D5470-95	Thermal Transmission Properties of Thin Thermally Conductive Solid Electrical Insulation Materials
E-228-95	Linear Thermal Expansion of Solid Materials with a Vitreous Silica Dilatometer
E457-72	Measuring Heat-Transfer Rate Using Thermal Capacitance (Slug) Calorimeter
E968-93	Heat Flow Calibration of Differential Scanning Calorimeters
E1142-93	Thermophysical Properties
E1269-95	Determining Specific Heat Capacity by Differential Scanning Calorimetry

E1461-92	Thermal Diffusivity of Solids by the Flash Method
F356-75	Beryllia Ceramics for Electronic and Electrical Applications
F394-84	Biaxial Flexure Strength of Ceramic Substrates
F635-85	Selection of Tests for Material Properties of Liquid Thermoset Encapsulating Compounds for Electronic and Microelectronic Encapsulation
F636-85	Selection of Tests for Material Properties of Transfer Molding Compounds for Electronic and Microelectronic Encapsulation
F1684-96	Iron-Nickel and Iron-Nickel-Cobalt Alloys for Low Thermal Expansion Applications

IPC (Institute for Interconnection and Packaging)

2215 Sanders Rd

Northbrook, IL 60062-6135

Phone: (847) 509-9700

Fax: (847) 509-9798

Internet: www.ipc.org

E-mail: webmaster@ipc.org

For over 38 years, the Institute for Interconnection and Packaging has been a trade association that brings together designers, board manufacturers, assembly companies, suppliers, and original equipment manufacturers. IPC provides standards, training guides, design guidelines, conferences, marketing, and environmental information for the electronic interconnection industry.

IPC-A-142	Specification for Finished Fabric Woven from Aramid for Printed Boards
IPC-CF-152	Composite Metallic Materials Specification for PWBs
IPC-TM-650	Test Methods Manual
IPC-EG-140	Specification for Finished Fabric Woven from 'E' Glass for Printed Boards
IPC-EG-141	Specification for Finished Fabric Woven from 'S' Glass for Printed Boards
IPC-FC-233	Adhesive Materials
IPC-FC-241	Flexible Metal Clad Dielectrics for Use in Fabrication of Flexible Printed Wiring
IPC-L-108	Specification for Thin Metal Clad Base Materials for Multilayer Printed Boards
IPC-L-109	Specification for Resin Impregnated Fabric (Prepreg) for Multilayer Printed Boards
IPC-L-112	Specification for Composite Metal Clad Base Materials for Printed Boards
IPC-L-115	Specification for Rigid Metal Clad Base Materials for Printed Boards
IPC-MC-324	Performance Specification for Metal Core Boards
IPC-MF-150	Metal Foil for Printed Wiring Applications

IPC-QF-143	Specification for Finished Fabric Woven from Quartz (Pure Fused Silica) for Printed Boards
IPC-RB-276	Qualification and Performance Specification for Rigid Printed Boards
IPC-RF-245	Performance Specification for Rigid Flex Printed Boards
IPC-TR-470	Thermal Characteristics of Multilayer Interconnection Boards
IPC-6011	Generic Performance Specification for Printed Boards

SEMI (Semiconductor Equipment and Materials International)

SEMI North America (Headquarters)
805 East Middlefield Road
Mountain View, California 94043-4080 USA
Phone: (415) 964-5111
Fax: (415) 967-5375
Internet: www.semi.org
E-mail: semihq@semi.org

SEMI is an international trade association that represents semiconductor and flat panel display equipment and materials suppliers.

G30-88	Test Method for Junction-to-Case Thermal Resistance Measurements of Ceramic Packages
G46-88	Test Method for Thermal Transient Testing for Die
G43-87	Test Method for Junction-to-Case Thermal Resistance Measurements of Molded Plastic Packages
G32-94	Guideline for Unencapsulated Thermal Test Chip
G42-87	Specification for Thermal Test Board Standardization for Measuring Junction-to-Ambient Thermal Resistance of Semiconductor Packages
G15-93	Standard Test Method for Differential Scanning Calorimetry of Molding Compounds
G4-94	Specification for Integrated Circuit Leadframe Materials Used in the Production of Stamped Leadframes
G22-86	Specification for Ceramic Pin Grid Array Packages
G33-90	Specification for Pressed Ceramic Pin Grid Array Packages
G38-0996	Test Method for Still- and Forced-Air Junction-to-Ambient Thermal Resistance Measurements of Integrated Circuit Packages
G34-89	Specification for Cer-Pack Package Constructions, Including Leadframes, Suitable for Automated Assembly by End Users
G1-96	Specification for Cer-Dip Package Constructions
G61-94	Specification for Cofired Ceramic Packages
G58-94	Specification for Cerquad Package Constructions
G50-89	Specification for Co-fired Ceramic Fine Pitch Leaded and Leadless Chip Carrier Package Constructions
G39-89	Specification for Brazed Lead Flatpack Constructions, Including Leadframes, Suitable for Automated Assembly

G2-94	Specification for Metallic Leadframes for Cer-Dip Packages
G13-88	Standard Test Method for Expansion Characteristics of Molding Compounds

JEDEC (Joint Electron Device Engineering Council)

Electronic Industries Association

2500 Wilson Boulevard

Arlington, VA 22201-3834

Phone: (703) 907-7560

Fax: (703) 907-7501

Internet: www.eia.org

E-mail: laurawal@eia.org

JEDEC is a part of EIA, dedicated to the standardization of electronic devices. Committees focus on particular aspects of electronic products to standardize the nomenclature and testing methods.

Proposed New Measurement Method—"High Effective Thermal Conductivity Test Board for Leaded Surface-Mount Packages"

Proposed New Publication—"Thermal Test Chip Guidelines (Wire Bond Type Chip)"

Proposed New JEDEC Standard—"Low Thermal Conductivity Test Card"

JESD9-A Metal Package Specification for Microelectronic Packages and Covers (April 1987). This standard establishes the general requirements and quality assurance provisions that can be specified and met in procuring microelectronic packages and covers, manufactured from matched seals with and without high thermal conductivity base materials, intended for use in fabricating hybrid microelectronic circuits. This document details those minimum requirements necessary for metal packages' use exclusively.

JESD26-A General Specification for Plastic Encapsulated Microcircuits for Use in Rugged Applications (April 1990). This standard establishes uniform procedures and defines the general requirements for the quality and reliability assurance of plastic encapsulated (noncavity) microcircuits used in ground, fixed, or benign applications. Detail requirements, specific characteristics of microcircuits, and other provisions that are sensitive to the particular use intended shall be specified in the applicable detail specification.

JESD27 Ceramic Package Specification for Microelectronic Packages (August 1993). The intent of this standard is to be a guide in the manufacture and procurement of ceramic packages, especially for the hybrid industry. Manufacturers of ceramic packages and procuring activities for these packages will now be able to use this document as the means for agreement in the imposition of minimum requirements in qualification, screening, and quality conformance.

IEEE (Institute of Electrical and Electronics Engineers)

345 E. 47th St.

New York, NY 10017

Phone: (212) 705-7999

Internet: www.ieee.org

E-mail: webmaster@ieee.org

IEEE is the world's largest technical professional society. It focuses on the advancement of theory and practice of electrical and electronic science through the sponsorship of conferences, symposia, and local meetings. Broad coverage of the industry is found in the many publications available.

Although Mil-Stds are not being kept current, they are still valuable references for many packaging issues.

MIL-STD 883C Method 1012.1 Thermal Characteristics

MIL-P-13949 General Specification for Plastic Sheet, Laminated, Metal Clad (for Printed Wiring)

MIL-P-55110 General Specification for Printed Wiring Boards

MIL-M-55627 Materials for Flexible Printed Wiring Boards

MIL-STD-2118 Flexible and Rigid Flexible Printed Wiring for Electronic Equipment, Design Requirements for

Method 1012.1 was created to determine the junction temperature, package thermal resistance, case and mounting temperature, and the thermal response time of microelectronic devices.

UL, Incorporated (Underwriters Laboratories)

333 Pfingsten Rd.

Northbrooks, IL 60062-2096

Phone: (708) 272-8800

Fax: (708) 509-1724

UL is chartered as a nonprofit organization to establish, maintain, and operate a laboratory to investigate the safety aspects of devices, systems, and materials. UL provides product safety testing, safety certification, product performance verification, product performance certification, and EMI evaluation/testing services for telecommunications equipment and accessories. It accomplishes this objective through many publications, meetings, and conferences.

REFERENCES

1. C. A. Harper, *Electronic Packaging & Interconnection Handbook,* McGraw-Hill, New York, 1991.
2. R. T. Howard, "Thermal Management Concepts in Microelectronic Packaging," *International Society for Hybrid Microelectronics Technical Monograph,* pp. 111–134, 1984.
3. "Physical Constants of IC Package Materials," in *INTEL Packaging Databook,* INTEL literature, Mt. Prospect, Illinois, 1995.
4. F. P. Incropera and D. P. DeWitt, *Fundamentals of Heat and Mass Transfer,* 3d ed., John Wiley & Sons, New York, 1983.
5. M. M. Gauthier, *Engineered Materials Handbook,* Desk Edition, ASM International, Materials Park, Ohio, 1995.
6. J. E. Sergent and C. A. Harper, eds., *Hybrid Microelectronics Handbook,* 2d ed., McGraw-Hill, New York, 1995.
7. Y. S. Touloukian et al., *Thermophysical Properties of Matter,* vols. 1–12, Plenum Publishing, New York, 1970.
8. W. M. Alvino, *Plastics for Electronics,* McGraw-Hill, New York, 1995.

9. J. Carvill, *Mechanical Engineers Data Handbook,* CRC Press, Boca Raton, Florida, 1993.
10. A. D. Kraus and A. Bar-Cohen, *Thermal Analysis and Control of Electronic Equipment,* McGraw-Hill, New York, 1983.
11. A. Bar-Cohen, T. Elperin, and R. Eliasi, "Theta JC Characterization of Chip Packages—Justification, Limitations, and Future," *IEEE Transactions on Components, Hybrids, and Manufacturing Technology,* **12**(4): 724–731, December 1996.
12. R. C. Weast, *CRC Handbook of Chemistry and Physics,* CRC Press, Inc., Boca Raton, FL, 1981.
13. K. Azar and J. E. Graebner, "Experimental Determination of Thermal Conductivity of Printed Wiring Boards," *12th IEEE SEMI-THERM Symposium,* pp. 169–182, May 1996.
14. D. Edwards, "Development of JEDEC Standard Thermal Measurement Test Boards," *12th IEEE SEMI-THERM Symposium,* pp. 183–190, May 1996.
15. "Thermal Considerations for Integrated Circuit Packaging," *Advanced Packaging Applications Manual,* rev 1.0, Symbios Logic, Fort Collins, Colorado, 1996.
16. C. A. Harper and R. M. Sampson, eds., *Electronic Materials & Processes Handbook,* 2d ed., McGraw-Hill, New York, 1996.

CHAPTER 4

HEAT GENERATION IN ELECTRONIC CIRCUITS

Al Krum
Jerry Sergent

4.1 RELATIONSHIP OF HEAT TO POWER

Heat is an unavoidable by-product of every electronic device and circuit. [1] In most instances designers attempt to minimize the generation of heat, as for every 10°C rise in temperature, the reliability of the circuit is decreased by approximately half. In addition, heat can cause parametric changes which may be detrimental to the performance of both active and passive components.

The intent of this chapter is to provide the reader with quantitative methods for determining heat generation in both active and passive components. With this information, the exact temperature rise in each component can be calculated, and the option of using various packaging techniques that can lower the temperature rise can be considered. Another option is to simply reduce the amount of heat generated through circuit and packaging design. Thermal management can be as simple as not generating the heat in the first place. [2]

4.2 POWER DISSIPATION

When current flows through a semiconductor device or through a passive element (except in superconductors), power is dissipated which results in a rise in the device or element temperature. This rise in temperature is a function of the device and element geometry and the thermal path to the heat sink.

4.3 GENERAL THEORY

Under DC excitation, the power dissipated in an active or passive device P can be calculated with the following equation

$$P = V \times \tag{4.1}$$

where P = power in watts or joules/second
I = DC current in amperes through the device
V = DC voltage drop in volts across the device

When the current or voltage are time-varying, periodic, and nonsinusoidal, the mean power loss (P_M) for that device is

$$P_M = \frac{1}{T} \int_{t_1}^{t_2} v(t)i(t)\,dt \tag{4.2}$$

where T = waveform period
$i(t)$ = instantaneous value of current through the element
$v(t)$ = instantaneous value of voltage through the element
t_1 = lower limit of conduction for current
t_2 = upper limit of conduction for current

The instantaneous power $p(t)$ can be graphically calculated for the switching waveforms of voltage and current as shown in Fig. 4.1 by multiplying the voltage and current at the various points of time.

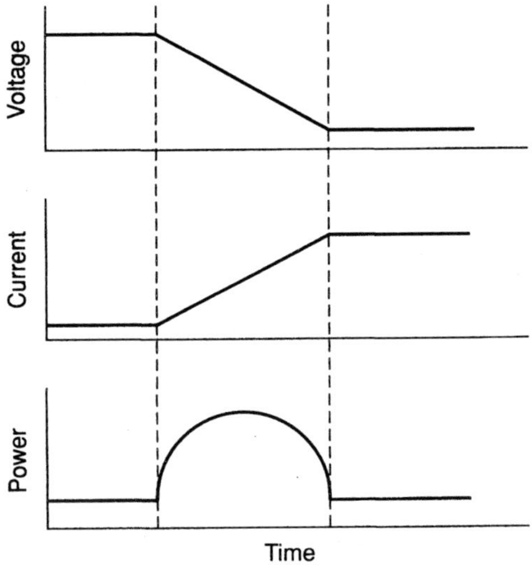

FIGURE 4.1 Graphical calculation of power for time-varying waveforms.

4.4 PASSIVE DEVICES

The family of passive devices consists of resistors, capacitors, inductors, and transformers.

4.4.1 Resistors

Joule's law states that the rate at which heat is produced in an electric circuit of resistance R is equal to the product of the resistance and the square of the current. [3] (See Fig. 4.2 for a schematic of a resistor.) This can be written as

FIGURE 4.2 Resistor schematic symbol.

$$P = I^2R \tag{4.3}$$

where P = steady state power in watts
I = steady state current in amperes
R = resistance in ohms

The instantaneous power $p(t)$ developed across a resistor R with a time-varying current $i(t)$ is

$$p(t) = i(t)^2R \tag{4.4}$$

For a sinusoidal steady state current, where $i(t)$ is described by

$$i(t) = I_M \sin(\omega t) \tag{4.5}$$

the average power (p_{av}) dissipated in a resistor R is

$$P_{AV} = \tfrac{1}{2} I_M^2 R \tag{4.6}$$

where I_M = peak value of sinusoidal current
ω = radian frequency ($2\pi f$)
f = frequency in hertz

A continuous current of 10 amps flowing through a 500-mΩ sense resistor would therefore dissipate

$$P = 10^2(0.5) = 50 \text{ watts}$$

All resistors exhibit some drift in resistance value over temperature. The amount of this drift, characterized as temperature coefficient of resistance (TCR), can range from just one or two parts per million per degree centigrade (ppm/°C) to several hundred. The value of TCR is a first-order function of the resistor material and a second-order function of deposition method and resistor geometry. Listed in Table 4.1 are typical TCRs for various types of microelectronic resistors (those typically found in hybrids and multichip modules).

A typical 1000-ohm thick film resistor fabricated with 1 KΩ/□ ink as shown in Table 4.1 might have a TCR of +50 ppm/°C. With a temperature change of 100°C, this change in resistance will cause a 0.5 percent change in power dissipated. Even the higher TCRs observed at the extreme ends of the resistance range have only a

TABLE 4.1 Resistivities of Microelectronic Resistor Materials

Film type	Film material	Resistivity, ohms/square	TCR, ppm/°C
Thin	Nichrome	50–400	50
Thin	Tantalum nitride	150	–75
Thick	Ruthenium oxide	<30 >100K	150
Thick	Ruthenium oxide	100–30K	50
Bulk foil	Nickel-chrome +*	0.25–1.0	2

* Proprietary composition from Vishay Resistor Products, Malvern, Pennsylvania. See Ref. 4.

1.5 percent change over a 100°C temperature change. Therefore, the temperature effect on power dissipation in resistors can usually be neglected.

4.4.2 Capacitors

The electrical symbol for a capacitor is shown in Fig. 4.3. In the ideal capacitor, no power is dissipated when DC is applied. The physical realization of capacitors has an equivalent circuit as shown in Fig. 4.4 and will dissipate heat under both DC and AC conditions.

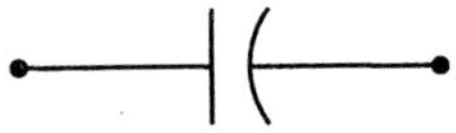

FIGURE 4.3 Capacitor schematic symbol.

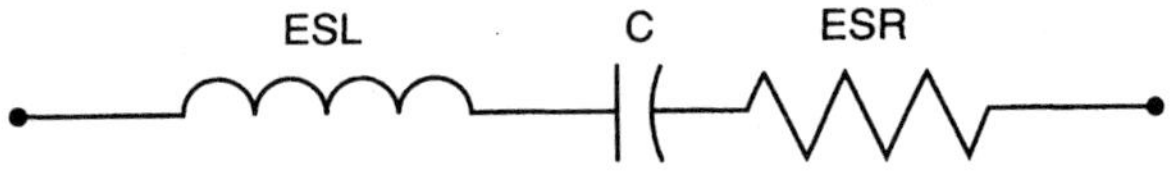

FIGURE 4.4 Capacitor equivalent circuit.

Capacitors under sinusoidal excitation dissipate power according to the formula

$$p(t) = \tfrac{1}{2}\,\omega C V_M^2 \sin 2\omega t \tag{4.7}$$

where C = capacitance in farads
V_M = peak sinusoidal voltage in volts applied to capacitor
ω = radian frequency = $2\pi f$
f = frequency in hertz

The equivalent series resistance (ESR) is an AC resistance of a capacitor reflecting both the series resistance R_s and the parallel resistance R_p at a given frequency so that the losses of these elements can be expressed in a single resistance R in the equivalent circuit. [5–7] The value of ESR varies over frequency as seen in Fig. 4.5.

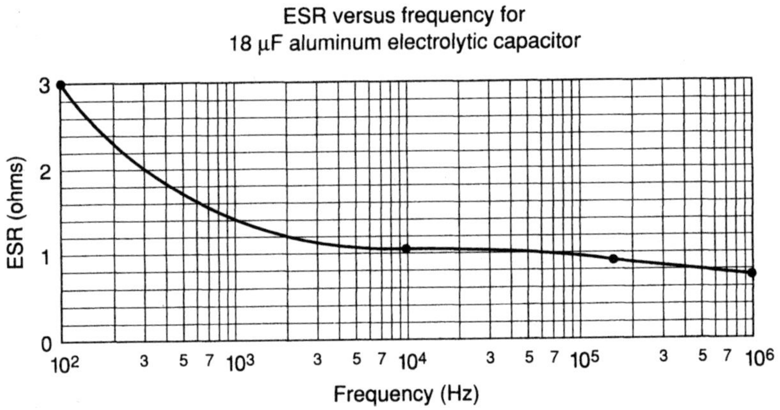

FIGURE 4.5 ESR versus frequency plot of electrolytic capacitor.

A 100-μF, 20-volt tantalum capacitor which conforms to MIL-C-39003 has an equivalent series resistance of 75 mΩ at 25°C and at a frequency of 100 kHz. [6] ESR can vary from one grade of capacitor to another, even within a particular manufacturer's line. Listed in Table 4.2 are typical ESRs for various capacitor dielectrics found in electronic circuits. [8–11]

TABLE 4.2 Typical ESR of Capacitors*

Dielectric material	Value, μF	ESR at 1 KHz, milli Ω
BX	0.1	19,000
X7R	0.1	16,000
X7R	0.18	10,000
BX	1.0	2,000
Z5U	3.3	600
Tantalum	2.2	1,000
Tantalum	22	200
Tantalum	33	200
Tantalum	33	260
Tantalum	68	168

* Based on Hopkins [32].

Capacitor ESR can contribute significant heat in switching mode power supplies if the proper dielectric for the output filters is not used. The current in the capacitor is a nonsinusoidal ripple current, making the calculation of the heat generated in capacitors due to ESR more complex. Depending on the application, the current waveform will take on different shapes, which can be determined analytically by circuit analysis. Once the current waveform is characterized, the average power can be calculated by

$$P_{\text{AV}} = \frac{1}{T} \int_{t_1}^{t_2} i(t)^2 \text{ESR}\, dt \tag{4.8}$$

where the appropriate value of ESR depending on conditions is inserted.

As shown in the capacitor equivalent circuit in Fig. 4.4, there is an equivalent series inductance, ESL, consisting of the effective series inductance contributed by the capacitor's wire bonds or leads and the body of the device. [12] In hybrid circuits and multichip modules, the capacitor is typically leadless and may have a series inductance in the order of 1 nH. At 1 MHz, the impedance of the capacitor due to the inductance is 6.28 mΩ. Therefore, the ESL contribution to power loss in capacitors can be considered negligible.

4.4.3 Magnetic Components: Inductors and Transformers

The electrical symbol for an inductor is shown in Fig. 4.6 and for a transformer in Fig. 4.7. There are two types of losses in inductors and transformers: winding resistance and core. With DC applied to the winding of an inductor or to the winding of a transformer, it dissipates power according to the following equation

FIGURE 4.6 Inductor schematic symbol.

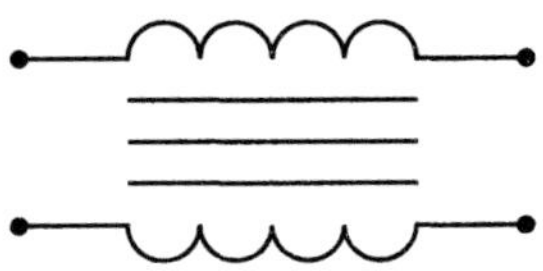

FIGURE 4.7 Transformer schematic symbol.

$$P = I^2 R_L \tag{4.9}$$

where R_L = DC resistance of the inductor or winding

If the excitation current has a significant high-frequency component, the winding resistance will increase due to skin effect as described in Sec. 4.7.

For a sinusoidal excitation of an inductor, the power dissipated is

$$p(t) = \tfrac{1}{2} L I_M^2 \omega \sin 2\omega t \tag{4.10}$$

where L = inductance in henrys
I_M = peak sinusoidal current in amperes applied to inductor
ω = radian frequency = $2\pi f$
f = frequency in hertz

When a ferromagnetic core is used, there are two components of the core losses: hysteresis and eddy current. *Hysteresis loss* is the energy dissipated into heat resulting from the reversal of magnetic domains in the core, while *eddy current loss* is the loss from the conductive current flowing in the metallic core induced by the changing flux. Core losses are typically obtained experimentally and consist of the combined hysteresis and eddy current losses. A mathematical expression for the combined core losses P_C per unit volume is

$$P_C = 6.51 f^n B_{\max}^m \tag{4.11}$$

where P_C = core loss in watts/kg
C, n, and m = constants depending on properties of coil material
f = switching frequency in kHz
$B_{\max}$ = maximum flux density in teslas

Example For the Allied Signal POWERLITE C-core AMCC-25 with a weight of 0.38 kg, the core loss relationship is [13–15]

$$P_C = 6.51 f^{1.51} B_{\max}^{1.74}$$

At a switching frequency of 50 kilohertz and a maximum flux density of 0.086 teslas, the core loss per kilogram is

$$P = P_C \times \text{mass} = 6.51 \times 50^{1.51} \times 0.086^{1.74} = 12.71 \text{ watts}$$

4.5 ACTIVE DEVICES

The power dissipation for all standard-product active integrated circuits can easily be obtained from device data sheets or calculated from laboratory measurements.

4.5.1 Frequency Effects

The power dissipation for bipolar devices is relatively constant with frequency, while, for CMOS devices, the power dissipation is a first-order function of frequency and a second-order function of device geometry. An analysis of a CMOS gate shows that its power consumption is composed of three components: switching power, dynamic short-circuit power, and DC leakage. The switching power components, resulting from the charging and discharging of the capacitance of transistor gates and interconnect lines during the changing of logic states, typically comprises 70 to 90 percent of the dissipated power. The switching power P can be calculated by

$$P = \frac{CV^2}{2} f$$

where C = input capacitance in farads
V = peak-to-peak voltage swing of signal in volts
f = switching frequency in Hz

A dynamic short-circuit power occurs when pull-up or pull-down transistors in a CMOS logic gate are briefly on during a change of state in the output node. This results in a momentary short circuit from V_{DD} to ground. This dynamic short-circuit power typically accounts for 10 to 30 percent of the dissipated power. Both the switching and dynamic short-circuit powers are a linear function of frequency and are typically categorized as a combined power loss in microwatts per megahertz per gate. For example, a CMOS chip with 0.5-micron geometry having 100,000 gates and a clock frequency of 100 megahertz might have a 1 μW/MHz per gate dissipation. If one-third of the gates were toggling simultaneously, the chip's power dissipation would be

$$P = 100{,}000 \times 0.33 \times 1 \times 100 = 3.3 \text{ watts}$$

The third component of a CMOS circuit's power dissipation is the DC leakage current, which typically accounts for less than 1 percent of the total dissipation. [16,17]

For standard-product CMOS devices, the three components of power dissipation are usually given in the devices' data sheets. However, for custom CMOS devices, the power dissipation must be modeled and subsequently verified with measurements on sample devices.

Several types of discrete active devices are reviewed for their power dissipation modes.

4.5.2 Power MOSFET

The power MOSFET, as depicted in Fig. 4.8, with the equivalent circuit in Fig. 4.9, has several sources of losses which result in a power dissipation of [18]

1. P_c: Conduction losses while the device is on
2. P_d: Reverse diode conduction and t_{rr} losses
3. P_L: Power loss due to drain-source leakage current (I_{DSS}) when the device is off

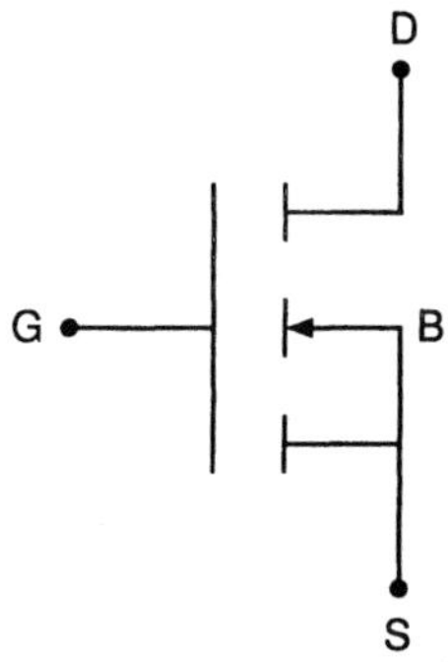

FIGURE 4.8 MOSFET schematic symbol.

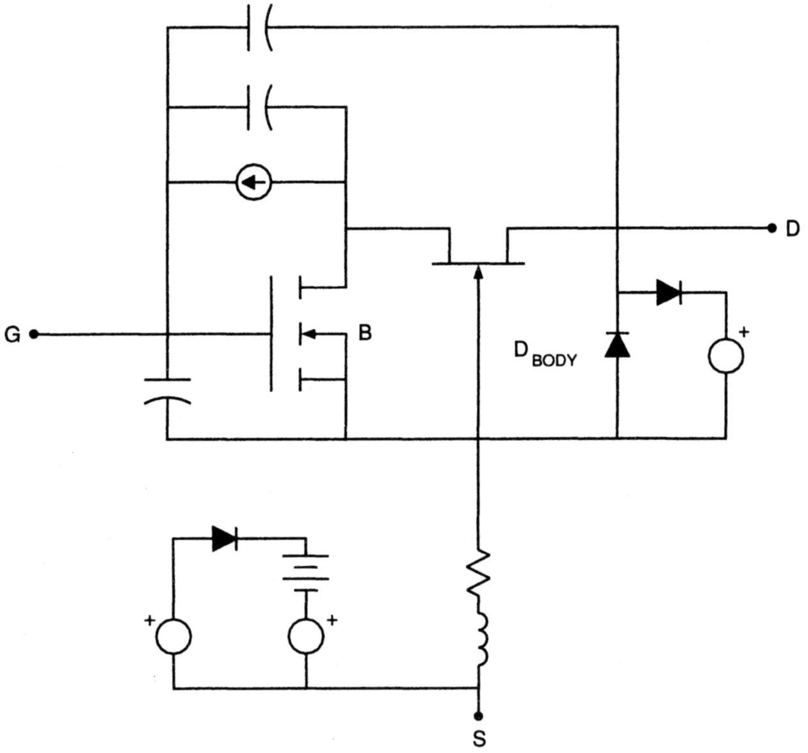

FIGURE 4.9 Power MOSFET equivalent circuit.

4. P_G: Power dissipated in the gate structure
5. P_S: Switching transition loss

1. *Conduction losses.* The conduction losses of a power MOSFET when switched on can be calculated by

$$P_c = I_D^2 r_{ds(\mathrm{ON})} \tag{4.13}$$

where I_D = drain current in amperes
$r_{ds(\mathrm{ON})}$ = drain to source on resistance in ohms

For the IRF450 power MOSFET, the conduction losses at 25°C with a drain current of 5 amps are

$$P_c = I_D^2 r_{ds(\mathrm{ON})}$$

$$P_c = 25 \times 0.40 = 10.0 \text{ watts}$$

In the linear region, the conduction loss of a power MOSFET can be calculated by

$$P_c = I_D V_{DS} \tag{4.14}$$

where V_{DS} equals drain to source voltage in volts.

2. *Leakage current losses.* When the MOSFET is off, a drain-source leakage current (I_{DSS}) flows. The power dissipated P_L in the device in the OFF condition is

$$P_L = I_{DSS} V_{DS} \tag{4.15}$$

where I_{DSS} equals drain to source leakage current in amperes.

The value of I_{DSS} increases with temperature and can be determined from manufacturers' data sheets.

Example For the IRF 450 at a junction temperature of 25°C, $I_{DSS} = 250$ microamps at $V_{DS} = 500$ volts

$$P_L = I_{DSS} \times V_{DS} = 250 \times 10^{-6} \times 500 = 0.125 \text{ watts}$$

For the same MOSFET with the same drain-to-source voltage operating at a junction temperature of 150°C, the value of I_{DSS} rises to 1000 microamps. The power becomes

$$P_L = I_{DSS} \times V_{DS} = 1000 \times 10^{-6} \times 500 = 0.50 \text{ watts}$$

3. *Switching transition loss* P_S. Switching transition losses occur because MOSFETs dissipate power during the transition from OFF to ON where they are in the linear region. As seen in the idealized waveforms for a MOSFET with resistive load shown in Fig. 4.10, switching transition losses can be calculated as the product of the drain-to-source voltage and the drain current. A general equation for P_S is

$$P_S = f_s\left(\int_0^{t_{S1}} V_{DS}(t)I_D(t)\,dt + \int_0^{t_{S2}} V_{DS}(t)I_D(t)\,dt\right) \tag{4.16}$$

where f_s = switching frequency in Hz
V_{DS} = MOSFET drain-to-source voltage in volts
I_D = MOSFET drain current in amperes
t_{S1} = first transition time (see Fig. 4.10)
t_{S2} = second transition time (see Fig. 4.10)

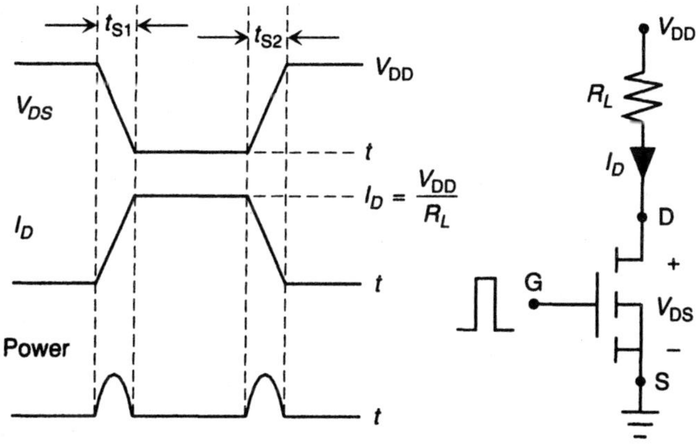

FIGURE 4.10 Power MOSFET switching losses for resistive load.

The switching waveform for an unclamped inductive load is shown in Fig. 4.11 with the graphical calculation of switching loss.

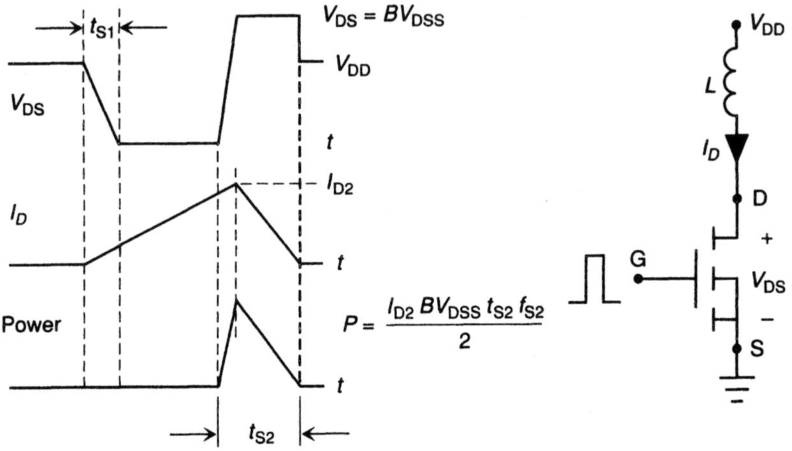

FIGURE 4.11 Power MOSFET switching loss for unclamped inductive load.

For a capacitive load C, the switching waveforms are shown in Fig. 4.12. For the case in which the turn-off time (t_{S2}) of the MOSFET channel is short and I_{DD} is small (or C large), then the rise time of V_{DS} is controlled by the capacitive load, not the t_{S2}. For the case in which I_{DD} is large (or C small), the charge time of the capacitive load C is short and the rise time is equal to t_{S2}.

4. *Gate losses* P_G. As shown in the MOSFET gate drive equivalent circuit in Fig. 4.13, the gate is a capacitive load with a series resistance. The total power dissipated in the gate circuit P_{GT} is

$$P_{GT} = V_{gs} Q_g f_s \tag{4.17}$$

where V_{gs} = gate-to-source voltage in volts
Q_g = peak charge in the gate capacitance in coulombs
f_s = switching frequency in Hz

The power dissipated internal to the MOSFET (P_G) is

$$P_G = V_{gs} Q_g \frac{R_G}{R_S + R_G} \tag{4.18}$$

The value of R_G is a function of the particular MOSFET and typically ranges from 0.05 to 4 ohms.

5. *Reverse diode losses* (P_D). The MOSFET equivalent circuit shown in Fig. 4.9 has an internal diode from the source to drain. A simplified circuit showing the integral diode is shown in Fig. 4.14. This diode dissipates power as in a normal PN junction. When conducting, this diode has a power dissipation (P_D) of

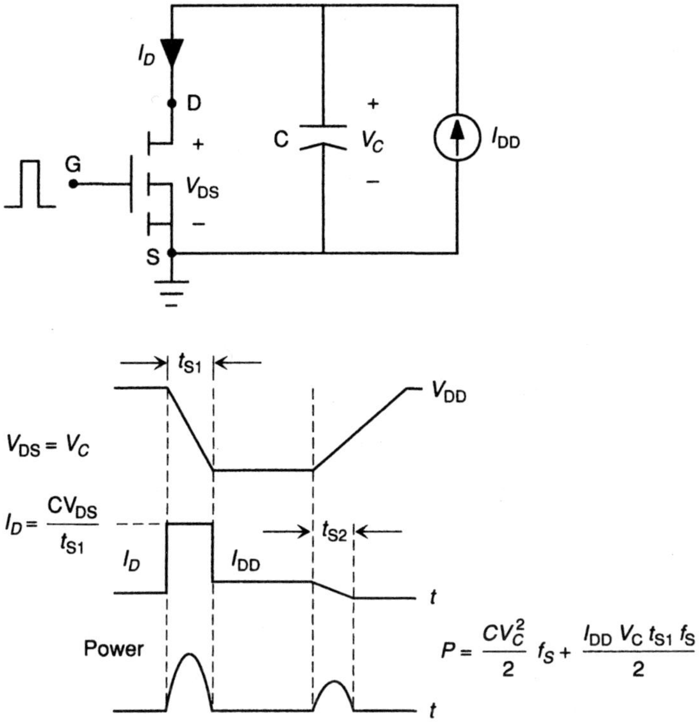

FIGURE 4.12 Capacitive switching load.

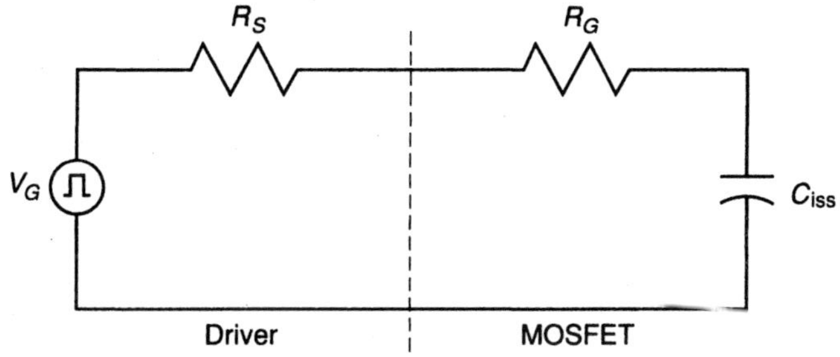

FIGURE 4.13 Gate drive equivalent circuit.

$$P_D = I_{RD(\text{AVG})} V_{f(\text{AVG})} \tag{4.19}$$

where $I_{RD(\text{AVG})}$ = average current through reverse diode
$V_{f(\text{AVG})}$ = average forward voltage drop across reverse diode

This reverse diode will dissipate additional power during the reverse turn-off interval (t_{rr}). This loss is usually small at low frequencies (less than 100 kHz). [19–21]

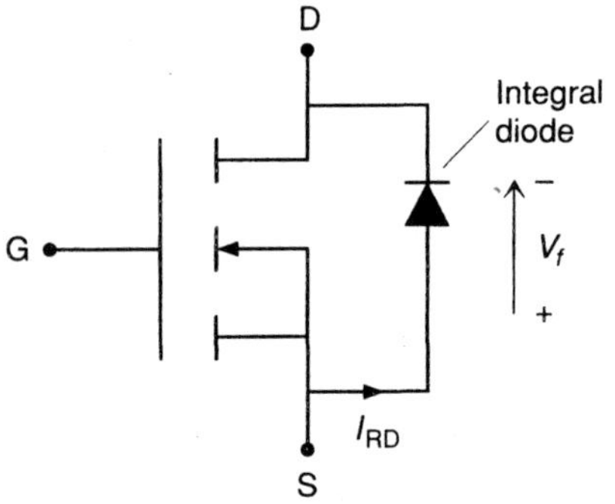

FIGURE 4.14 Integral diode in MOSFET.

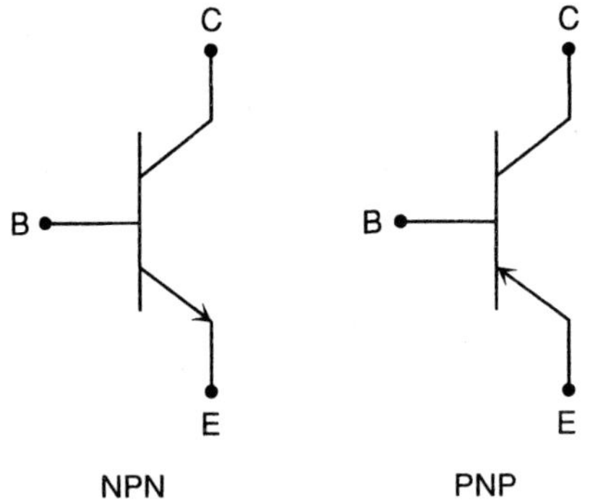

FIGURE 4.15 Schematic symbols for NPN and PNP transistors.

4.5.3 Bipolar Transistor

The bipolar transistor, shown in Fig. 4.15, can be in one of three different states at any time: ON, linear, and OFF. When the bipolar transistor is in the linear region, the majority of its power dissipation can be calculated from the following equation

$$P = I_C V_{CE} \tag{4.20}$$

where I_C = collector current in amperes
V_{CE} = collector-to-emitter voltage in volts

A small amount of power is dissipated in the base-to-emitter (P_{BE}) junction and is calculated by

$$P_{BE} = I_B V_{BE} \tag{4.21}$$

where I_B = base current in amperes
V_{BE} = base-to-emitter voltage in volts

Base current levels are typically in the microampere to milliampere level. Typical base-to-emitter voltages for silicon transistors are 0.6 volts at 25°C (with a temperature coefficient of approximately −2.2 millivolts/°C). With a 1-milliamp base current, the power dissipated in the base-to-emitter (P_{BE}) junction would be 0.7 milliwatts. This is small compared to the power dissipated in the collector to emitter and thus can be neglected when performing thermal analysis.

When the bipolar transistor is in the ON state, its power dissipation is

$$P = I_C V_{CE(\text{SAT})} \tag{4.22}$$

where $V_{CE(\text{SAT})}$ = collector-to-emitter saturation voltage in volts

The collector-to-emitter saturation voltage ($V_{CE(\text{SAT})}$) typically ranges from 0.2 to 1.5 volts. This value is a function of the die geometry, current levels, and processing parameters. $V_{CD(\text{SAT})}$ has a temperature variation of approximately 0.22 millivolts per degree centigrade. This variation is one-tenth that of the base-to-emitter or collector-to-base voltage variation.

When the bipolar transistor is in the OFF condition, the only current flowing through the collector is the collector cutoff current (I_{CBO}). This current is a function of transistor geometry, processing steps, and junction temperature. The power dissipated in the transistor in the OFF state (P_{OFF}) can be calculated by

$$P_{\text{OFF}} = I_{CBO} V_{CE} \tag{4.23}$$

where I_{CBO} = collector cutoff current or leakage current

V_{CE} = voltage applied from collector to emitter

This leakage current I_{CBO} approximately doubles with every 10°C rise in temperature. [22]

4.5.4 Junction FET

The junction FET, as shown in Fig. 4.16, is used in small signal applications as a switch and as an amplifier. Like other active devices, it has three modes of operation: ON, OFF, and linear. In the ON mode, its power dissipation (P_{ON}) is

$$P_{ON} = I_D^2 R_{DS(ON)} \tag{4.24}$$

where I_D = drain current
$R_{DS(ON)}$ = drain-to-source ON resistance

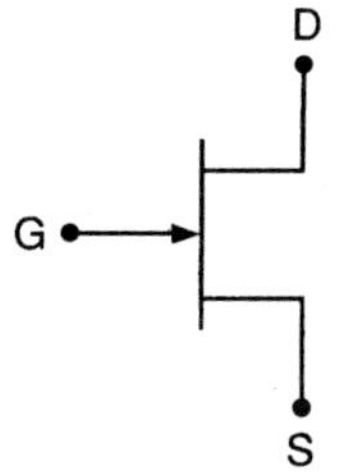

FIGURE 4.16 Junction FET symbol.

The $R_{DS(ON)}$ varies with temperature and the gate-to-source drive voltage. The 2N4392 n-channel junction FET has an $R_{DS(ON)}$ of 60 ohms at 25°C. With a drain current of 1 ma, the power dissipation at 25°C is

$$P_{ON} = I_D^2 R_{DS(ON)}$$

$$P_{ON} = (0.001)^2(60)$$

$$P_{ON} = 0.00006 \text{ watts} = 60\ \mu\text{W}$$

In the linear mode, the power dissipation may be expressed as

$$P = I_D V_{DS} \tag{4.25}$$

In the OFF mode, the power dissipation (P_{OFF}) is

$$P_{OFF} = I_{D(OFF)}\ V_{DS(OFF)} \tag{4.26}$$

where $I_{D(OFF)}$ = drain cutoff current in amperes
$V_{DS(OFF)}$ = drain-source off voltage

The magnitude of $I_{D(OFF)}$ is a function of temperature and approximately doubles with every 10 degrees centigrade rise in temperature. For the 2N4392 FET, the drain cutoff current at 25°C is 100 pA. At 150°C, this cutoff current rises to 200 nA.

The power dissipated at 25°C when the FET is off is [20,23–26]

$$P_{OFF} = I_{D(OFF)} V_{DS(OFF)} = 100 \times 10^{-12} \times 20 = 2 \times 10^{-9}\ \text{W} = 0.002\ \mu\text{W}$$

At 150°C, the power dissipated when the FET is off is

$$P_{OFF} = I_{D(OFF)} V_{DS(OFF)} = 200 \times 10^{-9} \times 20 = 4 \times 10^{-6}\ \text{W} = 4\ \mu\text{W}$$

4.5.5 Insulated Gate Bipolar Transistor (IGBT)

The *insulated gate bipolar transistor* (*IGBT*) is a minority carrier device that can be considered as a MOSFET driving a Darlington bipolar transistor as shown in Fig.

4.17. The ON-state output voltage drop, a collector-to-emitter saturated voltage, is considerably less than power MOSFETs and thus provides a lower power dissipation. The ON-state voltage drop is obtained from manufacturers' data sheets under the parameter "Collector-to-Emitter Saturation Voltage" (V_{CES}). It can range from approximately 1.5 to 3.0 volts and is a function of the device collector current as well as the device geometry. The power dissipated in the IGBT during the ON state is calculated by

$$P = V_{CES} I_C \tag{4.27}$$

where V_{CES} = collector-to-emitter saturation voltage
I_C = IGBT collector current

At 55 amps, the IRGPC50U from International Rectifier has a V_{CES} of 2.4 volts at 25°C and 1.9 volts at 150°C.

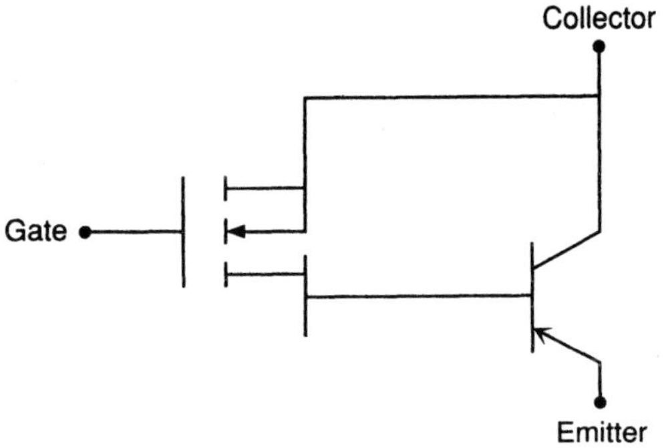

FIGURE 4.17 IGBT equivalent circuit.

The power dissipated in an IRGPC50U at 25°C with a 55-amp collector current is

$$P = V_{CES} I_C = 2.4 \times 55 = 132 \text{ watts}$$

This power decreases to 104.5 watts at 150°C.

As seen in the equivalent circuit, there is no integral reverse diode as seen in the power MOSFET. As a result, there are no associated diode losses.

When an IGBT is turned off, there is a finite turnoff time as seen in Fig. 4.18. The power dissipated during the turnoff time can be graphically calculated from the collector-to-emitter voltage waveform and the collector current waveform.

The tail in the current waveform at turnoff has the effect of increasing the power losses. The IGBT, like all minority carrier devices, will have increased power losses at higher temperatures.

When an IGBT is in the OFF state, the current flowing in the collector is I_{CES}, the zero gate voltage collector current. The power dissipated in the IGBT in the OFF state (P_{OFF}) is

$$P_{OFF} = V_{CE} I_{CES} \tag{4.28}$$

where V_{CE} = collector-to-emitter voltage
I_{CES} = zero gate voltage collector current

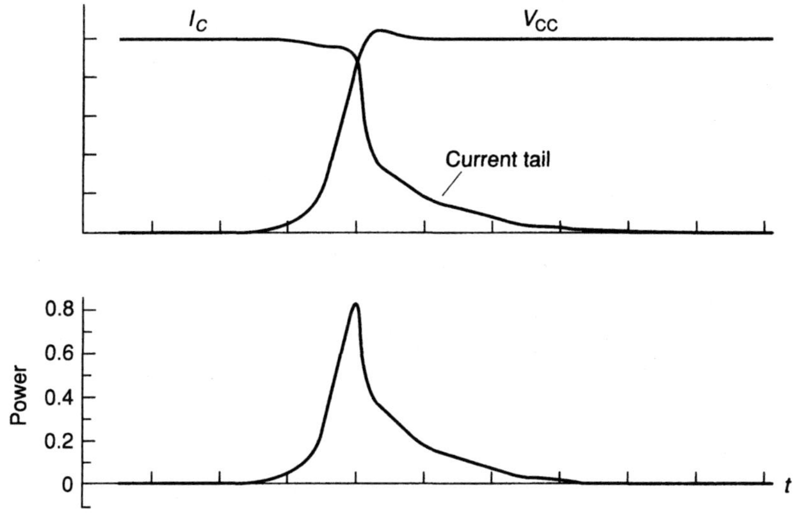

FIGURE 4.18 IGBT turnoff.

For the 55-ampere, 600-volt IRGPC50U IGBT from International Rectifier, the I_{CES} is 250 microamps at 25°C and 5.0 milliamps at 150°C. Designing to 80 percent of the collector-to-emitter breakdown voltage, the collector-to-emitter voltage is

$$V_{CE} = 0.8 \times 600 = 480 \text{ volts}$$

The power dissipated in the IGBT during the OFF state at 25°C is

$$P_{\text{OFF}} = V_{CE} I_{CES} = 480 \times 0.250 \times 10^{-3} = 120 \text{ mW}$$

At 150°C, the power dissipated in the IGBT during the OFF state is

$$P_{\text{OFF}} = V_{CE} I_{CES} = 480 \times 5.0 \times 10^{-3} = 2.4 \text{ W}$$

The IGBT, having a MOSFET input stage, also has gate losses as previously described in the power MOSFET section. [27–29]

4.5.6 Thyristor

The basic thyristor family of devices, as depicted in Fig. 4.19, includes the silicon-controlled rectifier (SCR) and the triac. The MOS-controlled thyristor (MCT) is a combination of a MOS transistor driver and a thyristor. [29–31]

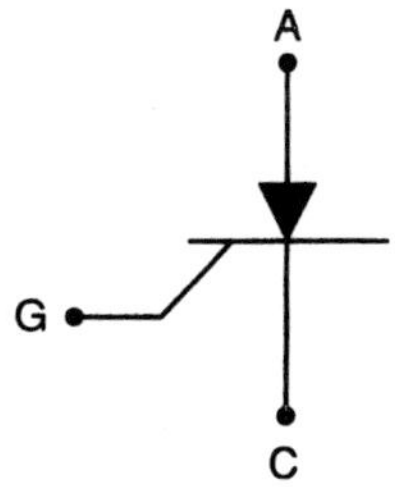

FIGURE 4.19 Thyristor schematic symbol.

Silicon Controlled Rectifier (SCR). The silicon-controlled rectifier (SCR) is a PNPN device with three terminals as shown in Fig. 4.20, with the voltage-current characteristics

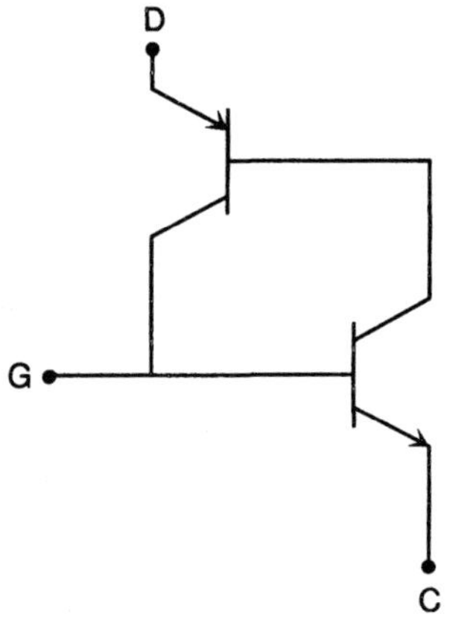

FIGURE 4.20 SCR equivalent circuit.

as shown in Fig. 4.21. When the SCR is triggered or switched ON (past the breakover region), it has a forward voltage drop V_F. Being a series of PN junctions, the magnitude of this forward drop is a function of both temperature and current. In the ON state (past its breakover region), the power dissipated by the SCR (P_{SCR}) is

$$P_{SCR} = V_F I_A \tag{4.29}$$

where V_F = anode-to-cathode voltage
I_A = anode current $\geq I_H$

V_F is typically about 1.7 volts (2 diode drops) at 25°C.

For the Motorola BRY55-100 SCR, the forward ON voltage is 1.7 volts maximum at 1 amp. Therefore, the power dissipated in the SCR would be

$$P_{SCR} = V_F \times I_A = 1.7 \times 1 = 1.7 \text{ watts}$$

When the SCR is in the OFF or untriggered state, there is a maximum current that can flow between the anode and cathode. For the forward direction, it is denoted I_{DRM} (peak forward-blocking current), while in the reverse direction it is I_{RRM} (peak reverse-blocking current). The power dissipated by the SCR in the OFF state (P_{OFF}) is several orders of magnitude less than that in the ON state and is calculated by either of the following equations

$$P_{OFF} = I_{RRM} V_{RRM} \tag{4.30}$$

$$P_{OFF} = I_{DRM} V_{DRM} \tag{4.31}$$

where V_{RRM} = peak repetitive reverse-blocking voltage from anode-to-cathode of the SCR
V_{DRM} = peak repetitive forward-blocking voltage

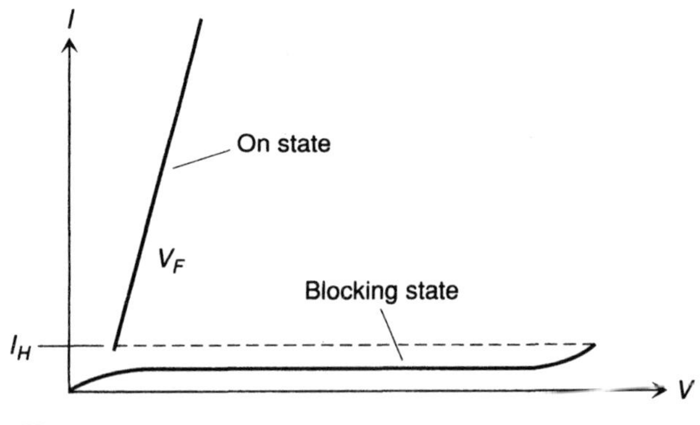

FIGURE 4.21 SCR voltage-current characteristics.

For the Motorola BRY55-100 SCR, $V_{RRM} = V_{DRM} = 100$ volts and $I_{RRM} = I_{DRM} = 100$ microamps. The power dissipated in the SCR in the forward-blocking mode is

$$P_{OFF} = I_{DRM} \times V_{DRM} = 100 \times 10^{-6} \times 100 = 0.01 \text{ W} = 10 \text{ mW}$$

The SCR can be turned ON by a single pulse or a group of several narrow pulses that meet the gate energy specification. The amount of power dissipated in the gate of an SCR can be considered negligible as the peak power is small and usually just one or several pulses.

During turn-on or turnoff of the SCR, there is a finite amount of time required for the switching to take place. The power dissipated during these transition times can be graphically calculated in the same manner described for the power MOSFET as described in the section on switching transition loss. [26,32]

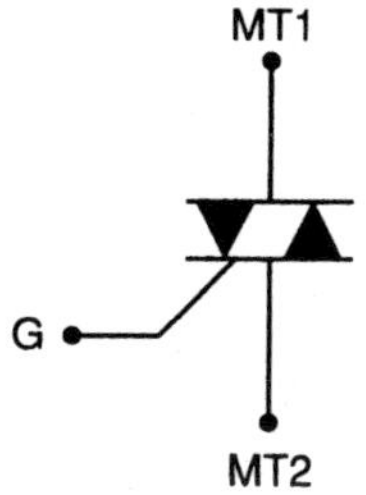

FIGURE 4.22 Triac schematic symbol.

Triac. The *triac,* a three-terminal semiconductor switch as shown in Fig. 4.22, is closely related to the SCR and provides the capability of being able to pass current in two directions. It may be considered as two parallel SCRs oriented in opposite directions.

As shown in the voltage-current characteristic in Fig. 4.23, the triac can be in one of two ON states. The power dissipated by the triac in either ON state is

$$P_{ON} = V_{ON} I \tag{4.32}$$

where V_{ON} equals ON-state voltage, typically 1.5 volts (≈two diode drops).

For the Motorola MAC15-10FP triac, the maximum power dissipated in the ON state with a current of 10 amps is

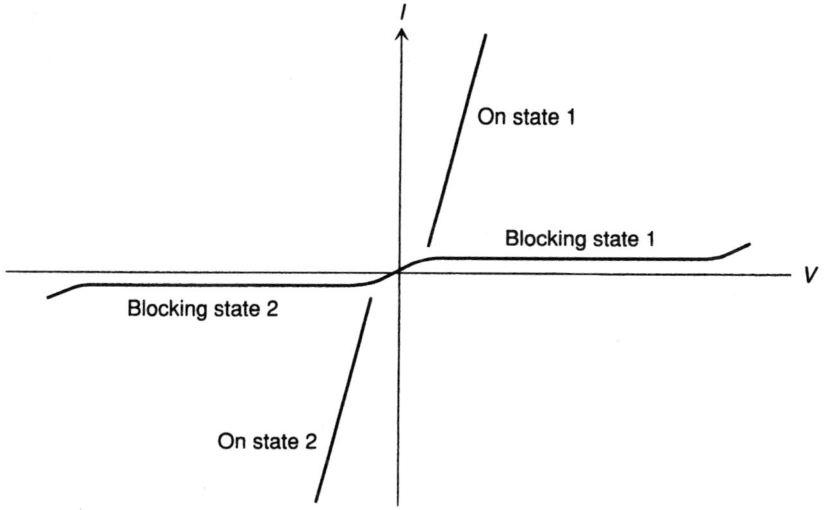

FIGURE 4.23 Triac voltage-current characteristic.

$$P_{\text{ON-state}} = V_{\text{ON-state}} \times I = [V_{\text{ON-state}}]_{\max} \times I = 1.6 \times 10 = 16 \text{ watts}$$

In the OFF state, the power dissipated by the triac is

$$P_{\text{OFF}} = V_{DRM} I_{DRM} \tag{4.33}$$

where V_{DRM} = peak blocking voltage (in either direction)
I_{DRM} = current in the OFF state

For the Motorola MAC15-10FP triac, the maximum power dissipated in the OFF state with a blocking voltage of 800 volts with $I_{DRM} = 10\ \mu\text{A}$ at 25°C is

$$P_{\text{OFF-state}} = V_{DRM} \times I_{DRM} = 800 \times 10\ \mu\text{A} = 0.008\ \text{W} = 8\ \text{mW}$$

The 10 μA at 25°C rises to 2 mA at 125°C. At this temperature, the power dissipated is 1.6 watts.

The turn-on of a triac can be accomplished with either a positive or a negative gate signal with very low energy requirements. Therefore, the gate power in a triac as in a SCR can be considered negligible.

During turn-on or turnoff of the triac, there is a finite amount of time required for the switching to take place. The power dissipated during these transition times can be calculated graphically by multiplying the current and the voltage at discrete time points. [32]

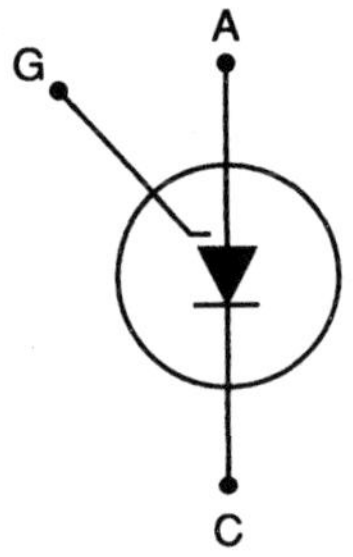

FIGURE 4.24 MCT schematic symbol.

4.5.7 MOS-Controlled Thyristor (MCT)

The MOS-controlled thyristor (MCT) circuit symbol is shown in Fig. 4.24, with the equivalent circuit shown in Fig. 4.25. The MCT exhibits power loss primarily in the forward mode where its conduction drops are

$$P = V_F I_F \tag{4.34}$$

where V_F = forward voltage drop from anode to cathode
I_F = forward current through the anode (or cathode)

Typical values of V_F are 1.4 volts at 25°C and 1.35 volts at 150°C. The magnitude of V_F increases only modestly at very high peak currents. For the Harris MCTG35P60F1 MCT with an ON current of 50 amps, the power dissipation is

$$P = V_F \times I_F = 1.35 \times 50 = 67.5 \text{ amps}$$

In the OFF state, the MCT has power dissipations of

$$P_{\text{OFF}} = I_{DRM} V_B \tag{4.35}$$

$$P_{\text{OFF}} = I_{RRM} V_B \tag{4.36}$$

where I_{DRM} = peak OFF-state blocking current
I_{RRM} = peak reverse-blocking current
V_B = blocking voltage applied across anode to cathode

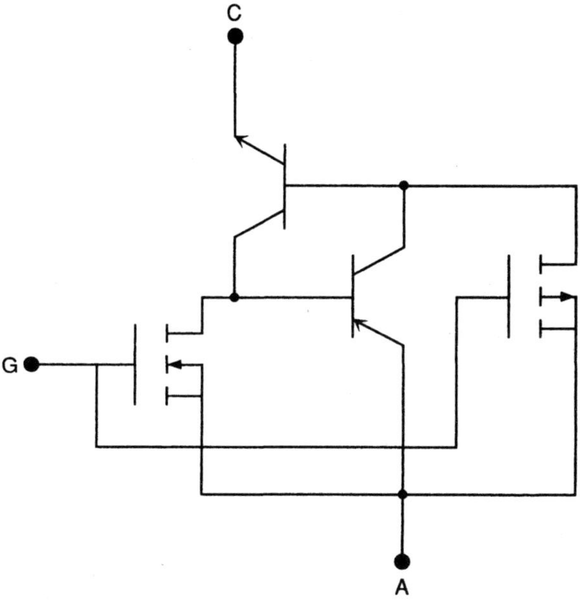

FIGURE 4.25 MCT equivalent circuit.

For the Harris MCTG35P60F1 MCT, rated at a maximum blocking voltage of 600 volts, the power dissipated in the device during the OFF state at 25°C is

$$P_{\mathrm{OFF}} = I_{DRM} \times V_B = 50 \times 10^{-6} \times 600 = 0.03 \text{ W}$$

At 150°C, I_{DRM} increases to 1.5 mA and the resulting power dissipation becomes

$$P_{\mathrm{OFF}} = I_{DRM} \times V_B = 1.5 \times 10^{-3} \times 600 = 0.9 \text{ W}$$

While the MCTG35P60F1 has the same I_{RRM} as I_{DRM} at 25°C the I_{RRM} rises to 2.0 mA at 150°C. Therefore, the power dissipated at 150°C due to reverse-blocking current is

$$P_{\mathrm{OFF}} = I_{DRM} \times V_B = 2.0 \times 10^{-3} \times 600 = 1.2 \text{ W}$$

Power losses in the gate circuit can be considered negligible. The MCT has switching losses during both turn-on and turnoff. [32]

4.5.8 Diodes

There are several types of diodes, as shown in Fig. 4.26, used in power-switching applications. One type, the Schottky barrier rectifier, is usually found in switched-mode power supplies as the output rectifier. [33] The low forward voltage drop, even at high currents, is beneficial in minimizing rectification loss. A 60HQ100 Schottky barrier diode from International Rectifier, [34] has a forward drop V_F of 0.50 volts at 100 amps. Its dissipation P_F in the forward mode is

FIGURE 4.26 Diode schematic symbol.

$$P_F = I_F V_F \tag{4.37}$$

where V_F = forward voltage drop of diode
I_F = forward current in diode

For this case,

$$P_F = 0.50 \times 100 = 50 \text{ watts}$$

The V_F in Schottky diodes generally decreases quite significantly as the junction temperature increases. The V_F also increases in Schottky diodes as the current density increases.

In the reverse mode, its dissipation P_R is

$$P_R = V_R I_R \tag{4.38}$$

where V_R = forward voltage applied to diode
I_R = reverse leakage current in diode

The reverse leakage current I_R is a first-order function of temperature and can be obtained from the manufacturer's data sheet. (See also Fig. 4.27.) For the 60HQ100 at 125°C with a reverse voltage V_R of 80 volts, the reverse current is 7 mA. Therefore, the power dissipated in the reverse mode at 125°C is

$$P_R = 0.007 \text{ (amps)} \times 80 \text{ (volts)} = 0.56 \text{ watts}$$

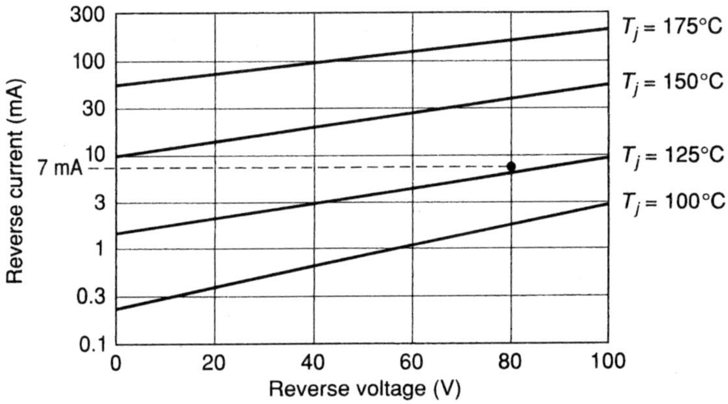

FIGURE 4.27 60HQ100 Schottky diode reverse characteristics.

An Ultrafast diode with similar forward-current handling capability (100 amps) and a reverse breakdown voltage of 600 volts, such as the Harris RURU10060, would have a forward voltage drop of 1.6 volts at 100 amps at 25°C and 1.4 volts at 150°C. In the forward mode with 100 amps flowing, its dissipation is

$$P = I_F V_F = 1.6 \times 100 = 160 \text{ watts} \qquad \text{(at 25°C)}$$

In the reverse mode, its leakage is 500 microamps at 600 volts. Therefore, the power dissipated in the reverse mode at 25°C is

$$P_R = 500 \times 10^{-6} \times 600 = 0.3 \text{ watts}$$

At 150°C, the leakage rises to 2.0 milliamps. This corresponds to a power dissipation of

$$P_R = 2.0 \times 10^{-3} \times 600 = 1.2 \text{ watts}$$

There are a variety of high-current diode types available to the circuit designer. Their power dissipations in the forward and reverse modes can be calculated as shown in the preceding examples.

4.5.9 Zener Diodes

FIGURE 4.28 Zener diode schematic symbol.

The schematic symbol for a Zener diode is shown in Fig. 4.28, with the V-I characteristic shown in Fig. 4.29. Zener diodes are typically operated in the reverse mode in which the voltage across the device is approximately equal to the nominal Zener voltage. The power dissipated in the Zener diode P_Z in this mode is

$$P_Z = I_Z V_Z \tag{4.39}$$

where I_Z = current in Zener in reverse mode
V_Z = voltage across Zener diode (Zener voltage)

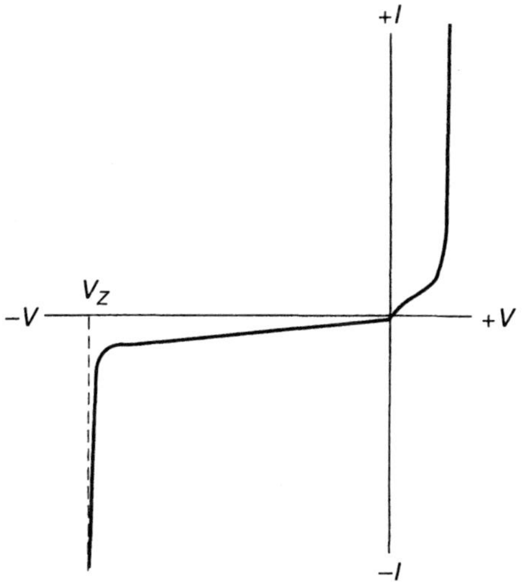

FIGURE 4.29 Typical Zener diode voltage-current characteristic curve.

The Zener voltage is defined at a specific reverse current, as it will vary slightly due to ohmic losses in the leads and bulk silicon. Zener voltages exhibit either positive or negative temperature coefficients that range from 0.005 percent/°C to over

0.1 percent/°C. A Zener diode with a 0.1 percent/°C temperature coefficient would have a 10 percent change in voltage over a 100°C change in temperature. This would equate to a 10 percent change in power dissipation. For Zener voltages above 6.2 volts, the temperature coefficient is positive while for voltages below 6.2 volts, the temperature coefficient is negative. There are Zener diodes with temperature coefficients of approximately 0.01 percent/°C or lower that are typically members of a family called *temperature-compensated* Zener diodes.

In the forward mode, the voltage across the Zener resembles that of any forward diode drop (approximately 0.7 volts). The power dissipated in the Zener still obeys Eq. 4.39.

4.6 INTERCONNECTIONS

4.6.1 Introduction

The amount of heat generated in the interconnections of microelectronic circuits is a first-order function of the electrical resistance and can be described by Joule's law:

$$P_I = I_I^2 R_I \tag{4.40}$$

where P_I = power dissipated in watts or joules/second
I_I = current in interconnection in amperes
R_I = resistance of interconnection in ohms

To calculate the power and the resultant heat generated, the electrical resistance must be known and is calculated using the basic resistance equation

$$R = \rho \frac{L}{A} \tag{4.41}$$

where R = resistance in ohms
ρ = resistivity of the material in ohm-cm
L = length of material in cm
A = cross-sectional area of material

This equation can be used to calculate the resistance of package pins, wire bonds, and substrate metallizations.

4.6.2 Wire Bonds

The power dissipated in wire bonds obeys Joule's law as previously described in Eq. 4.38. For low-power active devices such as logic and small signal analog, the typical wire bond is fabricated from either gold or aluminum having a diameter of 0.001 inch (see Fig. 4.30). If such a gold wire bond were 0.100 inch long, it would have a total resistance of

$$R = \rho_0 \frac{L}{A} \tag{4.42}$$

where R = resistance of wire bond in ohms
ρ_0 = resistivity of wire bond material in Ω-in

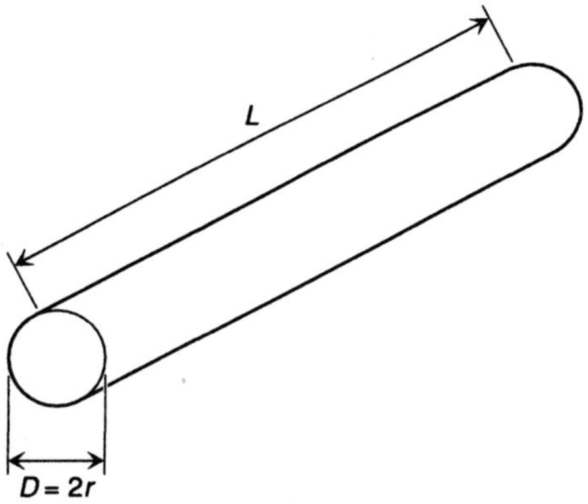

FIGURE 4.30 Wire bond dimensions.

L = length of wire bond in inches
A = cross-sectional area of wire bond = πr^2 in in^2
r = radius of wire bond in inches

From Table 4.3, the resistivity of gold is

$$\rho_{0\,\mathrm{Au}} = 2.44 \times 10^{-6}\ \Omega\text{-cm}$$

converting ρ_0 to ohm-inches

$$\rho_0 = 2.44 \times 10^{-6}\ (\Omega\text{-cm}) \times 1\ (\text{in})/2.54\ \text{cm} = 9.60 \times 10^{-7}\ (\Omega\text{-inch})$$

$$R = 9.6 \times 10^{-7}\ (\text{ohm-inch}) \times 0.10\ (\text{inch})/\pi(0.0005)^2$$

$$R = 0.122 \text{ ohms or } 122 \text{ milliohms}$$

TABLE 4.3 Resistance Properties of Selected Metals*

Material	Resistivity, μohm-cm	TCR, ppm/°C
Alloy 42	66.5	1,000
Alloy 52	43.0	2,900
Aluminum	2.83	3,400
Copper	1.72	3,900
Gold	2.44	3,400
Kovar	48.9	3,700
Nickel	7.80	6,000
Silver	1.63	3,800

* Based on CRC Handbook of Chemistry and Physics [18] and King [35].

The maximum current a 0.001-inch-diameter gold wire that is 0.100 inch long can handle at ambient temperature prior to fusing is 0.90 amps. Using a derated current of 0.65 amps, we find that the power dissipated in the wire is

$$P = I^2R$$

$$P = 0.65^2 \times 0.122$$

$$P = 0.052 \text{ watts} = 52 \text{ milliwatts}$$

Fifty-two milliwatts is a rather low power dissipation by itself, but most hybrids and multichip modules have a multitude of wire bonds. The heat generated from the wire bonds needs to be recognized when modeling heat-generating elements.

In high-power microcircuits where high currents are typically seen, the interconnections are typically aluminum wire with diameters ranging from 0.005 to 0.025 inch. A typical wire bond used on an IGBT might have a 0.020-inch diameter and a length of 0.25 inch. The resistance of such a wire would be 0.887 milliohms. If the currents in the wire bond were 50 amps (the fusing current is 60 amps), then the heat dissipated by that wire bond would be 2.22 watts. This is a significant amount of power and needs to be addressed in thermal modeling of the microcircuit.

4.6.3 Ribbons

Applications such as microwave circuits use ribbons, as shown in Fig. 4.31, for interconnections. The power generated in the ribbon is calculated in a manner similar to that of the wire bonds. Table 4.4 shows the characteristics of several hybrid interconnections. Since the ribbon has a rectangular cross section, the resistance is

$$R = \rho_0 \frac{L}{A} = \rho_0 \frac{L}{Wt} \tag{4.43}$$

where R = resistance in ohms
ρ_0 = resistivity of ribbon material in ohm-in
L = length of ribbon in inches
W = width of ribbon in inches
t = thickness of ribbon in inches

For a 0.002- by 0.005-inch gold ribbon 0.200 inch long, the resistance is

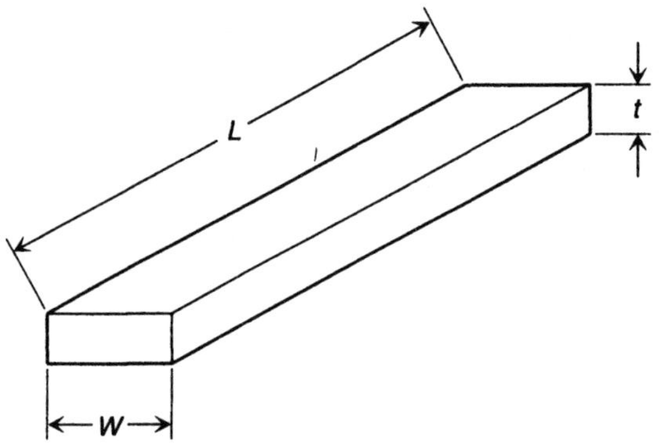

FIGURE 4.31 Ribbon interconnect dimensions.

TABLE 4.4 Characteristics of Hybrid Interconnections*

Type of interconnect	Resistance, milliohms/0.1 in	Max current before fusing, amps	Max. derated current,† amps
0.001 diameter Au	122	0.9	0.65
0.002 diameter Au	30	3.4	2.45
0.001 × 0.003 Au ribbon	30	3.4	2.45
0.002 × 0.005 Au ribbon	10	10.6	7.63
0.002 × 0.010 Au ribbon	5	21.6	15.55
0.005 diameter Al	5.8	5.5	3.96
0.010 diameter Al	1.4	22.0	15.84
0.020 diameter Al	0.35	88.0	63.36
0.025 diameter Al	0.23	137.0	98.64

* Based on Krum [37].
† Maximum derated current is 72 percent of fusing current.

$$R = \rho_0 \frac{L}{Wt} = 9.6 \times 10^{-7} \frac{0.2}{0.002 \times 0.005} = 19.2 \text{ m}\Omega$$

For a 5-amp current through the ribbon, the power dissipated is 0.48 watts.

4.6.4 Package Pins

There are two basic styles of packages used for microelectronic devices: hermetic and nonhermetic. In either style, the pin material is a low-resistance metal. This pin will dissipate power based upon its resistance and the applied current. For a hermetic package with a Kovar body, the pin is enclosed in a glass or ceramic bead as shown in Fig. 4.32. For low-power applications, a 0.018-inch-diameter Kovar pin is

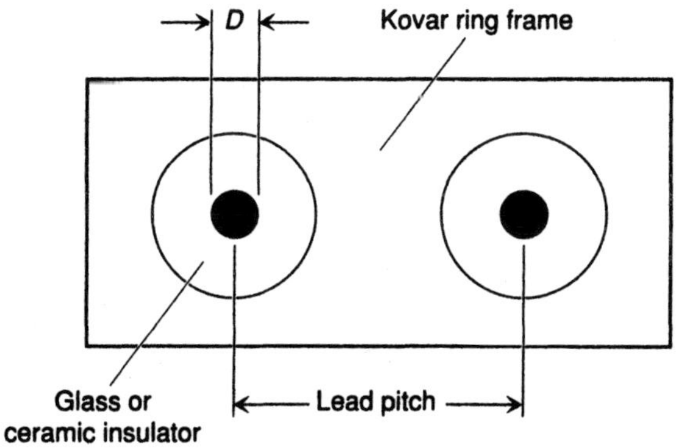

FIGURE 4.32 Axial leaded hermetic package pin configuration.

typically surrounded by a 0.040-inch-diameter glass bead in the package wall. The typical length for the pin would be 0.240 inch. The resistance of the pin is

$$R = \rho \frac{L}{A} = \frac{\rho}{\pi r^2} \tag{4.44}$$

$$R = 48.9\ \mu\Omega\text{-cm} \times \frac{1\ \text{in}}{2.54\ \text{cm}} \times \frac{0.240}{\pi(0.009)^2} = 0.018\ \Omega$$

This 0.018-inch-diameter pin has been shown [1] to be able to carry 3.2 amps without raising the temperature of the pin 40°C. If a current of 3.0 amps were flowing in the pin, the power dissipated in the pin would be

$$P = I^2R = 3^2(0.189) = 0.170\ \text{watts}$$

Larger-diameter pins made from higher electrical conductivity materials are used to carry higher currents. The maximum current for various pin materials and diameters is shown in Table 4.5.

TABLE 4.5 Maximum Design Current in Package Pins*

Material	Conductor diameter, inch	Max. design current,† amps
Kovar	0.018	1.3
	0.030	3.2
	0.040	5.0
	0.060	10.0
Alloy 52	0.020	1.5
	0.040	5.5
	0.060	11.0
Copper-core Kovar	0.040	7.5§
Kovar-copper butt weld‡ 0.070 Kovar, 0.050 Cu	0.050	10.3

* Based on Krum [1] and Krum [37].
† Current which raises temperature of pin 40°C.
‡ From end of Kovar on inside of package to 1.0 inch on end of Cu on outside of package.
§ Insulator is glass.

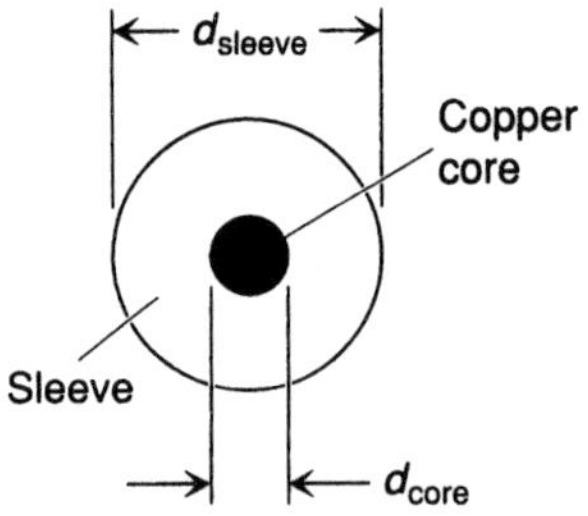

FIGURE 4.33 Cross section of copper-cored pin.

To achieve low resistance and thus low heat generation in a package pin, low-resistivity materials such as copper need to be used. For metal packages and ceramic packages with metal walls, pure copper is not typically used for pins due to its high TCE and its propensity for softening when subjected to brazing temperatures. One technique of minimizing resistance while maintaining the necessary mechanical properties of hardness and coefficient of expansion match is to use a copper-cored pin. Such pins, as shown in Fig. 4.33, have a copper core that is typically one-third the diameter of the overall pin. The outside material or sleeve is usually Kovar or Alloy 42. The resistance of a copper-cored pin would be modeled by the copper portion in parallel with the sleeve portion.

For the copper core with a diameter of d_{core} expressed in inches

$$R_{\text{Cu}} = \rho_0 \frac{L}{A} = \frac{1.72 \times 10^{-6}}{2.54} \frac{L}{\pi\left(\frac{d_{\text{core}}}{2}\right)^2} \tag{4.45}$$

For the sleeve with outside diameter of d_{sleeve} and inside diameter d_{core}:

$$A_{\text{sleeve}} = \pi\left[\left(\frac{d_{\text{sleeve}}}{2}\right)^2 - \left(\frac{d_{\text{core}}}{2}\right)^2\right] \tag{4.46}$$

$$R_{\text{sleeve}} = \frac{48.9 \times 10^{-6}}{2.54} \frac{L}{A_{\text{sleeve}}} \tag{4.47}$$

The equivalent resistance of the copper-cored pin is

$$R_{\text{Total}} = R_{\text{core}} \| R_{\text{sleeve}} = \frac{R_{\text{core}} \times R_{\text{sleeve}}}{R_{\text{core}} + R_{\text{sleeve}}} \tag{4.48}$$

Example For a copper-cored Kovar pin 0.250 inch long whose outside diameter is 0.060 inch and whose core is 0.020 inch diameter

$$R_{\text{Cu}} = \rho_{\text{Cu}} \frac{L}{A} = \frac{1.72 \times 10^{-6}}{2.54} \times \frac{0.250}{\pi(0.010)^2} = 0.542 \text{ m}\Omega$$

$$R_{\text{Kovar}} = \rho_{\text{Kovar}} \frac{L}{A} = \frac{48.8 \times 10^{-6}}{2.54} \times \frac{0.250}{\pi[(0.03)^2 - (0.01)^2]} = 1.91 \text{ m}\Omega$$

$$R_{\text{Total}} = \frac{0.542 \times 1.91}{0.542 + 1.91} = 0.422 \text{ m}\Omega$$

For a current of 50 amps flowing in the pin, the power dissipated would be

$$P = I^2R = 50^2(0.00042) = 1.05 \text{ watts}$$

Another type of package pin used in high-current applications requiring a hermetic package is the *butt-weld* pin as shown in Fig. 4.34. This pin consists of a short Kovar section which provides the hermeticity between the insulating glass or ceramic and a longer section of copper wire that is butt-welded or brazed to the Kovar section. Kovar is used to obtain a high-reliability glass-to-metal seal while the copper is used to minimize the overall pin resistance.

Example For a 20-amp current flowing in this butt-welded pin, the power dissipated is

$$P = I^2R = 20^2(0.00271) = 1.08 \text{ watts}$$

See Fig. 4.35.

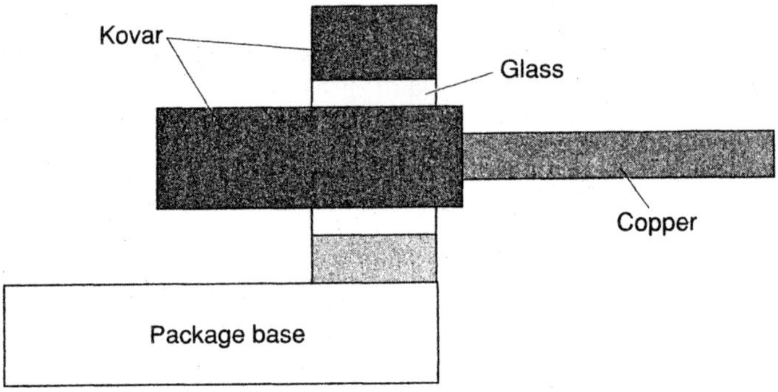

FIGURE 4.34 Butt-welded pin assembly.

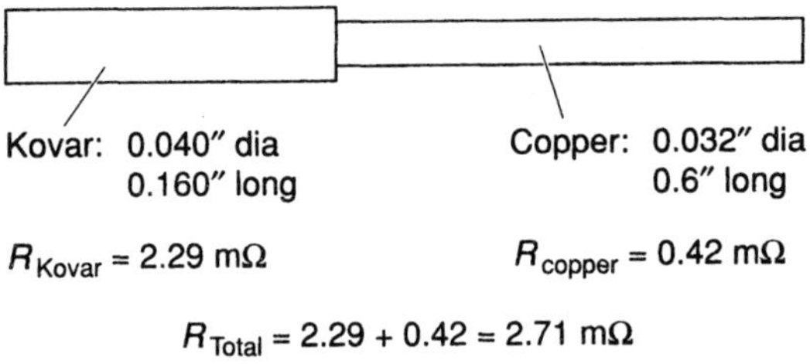

FIGURE 4.35 Butt-welded pin resistance calculation.

4.6.5 Pins on Nonhermetic Packages

Nonhermetic packages typically have stamped or etched lead frames fabricated from nickel-iron, clad strip, or copper-based. The most widely used material for lead frames is Alloy 42, a 42 percent nickel, 58 percent iron alloy which has a resistivity of 57.33 μΩ-cm. [38] Copper-clad materials have composite resistivities that are proportional to the thickness and resistivities of their constituent layers. Copper-based lead frame materials have a high copper content (typically greater than 90 percent) with some impurities added to modify either their TCE or hardness. The power dissipated in lead frames is calculated using the basic resistivity equation previously described for the material and geometry and then multiplying by the square of the current.

4.6.6 Losses in Substrates

Since any type of metallization used for interconnections on substrates will have some finite resistance, the metallization will have losses which produce heat. These losses are discussed for the various types of metallizations shown in Fig. 4.36.

The resistance R of any conductor can be calculated using a variation of Eq. 4.43. For film conductors, where the width is much greater than the thickness, it is usually more convenient to use a term called the *sheet resistivity*, as defined in Eq. 4.49. While the width of the conductor is defined in the design, the thickness is a function of the manufacturing process and can vary.

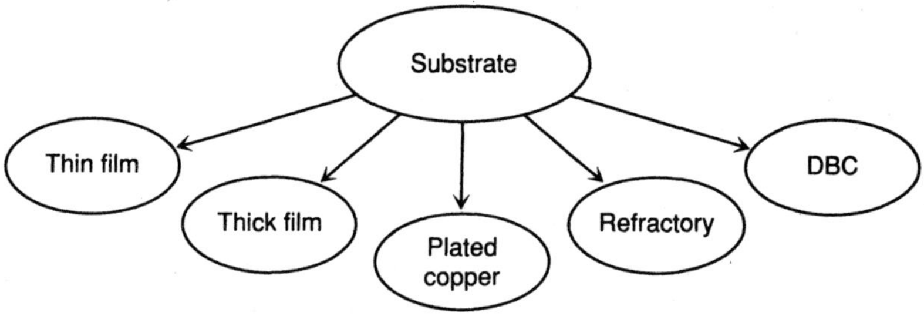

FIGURE 4.36 Substrate metallizations.

$$\rho_S = \frac{\rho_B}{t} \tag{4.49}$$

where ρ_S = sheet resistivity in Ω/□ (ohms/square)
ρ_B = bulk resistivity in Ω-length
t = thickness of the film

In many applications, the resistivity is normalized by dividing by a standard conductor thickness (t_0) for that process.

$$\rho_{SN} = \frac{\rho_B}{t_0} \tag{4.50}$$

where ρ_{SN} = normalized sheet resistivity in Ω/□
t_0 = normalized thickness (25 μ for thick film)

In terms of the sheet resistivity,

$$R = \rho_S \frac{L}{W} \tag{4.51}$$

The resistivity of conductors varies considerably with temperature. The temperature coefficient of resistance (TCR) for various conductor metallizations is shown in Table 4.3. Gold, while having a low 2.44-μΩ-cm resistivity, has a 3400 ppm/°C TCR. For a 100°C change in temperature, the resistance will change by 34 percent. Therefore, the power dissipated in a gold conductor will increase by 34 percent for a 100°C rise in temperature (assuming the current remains constant).

Presently there is no specification for maximum current allowed in thin film conductors. The new performance specification MIL-PRF-38534 C [39] does not give a quantitative value for maximum currents on any substrate. Its predecessor, MIL-H-38534 B, [40] specified in paragraph 3.5.6 a maximum current density on an unglazed alumina substrate of 6×10^5 amps/cm^2. While many have argued over the correctness of this current density, it was a quantitative number that could be easily applied to substrate designs. It is used in this section to provide an order of magnitude for the power dissipation in a substrate.

Example A 0.001-inch-thick plated copper trace is 0.100 inch long and 0.010 inch wide and has a 1.0-amp current flowing through it. The power dissipated in this conductor is calculated as shown.

The resistance of the conductor is

$$R = \rho_{Cu} \frac{L}{Wt} = \frac{1.72 \times 10^{-6}}{2.54} \times \frac{0.100}{0.01 \times 0.001} = 6.77 \text{ m}\Omega$$

The power dissipated in the conductor is

$$P = I^2R = 1^2 \times 6.77 \times 10^{-3} = 6.77 \text{ mW}$$

Screen-Printed Substrate Metallizations. The classic thick film metallizations and refractory metallizations are usually applied using screen printing. Both types of metallizations are applied to the substrate in the same manner. Their power losses, while significantly different, can be calculated in the same manner.

As pointed out previously, the width of a conductor is defined in the design and is accurately reproduced in manufacturing. However, the conductor thickness is a function of the manufacturing process and will vary. For screen-printed conductors, the wet thickness is directly proportional to the screen's emulsion thickness and to screen-printer settings such as force, screen opening size, screen angle, screen tension, and squeegee speed. After drying and firing, the resulting dry thickness is what sets the sheet resistivity. It is a function of the wet thickness and the rheology of the conductor ink. In some applications, the designer needs to minimize power losses in substrate conductors and thus designs to minimize conductor resistance. One technique of reducing conductor resistance in thick film designs is to print the conductor twice, thus reducing the resistance by a factor of two. Other resistance-lowering techniques in thick film include using lower-resistivity inks and wider conductors.

Thick Film Metallization. As seen in Table 4.6, each type of thick film conductor metallization has a different sheet resistivity ρ_S.

TABLE 4.6 Resistivities of Thick Film Conductors*

Conductor metallization	Resistivity ρ, mΩ/□
Gold	3
Silver	1
Copper	2
Palladium silver	30
Platinum gold	50
Palladium gold	50

* Based on Greiner and Krum [41].

The maximum current allowed in a thick film conductor on an alumina substrate is specified in MIL-H-38534 B, paragraph 3.5.6 as 4 watts/cm^2. [40] This equates to 25.81 watts per square inch. This power density is based upon a maximum temperature rise for the conductor. Using the 0.010-inch-wide thick film gold conductor previously described on an alumina substrate, the maximum current allowed is

$$\frac{P}{A} = 25.81 \frac{\text{W}}{\text{in}^2} \tag{4.52}$$

from which

$$I_{\max} = W\sqrt{\frac{25.81}{\rho_S}} \tag{4.53}$$

where W equals the width of the conductor in inches.

For a gold conductor 0.010 inch wide,

$$I_{\max} = 0.010\sqrt{\frac{25.81}{0.003}} = 0.93 \text{ A}$$

Example Find the power dissipated in a thick film gold conductor, 0.100 inch long and 0.010 inch wide when a current of 0.5 amp is flowing. The sheet resistivity of the thick film gold is listed in Table 4.6 as 3 milliohms per square.

$$R = \rho_S \frac{L}{W} = 3 \times 10^{-3} \frac{0.100}{0.010} = 30 \text{ m}\Omega$$

The power dissipated in this thick film conductor is

$$P = I^2R = 0.5^2 \times 30 \times 10^{-3} = 7.5 \text{ mW}$$

For the 0.100-inch-long conductor, the resistance was previously found to be 30 milliohms. Therefore, the maximum specified power allowed in this thick film conductor is

$$P = I^2R = 0.93^2 \times 30 \times 10^{-3} = 25.9 \text{ mW}$$

Since beryllium oxide has a thermal conductivity approximately 10 times that of alumina, the temperature rise of a conductor on beryllium oxide can be significantly higher. The MIL-H-38534 B specification [40] allows the maximum current in a thick film conductor on beryllium oxide to be 80 watts per square centimeter. This equates to 516 watts per square inch.

For the same 0.100-inch-long, 0.010-inch-wide gold (0.003 mΩ/□) previously described on a beryllium oxide substrate, the maximum current is

$$\frac{P}{A} = 516 \frac{\text{W}}{\text{in}^2}$$

$$I = 0.010\sqrt{\frac{516}{0.003}} = 4.1 \text{ A}$$

The maximum specified power is

$$P = I^2R = 4.1^2 \times 30 \times 10^{-3} = 504 \text{ mW}$$

Refractory Metallizations. Refractory metallizations, deposited on ceramics prior to firing, include tungsten, molybdenum, moly-manganese, and tantalum. Chosen for their high melting points, they have relatively high electrical resistivities. Listed in Table 4.7 are some properties of refractory metals used for substrate metallizations.

The power dissipated in refractory metallization is calculated in a manner similar to that for thick film.

For a moly-manganese conductor, 0.100 inch long and 0.010 inch wide carrying 0.5 amps (using a 15-mΩ/□ sheet resistivity)

TABLE 4.7 Properties of Refractory Metals Used for Substrate Metallization*

Property	Tungsten	Molybdenum	Moly-manganese†	Tantalum
Melting point (°C)	3,380	2,610	2,300	3,000
Electrical Resistivity† (μ-ohm-cm at 20°C)	5.5	5.7	4.1	13.5
Typical sheet resistivity (milli Ω/□)	22–32	15	10–19‡	35‡
TCE (ppm/°C)	4.5	5.6	6.0	6.7

* Based on Licari and Enclow [42], Electro-Science Laboratories, Inc. [43,44], and Tummala [45].
† 10 percent Molybdenum, 90 percent Manganese.
‡ 12–15 μ fired thickness.

$$R = \rho_S \frac{L}{W} = 15 \times 10^{-3} \frac{0.100}{0.010} = 150 \text{ m}\Omega$$

$$P = I^2 R = 0.5^2 \times 0.150 = 37.5 \text{ mW}$$

Thin Film Metallization. As shown in Table 4.8, thin film metallizations have sheet resistivities ranging from 5 to 20 milliohms per square. (When the deposited gold, typically 2 μ-inch, is unplated, it has a sheet resistivity of 500 milliohms per square.) To achieve the lower sheet resistivities, the gold plating thickness is increased. At 5 milliohms per square, the typical process limit for plated gold, the gold thickness is 200 microinches. The maximum current (from MIL-H-38534 B) is calculated as follows:

$$J_{max} = 6 \times 10^5 \frac{\text{A}}{\text{cm}^2} \tag{4.54}$$

where J_{max} equals the maximum allowable current density in A/cm².

TABLE 4.8 Thin Film Conductor Resistivities*

Gold thickness, μ-inch	Resistivity ρ, mΩ/□
192	5.0
96	10.0
48	20.0
2	500

* Based on Krum [37] and Greiner and Krum [41].

For a conductor which is 0.010 inches wide and 2×10^{-4} inches thick,

$$A = 0.010 \text{ in} \times 0.0002 \text{ in}$$

$$= 2 \times 10^{-6} \text{ in}^2 \times 2.54^2 \text{ cm}^2$$

$$= 1.29 \times 10^{-5} \text{ cm}^2$$

$$I_{max} = 6 \times 10^5 \text{ amps/cm}^2 \times 1.29 \times 10^{-5}$$
$$= 7.74 \text{ amps}$$

The resistance of the conductor is

$$R = \rho_S \frac{L}{W} = 5 \times 10^{-3} \frac{0.100}{0.010} = 50 \text{ m}\Omega$$

The power dissipated in this thin film conductor with a current of 7.74 amps is

$$P = I^2R = 7.74^2 \times 0.050 = 2.995 \text{ watts}$$

Plated Copper. To reduce voltage drops and power dissipation in thin film, thick film, or refractory conductors, one may plate up to 0.005 inch of copper over the adhesion layer and metallization. After copper plating, the metallization is then plated with nickel and gold as shown in Fig. 4.37.

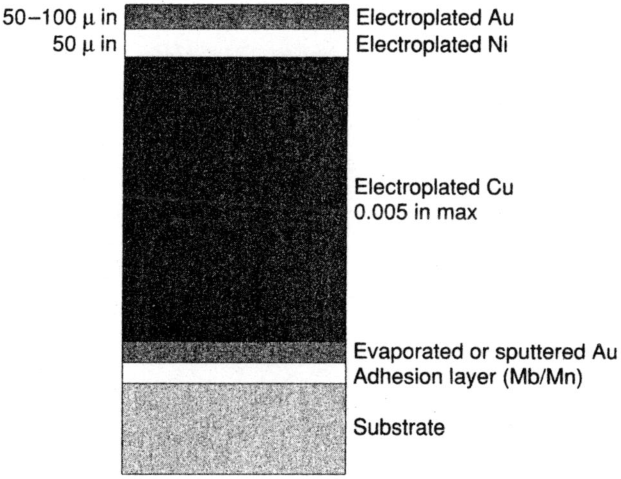

FIGURE 4.37 Plated copper process with thin film.

As with other substrate metallizations, there is no specification available for maximum current in the plated copper process. (*Note:* Plated copper was not covered in previous military specifications.)

Direct Bond Copper (DBC). *Direct bond copper* is a patented process [41,46] process in which copper foil is bonded directly to ceramic substrates without an intermediary bonding layer (Fig. 4.38). In this process, copper foil, from 0.001 to 0.020 inch thick, is chemically bonded to ceramic substrates such as alumina, beryllium oxide, and aluminum nitride. As shown in Fig. 4.39, there is a copper-oxygen eutectic composition at 0.39 percent oxygen which is liquidus at 1065°C where the eutectic will form a thin liquidus skin on the surface of the solid copper which remains mechanically stable. This copper-oxygen eutectic mixture then wets to the ceramic and, upon cooling, forms a Cu_2O-Cu complex which fuses the ceramic to the copper section. A cross section of a substrate metallized with direct bond copper is shown in Fig. 4.37. The circuit pattern can be etched after the fusing process or it can be stamped in the copper prior to the fusing. The resistance of the metallization in the direct bond copper process can be easily calculated knowing the conductor dimensions and the bulk resistivity of copper.

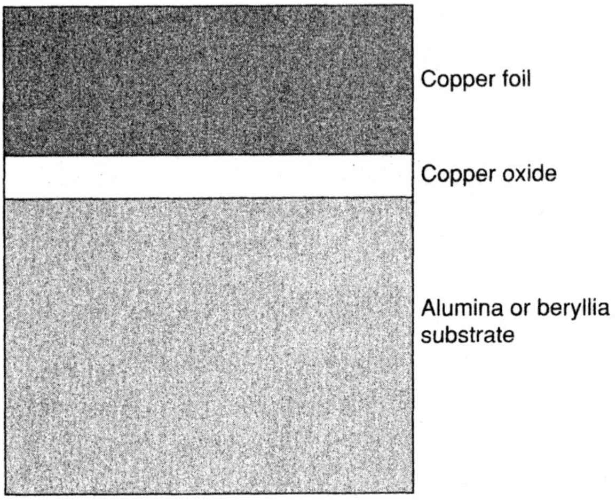

FIGURE 4.38 Cross section of direct bond copper metallization.

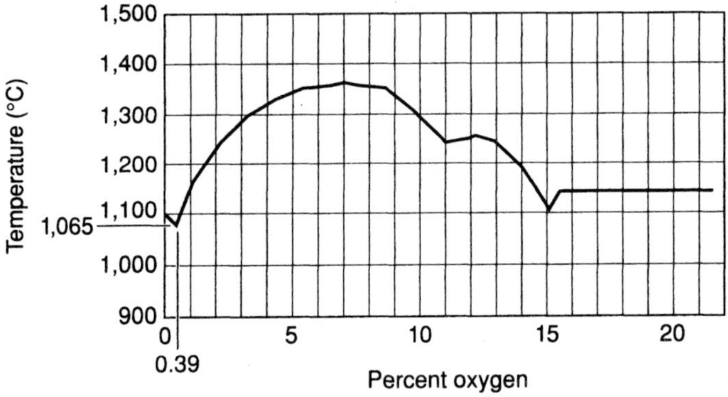

FIGURE 4.39 Copper-oxygen phase diagram.

Example A direct bond copper trace is 0.250 inch long and 0.100 inch wide with a foil thickness of 0.010 inch. The resistance of this trace is

$$R = \rho_{Cu} \frac{L}{Wt} = \frac{1.72 \times 10^{-6}}{2.54} \frac{0.250}{0.100 \times 0.010} = 0.169 \text{ m}\Omega$$

There is no specification for maximum current in direct bond copper. A typical current carried by 0.010-inch-thick copper in the DBC process might be 100 amps. The power dissipated in the preceding conductor is

$$P = I^2R = 100 \times 0.169 \times 10^{-3} = 0.0169 \text{ watts} = 16.9 \text{ mW}$$

4.7 HIGH-FREQUENCY LOSSES

At DC, charge carriers are evenly distributed though the cross section of a wire, pin, or conductor. As the frequency increases, the carriers move away from the center toward the edge and cause an apparent resistance increase. [33] This is due to the charge carriers finding it easier to move near the edge of the conductor. As shown in Fig. 4.40*a*, the charge carriers are evenly distributed as depicted by the even shading of the circular cross section of the conductor. At higher frequencies, the charge carriers are clustered near the peripheral as shown in Fig. 4.40*b*.

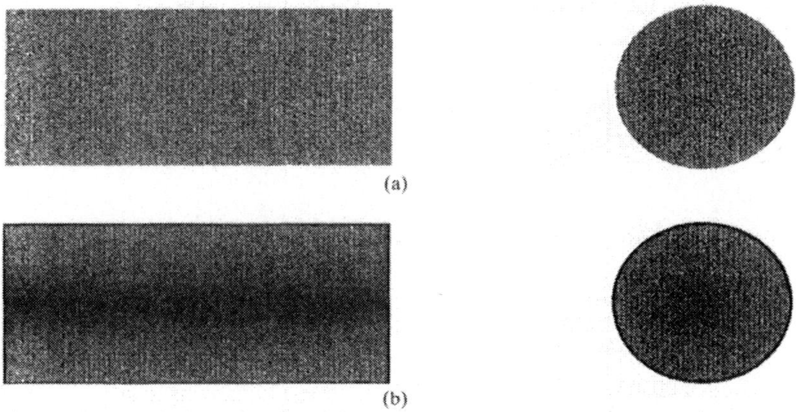

FIGURE 4.40 Skin effect: (*a*) charge distribution at DC; (*b*) charge distribution at high frequency.

The depth to which the current density drops off to 36.8 percent of the value at the surface is called *skin depth* (δ) and can be calculated for various conductor materials using Eq. 4.55. See Fig. 4.41.

$$\delta = \sqrt{\frac{\rho}{\pi \mu f}} \tag{4.55}$$

where δ = skin depth in meters
ρ = resistivity of conductor in ohm-meters
μ = permeability
f = frequency in Hz

With the frequency term in the denominator, the skin depth decreases as the frequency increases. Figure 4.42 shows a chart of skin depth for various conductor materials versus frequency. Table 4.9 shows the relative permeability for various conductor materials.

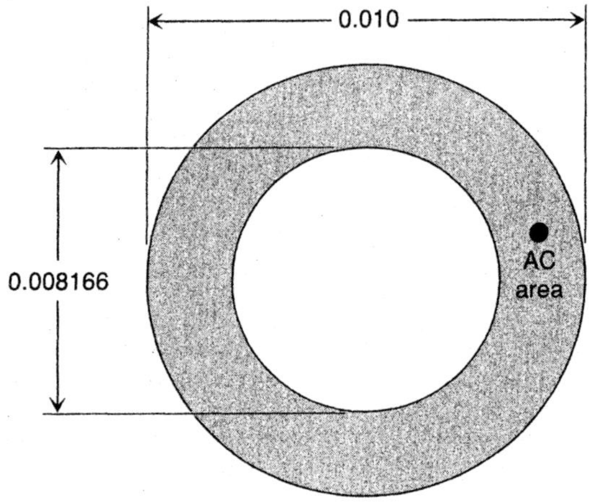

FIGURE 4.41 Effective AC area of conductor due to skin effect.

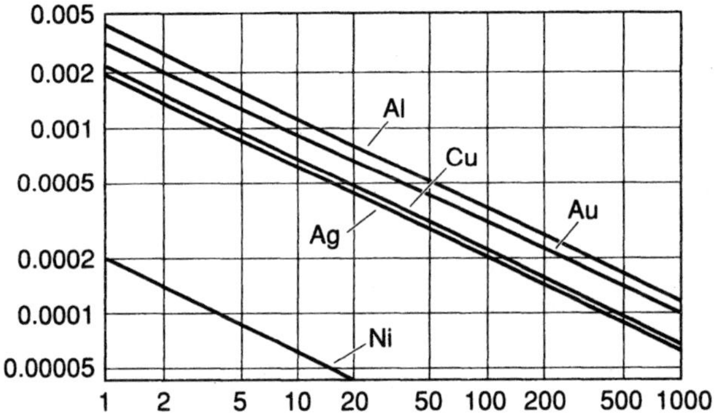

FIGURE 4.42 Skin depth of various conductor materials versus frequency.

Example A copper conductor has a signal operating at 10 megahertz. What is the skin depth?

$$\rho_{Cu} = 1.72 \times 10^{-8} \text{ ohm-m}$$

$$f = 10 \text{ MHz}$$

$$\mu_{Cu} = 4\pi \times 10^{-7}$$

$$\delta = \sqrt{\frac{1.72 \times 10^{-8}}{\pi \times 4\pi \times 10^{-7} \times 10 \times 10^{6}}} = 2.11 \times 10^{-5} \text{ m} = 0.829 \text{ mils}$$

For a circular conductor, such as a wire bond or a package pin, the AC resistance can be calculated as shown in the following example.

TABLE 4.9 Relative Permeability of Various Conductor Materials

Conductor material	Relative permeability
Silver	1
Copper	1
Gold	1
Aluminum	1
Nickel	50

Example An aluminum wire bond, 0.200 inches long with a diameter of 0.010 inches is operating at 10 megahertz. What is the effective resistance of the wire bond?

From Table 4.4, the DC resistance of the aluminum wire bond is 2.8 mΩ.

$$\text{DC area} = \pi r_1^2 = \pi(0.005)^2 = 7.854 \times 10^{-5}\ \text{in}^2$$

The skin depth is

$$\delta = \sqrt{\frac{2.83 \times 10^{-8}}{\pi \times 4\pi \times 10^{-7} \times 10 \times 10^6}} = 2.45 \times 10^{-5}\ \text{m} = 0.967\ \text{mils}$$

Effectively, most of the current is present in the outer 0.967 mils of the wire bond, the AC area. The AC area looks like the shaded ring in Fig. 4.41 in which the thickness of the ring is 0.967 mils.

$$\text{AC area} = \pi(0.005^2 - 0.004083^2) = 2.617 \times 10^{-5}$$

$$R_{AC} = \frac{\text{DC area}}{\text{AC area}} \times \text{DC resistance} \tag{4.56}$$

For this case,

$$R_{AC} = \frac{7.854 \times 10^{-5}}{2.617 \times 10^{-5}} \times 2.8 \times 10^{-3} = 8.4\ \text{m}\Omega$$

Once the AC resistance (R_{AC}) is known, then the power dissipation can be calculated. Assuming that the AC current is sinusoidal, then the power dissipated is

$$P_{AV} = \tfrac{1}{2} R_{AC} I_m^2 \tag{4.57}$$

REFERENCES

1. A. Krum, *Proceedings 1995 International Symposium on Microelectronics, Determination of Package Pin Current Ratings,* Reston, Virginia, pp. 223–228.
2. S. H. Leibson, "EDN Editorial," *EDN Magazine,* March 14, 1996, p. 11.

3. Donald G. Fink, *Electronic Engineers' Handbook,* McGraw-Hill, New York, 1975, pp. 1-20 to 1-22.
4. VMF-1900, Vishay Foil Resistors Data Sheet, Malvern, Pennsylvania.
5. Donald G. Fink, *Electronic Engineers' Handbook,* McGraw-Hill, New York, 1975, pp. 7-7 to 7-13.
6. Kemet Solid Tantalum and Multilayer Ceramic Catalog, Kemet Electronics Corporation.
7. Jeffrey D. Shepard, *Power Supplies,* Reston Publishing Company, Reston, Virginia, 1984, pp. 75, 159.
8. D. C. Hopkins, "Designing Power Hybrid Supplies," *Powertechniques Magazine,* June 1989, pp. 31–34.
9. D. C. Hopkins, M. M. Jovanovic, F. C. Lee, and F. W. Stephenson, "Off-line ZCS-QRC Thick-Film Hybrid Circuit," Virginia Power Electronics Center, *Sixth Annual Power Electronics Seminar,* September 1988, pp. 71–83.
10. *Olean Advanced Products Data Book,* S-OAP10M295-N, AVX Corporation.
11. Kemet Surface Mount Catalog F-3102, September 1994, p. 20.
12. Gary Breed, "Capacitor Behavior at Radio Frequencies," *RF Design,* January 1996, pp. 58–61.
13. W. H. Buchsbaum, *Buchsbaum's Complete Handbook of Practical Electronic Reference Data,* Second Edition, Prentice Hall, Englewood Cliffs, New Jersey, 1973, p. 127.
14. J. W. McClean, "Downsizing PFC Inductors in Switchers," *Electronic Design,* May 1, 1996, pp. 123–132.
15. MIT Staff, *Magnetic Circuits and Transformers, A First Course for Power and Communication Engineers,* John Wiley & Sons, New York, 1962, pp. 124–129, 148–154.
16. Lipman, "EDA Tools Let You Track and Control CMOS Power Dissipation," *EDN,* pp. 65–76, November 23, 1995.
17. Francine Ferguson and Ashutosh Mauskar, "High Level Design for Low Power," *The Electronic Design Automation & Test Conference,* Asia Pacific (Korea, Taiwan, China), August 1995, p. 364 (this conference was held in the 3 countries).
18. *CRC Handbook of Chemistry and Physics,* CRC Press, Boca Raton, Florida, 1984, p. E-78.
19. Rudy Severns and Jack Armijos, eds., *MOSPOWER Applications Handbook,* Siliconix, Santa Clara, California, 1984, pp. 4-4 to 4-9.
20. *HEXFETâ Power MOSFET Designer's Manual,* International Rectifier, 1993, pp. 833–838.
21. Edwin S. Oxner, *Power FETs and Their Applications,* Prentice Hall, Englewood Cliffs, New Jersey, 1982, pp. 9–11, 21–26.
22. J. Millman and C. Halkias, *Integrated Electronics, Analog and Digital Circuits and Systems,* McGraw Hill, New York, 1972, pp. 129–142.
23. C. F. Wheatley, H. R. Ronan, and G. M. Dolny, *Spicing-Up SPICE II Software for Power MOSFET Modeling,* Harris Semiconductor Application Note AN8610.1, August 1991.
24. J. Watson, *An Introduction to Field Effect Transistors,* Siliconix, Santa Clara, California, 1970, pp. 20–25, 88–90.
25. FET Design Catalog, Siliconix, Santa Clara, California, November 1979, pp. 3–26.
26. M. E. Van Valkenberg, *Network Analysis,* Prentice Hall, Englewood Cliffs, New Jersey, 1964, pp. 375–393.
27. *IGBT Designer's Manual,* International Rectifier, 1991, pp. 93–106.
28. C. S. Mitter, *Introduction to IGBTs, PCIM,* **21**(12): 32–39, December 1995.
29. MCT/IGBTS/DIODES, Harris Semiconductor, 1995, pp. 2-3 to 2-7, 5-118 to 5-120.
30. *MCT User's Guide,* Harris Semiconductor.
31. *Power Modules Designer's Guide,* International Rectifier Corporation, p. E-14, 1991.

32. Motorola Thyristor Device Data Book, 1995.
33. George Chryssis, *High Frequency Switching Power Supplies,* McGraw-Hill, New York, 1984, pp. 111, 128.
34. *Schottky Diode Designer's Guide,* International Rectifier Corporation, 1992, pp. D251–D254, E4–E8.
35. J. A. King, *Materials Handbook for Hybrid Microelectronics,* Artech House, Boston, Massachusetts, 1988, p. 353.
36. James Hardy, *High Frequency Circuit Design,* Reston Publishing Company, Reston, Virginia, 1979, pp. 26–28.
37. A. Krum, "Design & Processing of Power Hybrids," *Hybrid Circuit Technology,* January 1987.
38. P. V. Robock and L. T. Nguyen, "Plastic Packaging," in *Microelectronics Packaging Handbook,* R. R. Tummala and E. J. Rymaszewski, eds., Van Nostrand Reinhold, New York, 1989, pp. 544–548.
39. MIL PRF 38534C, Performance Specification, General Specification for Hybrid Microcircuits, 22 August 1995.
40. MIL-H-38534B, Military Specification, General Specification for Hybrid Microcircuits, 26 August 1992.
41. S. Greiner, and A. Krum, "Selecting a Substrate Technology for Power Hybrid Applications," *Electronic Packaging and Production,* pp. 54–57, May 1992.
42. James Licari and Leonard Enclow, *Hybrid Microcircuit Technology Handbook,* Noyes Publications, Park Ridge, New Jersey, 1987, pp. 79–83, 91.
43. Data sheets D-2170 and D-2180, Electro-Science Laboratories, Inc.
44. Private conversation, Cornelius Huang of Electro-Science Laboratories, June 10, 1996.
45. R. R. Tummala, "Ceramic Packaging," in *Microelectronic Packaging Handbook,* R. R. Tummala and E. J. Rymaszewski, eds., Van Nostrand Reinhold, New York, 1989, p. 489.
46. US Patent 1,563,383, Direct Bond Copper Ceramic Substrate for Electronic Applications, J. Dixon and Kuneman, issued 1/7/86.

CHAPTER 5
BASIC THERMAL ANALYSIS

Al Krum

5.1 HEAT FLOW

The second law of thermodynamics states that heat always flows spontaneously from a hotter region to a cooler region. All active and passive devices are sources of heat as described in Chap. 4. These devices are always hotter than the average temperature of their immediate surroundings. There are three mechanisms for heat transfer: conduction, convection, and radiation as depicted in Fig. 5.1. These mechanisms and the temperature distribution from the heat source to the surroundings are discussed in the following sections.

5.2 CONDUCTION

Thermal conduction is a process in which heat flows through a solid, liquid, or gas, or between two media that are in intimate contact with each other. Conduction, the dominant mechanism for heat transfer within solids, involves the transfer of kinetic thermal energy from one electron to another, causing no visible motion of the particles of the body. Conduction through dielectric solids is almost entirely due to lattice vibrations while conduction through metallic solids has the added energy transport by free electrons. This thermal energy transfer via electrons is similar to that of electric charge. It is not surprising to find that good electrical conductors such as copper and silver are good thermal conductors.

Liquids also conduct thermal energy, but to a significantly lesser extent than solids. When a material changes phase from a solid to a liquid, there is a lessening of its intermolecular bonds and a deterioration of the ordered state of the solid results. There is more freedom for thermal motion of the molecules. As a result, the material's thermal conductivity is lowered.

When a liquid changes to a gas, there is a further loosening of molecular bonds. The gas molecules are free to move in any direction and the only restraints are random collisions. This random motion of the gas molecules results in a significantly lower probability that contact areas will exist. Therefore, the thermal conductivity of gases is quite low.

5.2.1 Fourier's Law

Fourier's law of heat conduction, named after the French mathematician, Jean Fourier, states that the rate of heat flow equals the product of the area normal to the heat flow path, the temperature gradient along the path, and the thermal conductivity of the medium. Mathematically, Fourier's law can be expressed as

$$\frac{dq}{dt} = -KA\,\frac{dT}{dX} \tag{5.1}$$

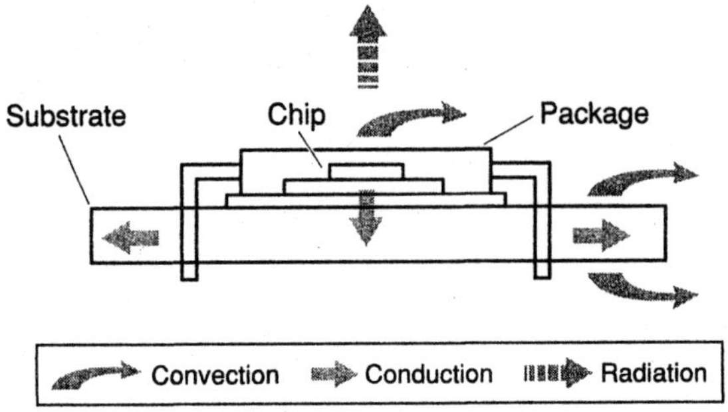

FIGURE 5.1 Heat transfer from packaged chip on a PWB.

where K = thermal conductivity of medium in watts/m-K or watts/in-°C
A = cross-sectional area of medium normal to the heat flow path in in^2 or cm^2
T = temperature of medium in °C
X = position along the medium in inch or cm
t = time in seconds
q = heat generated per unit volume in joules/cm^3
Q = heat flow in watts normal to the cross-sectional area of heat transfer

$$\frac{dQ}{dt} = \text{power in watts or calories/second} \tag{5.2}$$

$$\frac{dT}{dX} = \text{temperature gradient in °C/in or °C/cm} \tag{5.3}$$

The temperature gradient and the cross-sectional area are defined at the same point x. Heat flow is considered positive when the temperature is decreasing. As noted in Chap. 3, the thermal conductivity of many materials varies with temperature. In most cases this variation with temperature can be considered a second-order effect and contributes only a minor source of error.

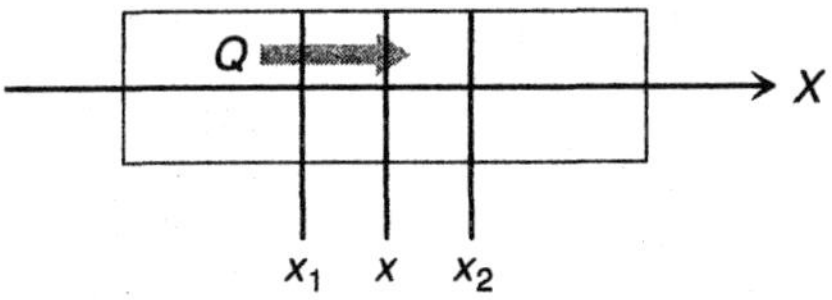

FIGURE 5.2 Heat conduction in a one-dimensional solid element.

Fourier's law can be extended from the preceding one-dimensional law previously described (Fig. 5.2) to three dimensions. For a volume element (ΔV) as shown in Fig. 5.3, Fourier's law can be rewritten as

$$q_x = -k_x \Delta y \Delta z \frac{\partial T}{\partial X} \quad \text{at point 3} \tag{5.4}$$

$$q_{x+\Delta x} = -k_x \Delta y \Delta z \frac{\partial T}{\partial X} \quad \text{at point 4} \tag{5.5}$$

$$q_y = -k_y \Delta x \Delta z \frac{\partial T}{\partial Y} \quad \text{at point 1} \tag{5.6}$$

$$q_{y+\Delta y} = -k_y \Delta x \Delta z \frac{\partial T}{\partial Y} \quad \text{at point 2} \tag{5.7}$$

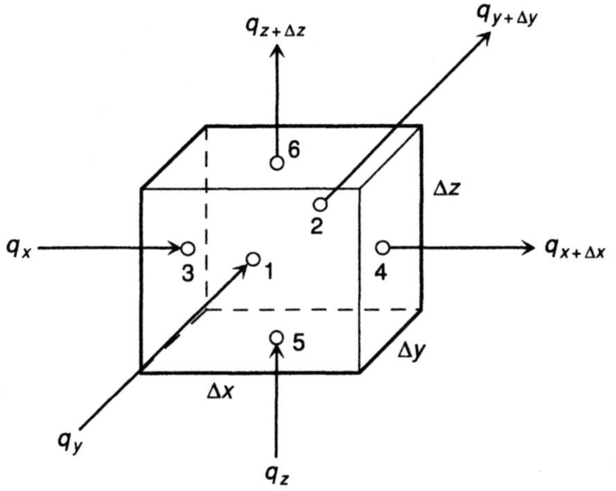

FIGURE 5.3 Volume element (ΔV).

$$q_z = -k_z \Delta x \Delta y \frac{\partial T}{\partial Z} \qquad \text{at point 5} \tag{5.8}$$

$$q_z = -k_z \Delta x \Delta y \frac{\partial T}{\partial Z} \qquad \text{at point 6} \tag{5.9}$$

The elemental volume is an internal heat-generating source of Q_V watts/unit volume.

5.2.2 Conservation of Energy

In a steady state condition, that is, no energy storage, heat in equals heat out (Fig. 5.4), so that for the elemental volume ΔV (with *heat in* being defined as positive)

$$\left[\left(-k_x \Delta y \Delta z \frac{\partial T}{\partial x}\right)\bigg|_3 - \left(-k_x \Delta y \Delta z \frac{\partial T}{\partial x}\right)\bigg|_4\right] + \left[\left(-k_y \Delta x \Delta z \frac{\partial T}{\partial y}\right)\bigg|_1 - \left(-k_y \Delta x \Delta z \frac{\partial T}{\partial y}\right)\bigg|_2\right]$$
$$+ \left[\left(-k_z \Delta x \Delta y \frac{\partial T}{\partial z}\right)\bigg|_5 - \left(-k_z \Delta x \Delta y \frac{\partial T}{\partial z}\right)\bigg|_6\right] + Q_v \Delta x \Delta y \Delta z = 0 \tag{5.10}$$

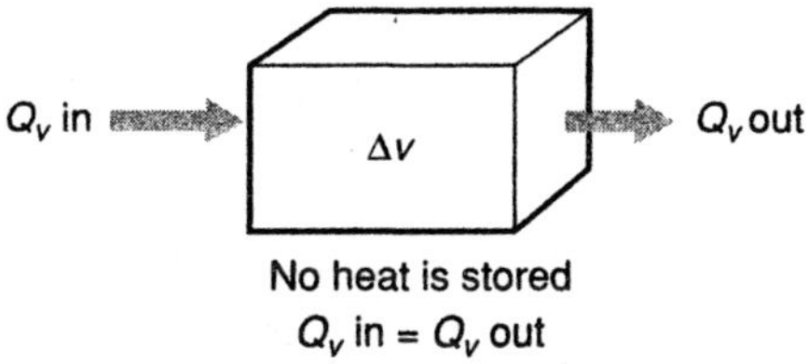

FIGURE 5.4 Conservation of energy.

Dividing both sides of the equation by the volume $\Delta V = \Delta x \Delta y \Delta z$:

$$\left(\frac{1}{\Delta x}\right)\left(k_x \frac{\partial T}{\partial x}\right)\Bigg|_3^4 + \left(\frac{1}{\Delta y}\right)\left(k_y \frac{\partial T}{\partial y}\right)\Bigg|_1^2 + \left(\frac{1}{\Delta z}\right)\left(k_z \frac{\partial T}{\partial z}\right)\Bigg|_5^6 = -Q_v \tag{5.11}$$

$$\lim_{\Delta x \to 0}\left(\frac{1}{\Delta x}\right)k_x \frac{\partial T}{\partial x}\Bigg|_3^4 = \lim_{\Delta x \to 0}\left(\frac{1}{\Delta x}\right)k_x \frac{\partial T}{\partial x}\Bigg|_{x,y,z}^{x+\Delta x,y,z} \frac{\partial}{\partial x}\left(k_x \frac{\partial T}{\partial x}\right) \tag{5.12}$$

taking the limit as $\Delta V \to 0$ in the x direction.

Repeating this process in the y and z directions, the result is the three-dimensional heat conduction for the steady state condition

$$\frac{\partial}{\partial x}\left(k_x \frac{\partial T}{\partial x}\right) + \frac{\partial}{\partial y}\left(k_y \frac{\partial T}{\partial y}\right) + \frac{\partial}{\partial z}\left(k_z \frac{\partial T}{\partial z}\right) = -Q_v \tag{5.13}$$

Equation 5.1 can be written as

$$dT = -\left(\frac{Q_K}{kA}\right) dx \tag{5.14}$$

Integrating both sides of Eq. 5.14,

$$\int_{T_1}^{T_2} dT = -Q_k \int_{x_1}^{x_2} \frac{dx}{kA_k} \tag{5.15}$$

Assuming that the thermal conductivity k does not vary over the length $L = x_2 - x_1$, Eq. 5.15 reduces to

$$\Delta T = T_2 - T_1 = Q_k \frac{L}{kA} \tag{5.16}$$

where $\Delta T = T_2 - T_1$ is the temperature difference along the length L.

The thermal resistance θ can be defined as

$$\theta = \frac{L}{kA} \tag{5.17}$$

$$= \frac{\Delta T}{Q_k} \tag{5.18}$$

As shown in Fig. 5.5, a heat source producing Q_k watts is mounted on a block attached to a heat sink which is at a constant, uniform temperature of $T_{\text{heat sink}}$. The temperature of the heat source in this configuration can be calculated from the following:

$$T_{\text{heat source}} = T_{\text{heat sink}} + \theta \times Q_k \tag{5.19}$$

The temperature difference ΔT is the difference between the heat source and heat sink temperatures

$$\Delta T = T_{\text{heat source}} - T_{\text{heat sink}} \tag{5.20}$$

Example A copper rod, 6 inches long, 0.5 inch wide, and 0.5 inch thick as shown in Fig. 5.6 has a 40°C temperature difference over its length. One end is fixed at 100°C. The thermal conductivity of the copper rod is 10 watt/°C-in.

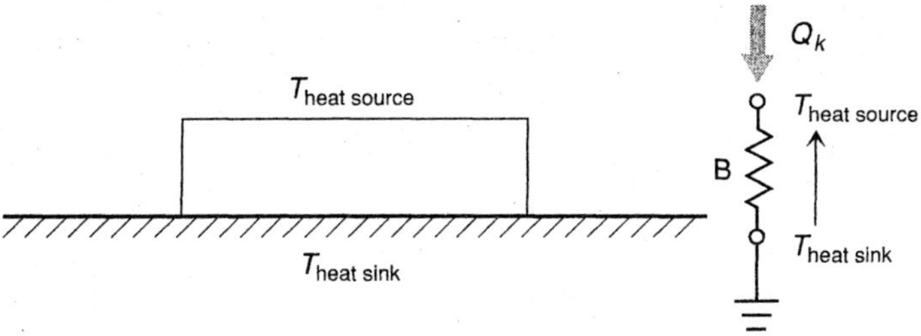

FIGURE 5.5 Temperature distribution.

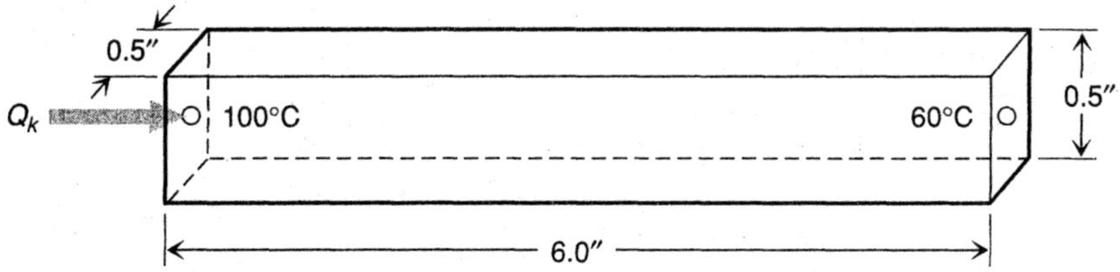

FIGURE 5.6 Copper rod with 40°C temperature difference.

$$\theta = \frac{L}{kA} = \frac{6}{(10\ \text{W/°C-in})(0.5\ \text{in})(0.5\ \text{in})} = 2.4\text{°C/W}$$

The power transferred Q_k can be calculated from Eq. 5.21:

$$\theta = \frac{\Delta T}{Q_k}$$
$$Q_k = \frac{\Delta T}{\theta} = \frac{40\text{°C}}{2.4\text{°C/W}} = 16.67\ \text{W} \tag{5.21}$$

5.2.3 Electrical Analogies

Thermal characteristics can be simplified by considering the thermal properties analogous to electrical properties. The temperature drop ΔT is analogous to the voltage drop V. The rate of heat flow P (or Q) is analogous to current flow. Thermal resistance θ is analogous to electrical resistance R. Thermal conductivity K is analogous to electrical conductivity σ.

When several materials are stacked in series, such as a die attached with epoxy to a substrate which is soldered to a package base, the equivalent or total thermal resistance becomes the sum of the individual thermal resistances. This is analogous to several resistors in series. For N thermal resistances in series, the equivalent thermal resistance θ_{equiv} is

$$\theta_{equiv} = \theta_1 + \theta_2 + \theta_3 + \cdots + \theta_N \tag{5.22}$$

The temperature at a particular interface may be calculated as follows:

$$T_{j,j-1} = T_{\text{heat sink}} + Q_K \sum \theta_{j-\text{hs}} \tag{5.23}$$

where $T_{j,j-1}$ = temperature at interface of layers j and $j-1$
$\Sigma\,\theta_{j,j-1:hs}$ = sum of thermal resistances from interface of layers j and $j-1$ to the heat sink

Figure 5.7 illustrates the stackup of thermal resistances using the electrical circuit analogy with the legends listed as follows:

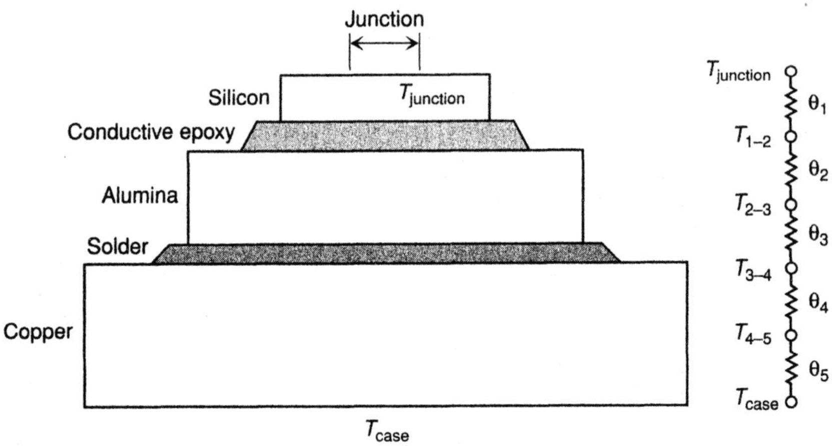

FIGURE 5.7 Stackup of five materials with electrical analog.

θ_1 = thermal resistance of die

θ_2 = thermal resistance of die attach

θ_3 = thermal resistance of substrate

θ_4 = thermal resistance of substrate attach

θ_5 = thermal resistance of package

T_j = junction temperature

T_{case} = case temperature

The temperature at the interface of two layers cannot change discontinuously. For example, the temperature at the bottom of the die will be the same as the temperature at the top of the die attach material and is denoted in Fig. 5.7 as $T_{1\text{-}2}$.

Example In Fig. 5.7, the heat dissipated in a junction is 10 watts. The thermal resistance of the copper is 0.5°C/watt. The case temperature is 100°C. Find the temperature of the top of the copper base.

From Eq. 5.23,

$$T_{j,j-1} = T_{\text{heat sink}} + Q_k \sum \Theta_{j-\text{hs}}$$

$$T_{4,5} = 100°\text{C} + 10\text{ W} \times \Theta_5 = 100 + 10 \times 0.5 = 105°\text{C}$$

When there is more than one heat path from the dissipating element to ambient, then the equivalent thermal resistance can be considered to be the parallel equivalent of the individual thermal resistances. This is analogous to electrical resistors in parallel. For N thermal paths, the equivalent thermal resistance θ_{equiv} is calculated as

$$\frac{1}{\theta_{equiv}} = \frac{1}{\theta_1} + \frac{1}{\theta_2} + \frac{1}{\theta_3} + \cdots + \frac{1}{\theta_N} \tag{5.24}$$

In Fig. 5.8, there are two heat paths from the die. The first path, θ_1, is from the die through the substrate. The second, θ_2, is from the top of the die through the lead frame. The equivalent thermal resistance for the configuration in Fig. 5.8 is

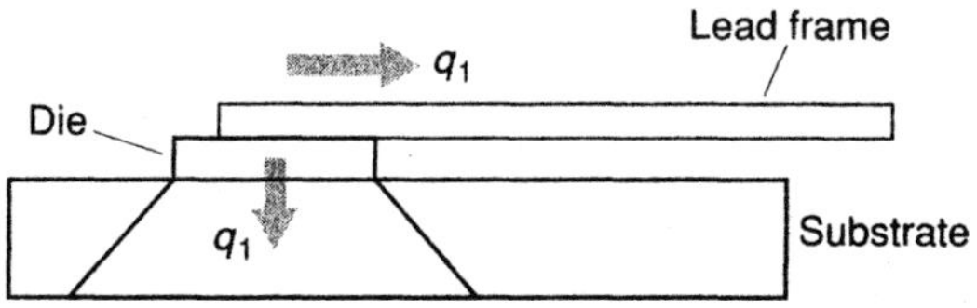

FIGURE 5.8 Parallel heat paths.

$$\frac{1}{\theta_{equiv}} = \frac{1}{\theta_1} + \frac{1}{\theta_2} \tag{5.25}$$

where θ_1 = thermal resistance of the die and substrate
θ_2 = thermal resistance through the lead frame

5.3 CONVECTION

Convection is the transfer of thermal energy between two surfaces as a consequence of a relative velocity between them. [1] It occurs only in fluids where the transfer mechanism is the mixing of the fluids. While each of the surfaces may be a fluid, the most practical application is where one is a solid surface and the other is a fluid. The heat loss due to Newtonian cooling (named after Sir Isaac Newton) or convection cooling is proportional to the temperature difference between them. In mathematical terms, this can be written as

$$Q_c = h_c A_s (T_s - T_A) = h_c A_s \Delta T \tag{5.26}$$

where Q_c = heat transferred from a surface to ambient by convection in watts
A_s = surface area in cm^2 or in^2
T_s = surface temperature in °C
T_A = ambient temperature in °C (temperature to which the heat is being transferred)
h_c = convection heat transfer coefficient in watts/cm^3-°C or watts/in^3-°C

Since h_c is both position and temperature dependent, convection heat transfer solutions are more complex than conduction. Many analysts use a simplifying approximation in which the average surface temperature is used in conjunction with an average heat transfer coefficient h_c.

Equation 5.26 can be rewritten as

$$\Delta T = \frac{1}{h_c A_s} Q_c \tag{5.27}$$

The term θ_s can be defined as the convective surface thermal resistance.

$$\theta_s = \frac{1}{h_c A_s} \tag{5.28}$$

It should be noted that the preceding equation is not a law of heat transfer as seen in Fourier's law, but is a definition of the heat transfer coefficient. This definition simply states the quantity of heat transferred through a temperature difference. This temperature coefficient actually depends on the surface and surrounding temperatures, fluid velocity (for forced convection), fluid viscosity, fluid density, and surface geometry.

There are two types of convection cooling: natural (or free) and forced. In *natural* convection cooling, heat flows by conduction or contact from the surface to the fluid particles in intimate contact with the surface. The fluid particles increase in internal energy causing the density of the nearby fluid to decrease. Buoyant forces then cause the particles to move to a region lower in temperature where further energy transfer takes place by conduction. There is a resulting boundary layer of hot air immediately adjacent to the surface. Natural convection is due entirely to differences in density within the fluids resulting from different temperatures and does not use externally forced air movement.

In *forced* convection, the thermal energy is transferred from the solid to the adjacent fluid particles in the same manner as in natural convection. However, the subsequent fluid action occurs through artificially induced fluid motion generated by fans, pumps, or blowers. There are three types of air-moving devices: centrifugal, propeller, and axial flow. Centrifugal fans are designed to move small volumes of air at high velocities and are capable of working against a high resistance. Propeller types are designed to move large volumes of air at low velocities. Axial flow fans are an intermediate type of air mover between the centrifugal and propeller types.

For forced convection, the convection heat transfer coefficient is calculated from

$$h = B \frac{V^{0.75}}{L^{0.25}} \tag{5.29}$$

where B = constant of air properties and surface configuration
V = linear velocity of air in cm/sec or in/sec
L = characteristic length of surface in direction of flow in cm or in

As can be seen in Eq. 5.29, the linear velocity of the air current as it passes the dissipating element is the key factor to the amount of heat that can be removed.

For natural or free convection, the convection heat transfer coefficient is calculated by

$$h = DE \frac{\Delta T^{0.25}}{L^{0.25}} \tag{5.30}$$

where D = constant for air properties (see Fig. 5.9)
E = constant for surface configuration ($E = 1.9 \times 10^{-4}$ for a flat plate)
L = characteristic length in cm or inches of dissipator surface with area factor
ΔT = temperature difference in °C between dissipator and ambient air

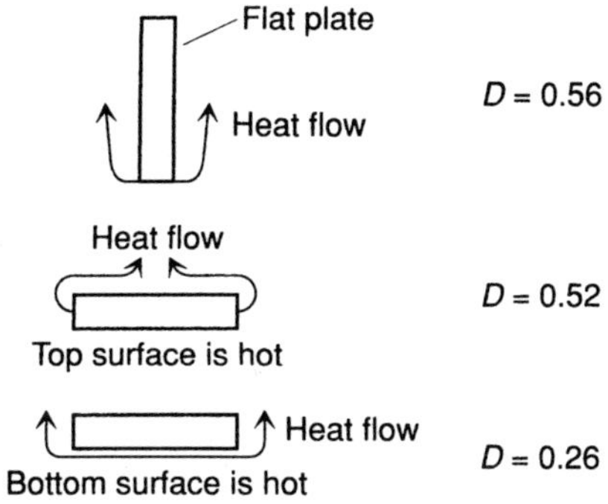

FIGURE 5.9 The value of D in Eq. 5.30.

Forced convection can be divided into laminar flow and turbulent flow. For air, the transition from laminar to turbulent flow usually occurs at a velocity of 180 feet per minute (180 LFM). In laminar (or streamline) flow, the fluid particles follow a smooth, continuous path. The velocity vectors of the particles are always parallel and never intersect. The heat is transferred by molecular conduction in the fluid and by the solid-fluid interface. Turbulent flow, characterized by the irregular motion of fluid particles, has eddies in the fluid in which the particles are continuously mixed and rearranged. The heat is transferred from the eddies back and forth across the streamlines. The greater heat transfer occurs for turbulent flow.

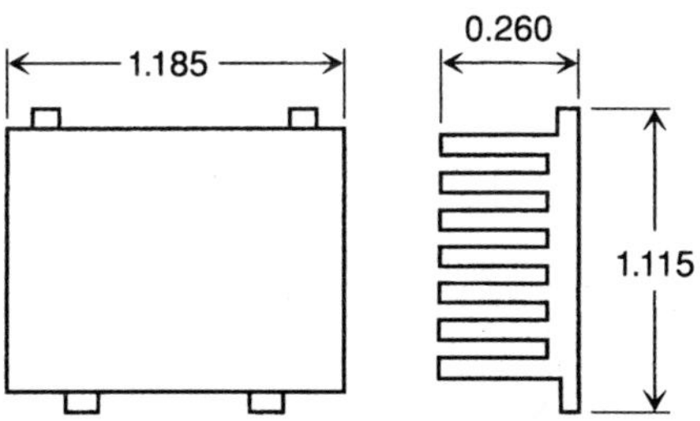

FIGURE 5.10 Aavid 583068 heat sink. *(From Aavid Thermal Technologies, Ref. 2.)*

While individual components, packages, and heat sinks can be modeled with the preceding equations, the experienced thermal designer will find that the most significant amount of convection heat flow will be from the attached heat sink to ambient. Most commercially available heat sinks have been thoroughly characterized for their natural and forced convection parameters. An example is the Aavid Thermal Technologies' 583068 heat sink. See Figs. 5.10 through 5.12.

5.3.1 Convection Cooling Examples

Natural Convection. A flat plate with a characteristic length of 2.0 inch (assumed value for this example), shown in Fig. 5.13, can be considered a model for a flat heat sink. The plate size is 2.0 × 2.0 inches. The plate bottom is at 125°C while the ambient air is 25°C. The value of D for this configuration is 0.26 (from Fig. 5.9). Calculating the value of the connective transfer heat coefficient h_c, from Eq. 5.30

$$h_c = DE\frac{\Delta T^{0.25}}{L^{0.25}} = 0.26 \times 1.9 \times 10^{-4}\,\frac{(125 - 25)^{0.25}}{2.0^{0.25}}$$

$$h_c = 1.31 \times 10^{-2}$$

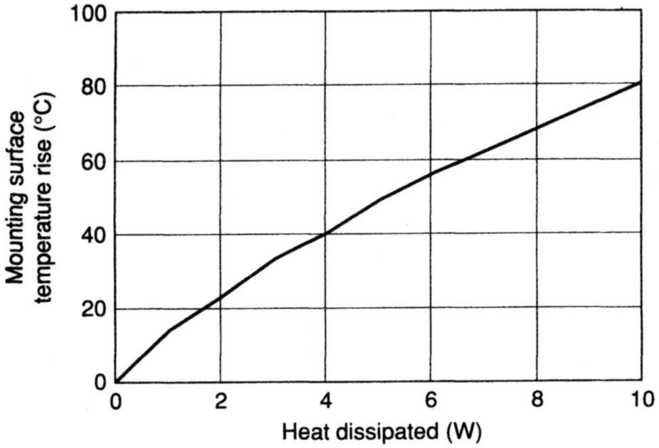

FIGURE 5.11 Aavid 583068 heat sink with natural convection. *(From Aavid Thermal Technologies, Ref. 2.)*

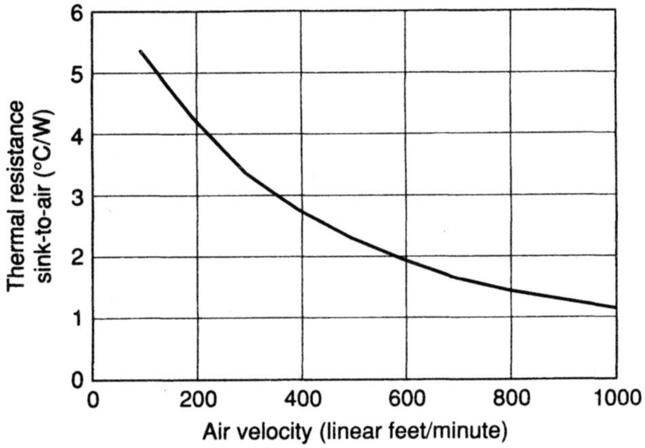

FIGURE 5.12 Aavid 583068 heat sink with forced convection. *(From Aavid Thermal Technologies, Ref. 2.)*

The convection surface thermal resistance is

$$\theta_s = \frac{1}{h_c A_s} = \frac{1}{1.31 \times 10^{-2} \times 4} = 1.9°\text{C/watt}$$

Forced Convection. Air with a velocity V of 500 feet/minute is blown across a plate with a characteristic length L of 2.0 inches. The surface area A is 4.0 square inches and the air property/surface configuration constant B is 1.0×10^{-3}.

From Eq. 5.29

$$h = B\frac{V^{0.75}}{L^{0.25}} = 1.0 \times 10^{-3}\frac{(500 \times 12)^{0.75}}{2^{0.25}} = 0.482\frac{\text{W}}{\text{in}^2\text{-°C}}$$

The thermal resistance is

$$\theta_s = \frac{1}{hA_s} = \frac{1}{0.482 \times 4} = 0.518°\text{C/watt}$$

5.4 RADIATION

Radiation cooling is the transfer of heat by electromagnetic emission, primarily in the infrared wavelengths. Since radiation cooling does not require a transport medium, it is the only means of cooling in a complete vacuum. Specifically, radiation cooling is maximized when there is no intervening material.

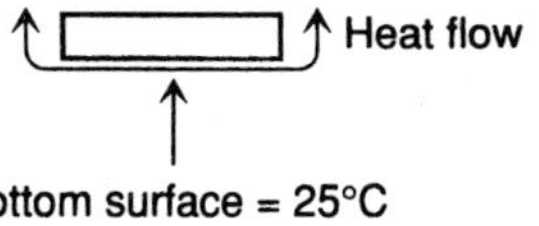

FIGURE 5.13 Example of natural convection.

Radiation from solid objects may be considered a totally surface-related phenomenon that is electromagnetic in character.

A visibly black surface absorbs all of the visible light which falls upon it. Its thermal analogy is called a *black body* and is defined as a surface which absorbs all of the thermal radiation incident upon it. The emissive power of a black body or surface is defined as E_b. Good absorbing materials are also good emitting materials.

Real surfaces do not radiate precisely as described by black body equations as no surface is ideally black. The actual monochromatic emissive power of a real surface is always less than E_b. The emissivity of a body or surface (ε) is defined as the ratio of the radiated flux (E) emitted by a body to that (E_b) emitted by a black body at the same temperature.

$$\varepsilon = \frac{E}{E_b} \tag{5.31}$$

A black body or perfect emitter would have an emissivity ε of 1.0. A perfect reflector would have an emissivity of 0. Listed in Table 5.1 are the emissivities for various materials used in microelectronic applications. Aluminum, a low-cost material with a high thermal conductivity, has a low emissivity of 0.04 in a polished state. After black anodization, the emissivity is significantly increased to 0.80.

The rate of emission of radiant energy from the surface of a body R can be expressed by Eq. 5.32.

$$R = \varepsilon\sigma T^4 \tag{5.32}$$

with R defined as

$$R = \frac{Q}{A} \qquad \text{W/m}^2 \tag{5.33}$$

where ε = surface emissivity in joules/sec-cm^2
σ = Stefan-Boltzmann constant (3.65×10^{-11} watts/in^2-K^4)
Q = heat transferred in watts
A = radiating surface area in m^2
T = temperature of surface in K

TABLE 5.1 Emissivities of Various Materials

Material	Emissivity
White alumina (96%)	0.88
Beryllia	0.87
Polished aluminum	0.04
Rough aluminum	0.06
Anodized aluminum	0.80
Polished copper	0.03
Machined copper	0.07
Thick oxide-coated copper	0.78
Gold	0.04
Kovar	0.05
Silver	0.02
Rolled-sheet steel	0.55
Oxidized steel	0.78
Stainless steel—alloy 316	0.28
Dull nickel plate	0.11
Polished silver	0.02
Bright tin	0.04

Combining Eqs. 5.32 and 5.33,

$$Q = \varepsilon\sigma A T^4 \tag{5.34}$$

The heat transferred via radiation between two black body surfaces ($\varepsilon = 1$) in which one body is completely enclosed by the other (and the internal body cannot see any part of itself) may be calculated by

$$Q = A\sigma(T_1^4 - T_2^4) \tag{5.35}$$

where T_1 = temperature of *hot* body in K
T_2 = temperature of *cold* body in K (air molecules or other absorbing body)

For non–black body surfaces, the heat transferred via radiation is

$$Q = SA\varepsilon\sigma(T_1^4 - T_2^4) \tag{5.36}$$

where S is the shielding factor or view factor.

The *shielding factor* (or *view factor*) S, whose value ranges from 0 to 1, is a measure of how well the emitter sees the absorber. Typical values are shown in Table 5.2.

TABLE 5.2 Shielding Factors for Various Configurations

Configuration	Shielding factor
Infinite parallel plates	1.0
Body completely enclosed by another body; internal body cannot see any part of itself	1.0
Two squares in perpendicular planes with a common side	0.20
Two equal, parallel squares separated by a distance equal to the side	0.19
Two equal, parallel circular disks separated by a distance equal to the diameter	0.18

5.4.1 Radiation Example: Non-black Body

The bottom of a heat sink is at 150°C while the ambient air is at 25°C. The heat sink is nickel-plated with a surface area of 4.0 square inches. The shielding factor is 1.0. The amount of heat transferred due to radiation is calculated from Eq. 5.36 as follows:

$$Q = P = SA\varepsilon\sigma(T_1^4 - T_2^4)$$

$$P = (1)(4)(0.11)(3.65 \times 10^{-11})[(150 + 273)^4 - (25 + 273)^4]$$

$$P = 0.388 \text{ watts}$$

If the heat sink bottom temperature is changed to 125°C, then

$$T_2 = 125°\text{C} \qquad Q = 0.111 \text{ watts}$$

5.5 TRANSIENT HEAT FLOW

$$\frac{\partial}{\partial x}\left(k_x \frac{\partial T}{\partial x}\right) + \frac{\partial}{\partial y}\left(k_y \frac{\partial T}{\partial y}\right) + \frac{\partial}{\partial z}\left(k_z \frac{\partial T}{\partial z}\right) = -Q_v + \rho C_\Theta \frac{\partial T}{\partial t} \tag{5.37}$$

The general differential equation for heat flow under time-varying or transient conditions is shown in Eq. 5.37. The last term, $\rho C_\Theta(\partial T/\partial t)$, describes the time-varying portion of the heat flow. The solution of this differential equation leads to an equation for thermal capacitance.

$$C_\Theta = C_T M = C_T \rho \int_0^L A\, dl = C_T \rho V \tag{5.38}$$

where C_Θ = thermal capacitance in W-sec/°C
ρ = density of the material in kg/m^3
C_T = specific heat of the material in W-sec/kg-°C
V = volume of the material in m^3
l = length of which the heat is conducted in m
A = cross-sectional area of the material in m^2

The densities (ρ) and specific heats (C_T) of various materials used in microelectronic packaging are shown in Table 5.3.

5.5.1 Thermal Capacitance Example

Find the thermal capacitance of the Kovar plate shown in Fig. 5.14.

$$C_\Theta = C_T \rho V$$

$$C_\Theta = 0.110 \times 8.2 \frac{\text{cal}}{\text{in}^3} \times \frac{\text{cal}}{\text{gm-°C}} \times \frac{1 \text{ watt-sec}}{0.2398 \text{ cal}} \times 0.3 \text{ in} \times 0.4 \text{ in} \times 0.050 \text{ in}$$

$$C_\Theta = 0.023 \text{ watt-sec/°C}$$

TABLE 5.3 Material Property Values for Thermal Capacitance Calculations

Material	Specific heat, cal/gm-°C	Density, gm/cm³
Air	0.240	0.0013
Alumina (96%)	0.210	3.7
Alumina (99%)	0.210	3.9
Aluminum nitride	0.167	3.3
Aluminum	0.220	2.7
Beryllia	0.250	2.88
Copper	0.093	8.9
Epoxy	0.200	2.2
Gold	0.030	19.3
Kovar	0.110	8.2
Silicon	0.180	2.4
Silver	0.056	10.5

In an electrical resistor-capacitor (RC) network as shown in Fig. 5.15, the voltage across the capacitor V_c charges exponentially from zero to a final value of V_0 according to the equation

$$C = V_0(1 - \varepsilon^{-(t/RC)}) \tag{5.39}$$

The rate at which the capacitor charges to the equilibrium voltage is the time constant τ.

$$\tau = RC \tag{5.40}$$

The thermal analog to this circuit is shown in Fig. 5.16.

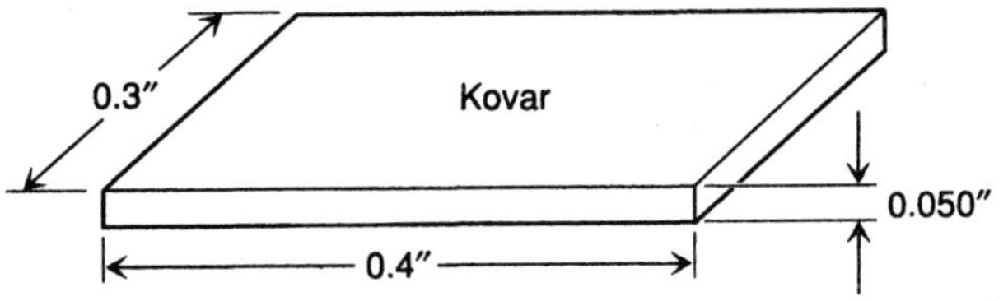

FIGURE 5.14 Thermal capacity of a Kovar plate.

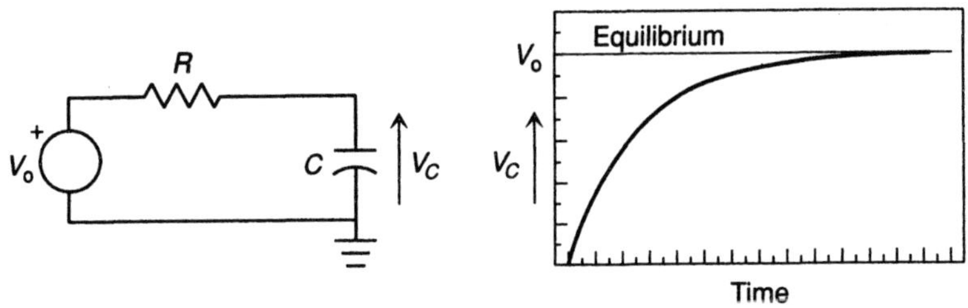

FIGURE 5.15 Resistor-capacitor network and time response.

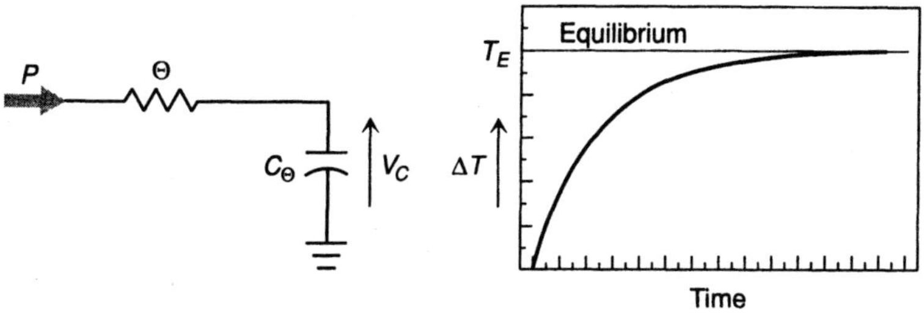

FIGURE 5.16 Thermal network and time response.

The temperature ΔT across the material can be calculated as a function of time by

$$\Delta T = T_E(1 - \varepsilon^{t/\Theta C_\theta}) \tag{5.41}$$

where ΔT = temperature difference
T_E = equilibrium temperature

The rate at which the material approaches temperature equilibrium is the thermal time constant τ.

$$\tau = \Theta C_\Theta \tag{5.42}$$

Using the material values for the Kovar plate shown in Fig. 5.14 (with heat flowing from top to bottom through the 0.050-inch thickness), we calculate a thermal resistance Θ using Eq. 5.17 with the length being the thickness of the plate.

$$\Theta = \frac{t}{kA} = \frac{0.050 \text{ in}}{(0.34 \text{ W/°C-in})(0.3 \text{ in})(0.4 \text{ in})} = 1.22\text{°C/watt}$$

The thermal time constant is calculated from Eq. 5.42.

$$= \Theta C_\theta = 1.22\text{°C/watt} \times 0.023 \text{ watt-sec/°C} = 0.921 \text{ sec}$$

For the exponential curve in Fig. 5.16 that describes the Kovar block, the time to reach 99.9 percent of final value is calculated as shown in the following equation:

$$\frac{\Delta T}{T_E} = 1 - \varepsilon^{-t/\tau} = 0.999 \tag{5.43}$$

Solving Eq. 5.43 for t yields 6.359 seconds. Therefore, the Kovar block will reach 99.9 percent of its equilibrium temperature in 6.359 seconds.

5.6 SUPERPOSITION

Superposition can be defined as a condition, irrespective of the duration of separate events, that their independent effects are additive. In an electrical circuit, superposition means that the voltage at any point is equal to the sum of the various voltages when each of the current sources acts on its own.

This principle can be extended to thermal circuits using the thermal analogy where the temperature at any point in a thermal circuit is equal to the sum of the

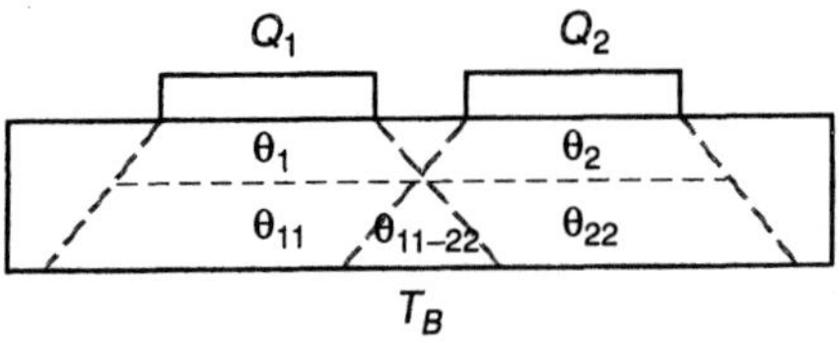

FIGURE 5.17 Multiple heat sources. *Note:* This assumes heat sources Q_1 and Q_2 have equal areas.

temperatures which would arise if each of the heat sources acted on its own. As shown in Fig. 5.17, there are two heat sources, Q_1 and Q_2.

Many thermal problems can be solved easily by combining several thermal resistances using series/parallel combinations and applying the principle of superposition. An example is the multiple heat sources shown in Fig. 5.17 with the electrical analog shown in Figs. 5.18 and 5.19. Assuming a substrate base at a constant temperature T_B, we need to solve for the temperature rises ΔT_1 and ΔT_2 in heat sources Q_1 and Q_2.

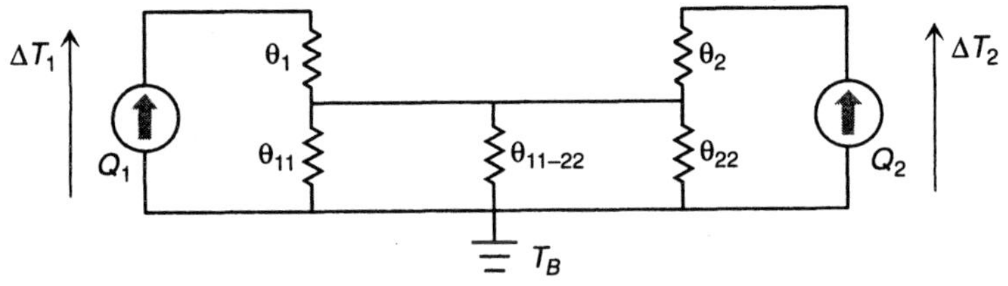

FIGURE 5.18 Electrical analog of multiple heat sources.

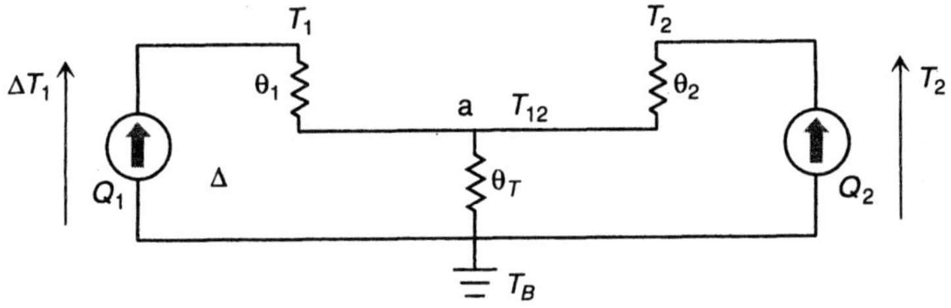

FIGURE 5.19 Electrical analog reduced.

Using the electrical analog, we find that the resistances θ_{11}, θ_{22}, and $\theta_{11\text{-}22}$ are in parallel and have an equivalent thermal resistance θ_T.

$$\frac{1}{\theta_T} = \frac{1}{\theta_{11}} + \frac{1}{\theta_{22}} + \frac{1}{\theta_{11\text{-}22}} \tag{5.44}$$

With heat source Q_2 off (open circuit), the temperature T_{12} at node a is calculated as

$$T_{12} = Q_1 \times \theta_T + T_B$$

With heat source Q_1 off, the temperature at node a is T'_{12}:

$$T'_{12} = Q_2 \times \theta_T + T_B$$

The temperature at node a with both heat sources Q_1 and Q_2 on is the sum of the individual temperature rises:

$$\Delta T = T_{12} + T'_{12} = (Q_1 + Q_2)\theta_T$$

The temperature rise from Q_1 to the heat sink is

$$\Delta T_1 = (Q_1 + Q_2)\theta_T + Q_1\theta_1$$

The absolute temperature T_1 at heat source Q_1 is

$$(Q_1 + Q_2)\theta_T + Q_1\theta_1 + T_B$$

The temperature rise from Q_2 to the heat sink is

$$\Delta T_2 = (Q_1 + Q_2)\theta_T + Q_2\theta_2$$

The absolute temperature T_2 at heat source Q_2 is

$$(Q_1 + Q_2)\theta_T + Q_2\theta_2 + T_B$$

5.7 RECIPROCITY

The principle of *reciprocity* states that the effect at position A due to an event at position B is the same as the effect at B should the event have occurred at position A. The electrical analog of reciprocity occurs for network N when the voltage at point A due to a current source at B is the same as the voltage at B should the same source act at A.

In Fig. 5.20, the first network N_1 has a current I at point A which produces a voltage V at point B. If the two networks are identical ($N_1 = N_2$), then a current I at point B will produce a voltage V at point A.

In the thermal case, reciprocity can be stated as "the temperature at any point A due to a heat dissipation Q at point B is equal to the temperature at point B due to a heat dissipation Q at point A." [3] See Fig. 5.21.

5.8 THERMAL SPREADING

In microelectronic packaging, conductive heat flows from the relatively small heat-dissipating element (the active or passive device) to a much larger heat sink, spreading in a complicated manner. As previously described in the second law of thermodynamics, heat flows from the hotter region to the cooler region(s). This heat flow is three-dimensional as described by Fourier's law. By placing a high thermal conductivity material in the heat path, the heat flow in the x and y horizontal directions is greater than that in the z or vertical direction. Therefore, the heat will spread with the result of increasing the effective thermal cross-sectional area of a relatively poor thermally conducting material. This provides an overall thermal resistance

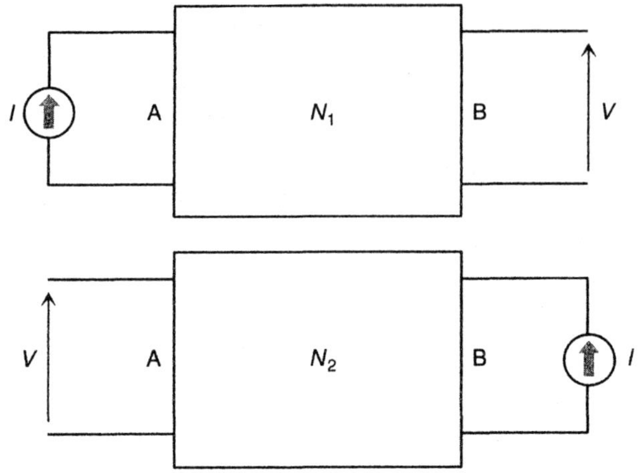

FIGURE 5.20 Reciprocal networks.

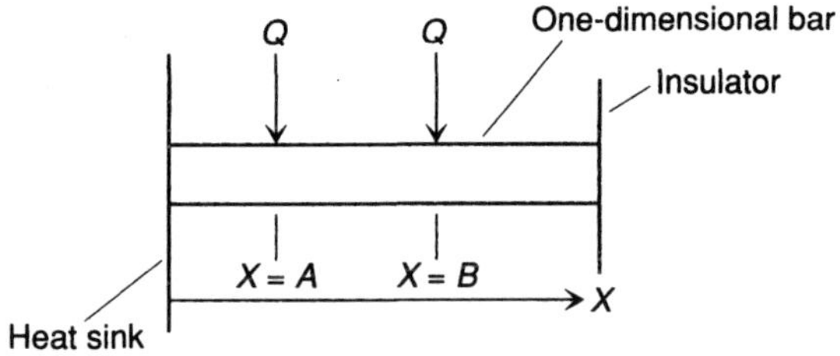

FIGURE 5.21 Thermal reciprocity.

lower than that without the spreader. The importance of increasing the cross-sectional area can be seen in Eq. 5.16 where the thermal resistance is inversely proportional to the area. Heat spreaders, typically high thermal conductivity materials, are placed between the dissipating element and the heat sink (or next layer) as shown in Fig. 5.22.

5.8.1 Spreading Angle

The *spreading angle* α is defined in Fig. 5.23. If the thermal conductivity K of the lower layer(s) is significantly higher than the current layer, spreading will be at a minimum. There will always be some lateral heat spreading due to the randomness of the thermal energy. Therefore, the spreading angle will be greater than 0°.

For the condition where the lower material's thermal conductivity layer(s) is very low (a thermally insulating material) compared to the current layer, the spreading angle will be very large, approaching, but never reaching, 90°.

N. Nguyen [4] gives an approximation of the spreading angle based upon the ratio of thermal conductivities of the current layer versus the underlying layer.

$$\alpha = \tan^{-1} \frac{K_1}{K_2} \tag{5.45}$$

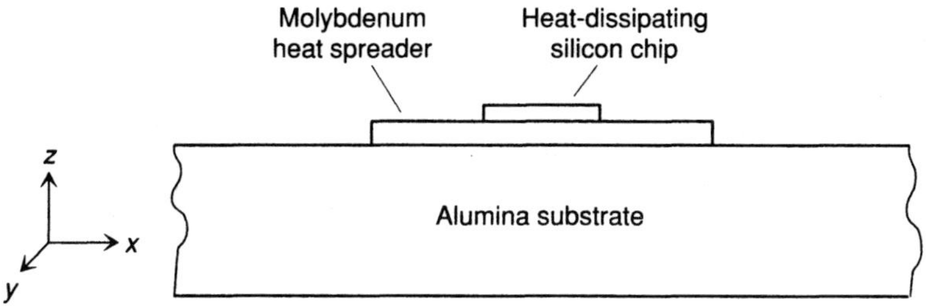

FIGURE 5.22 Heat spreader.

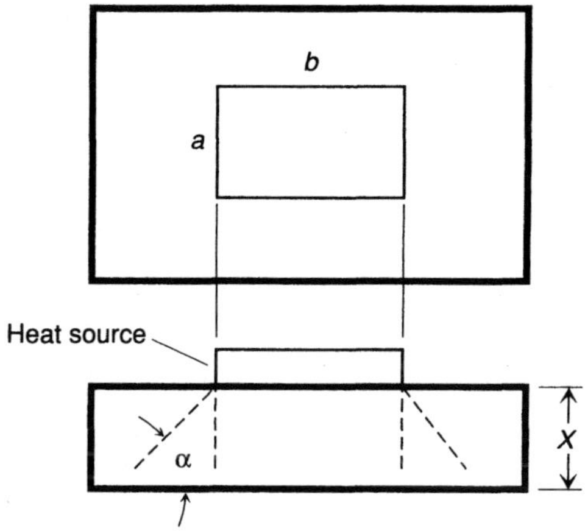

FIGURE 5.23 Spreading angle α for a rectangular heat source.

where α = spreading angle (degrees)
K_1 = thermal conductivity of current layer
K_2 = thermal conductivity of underlying layer

The heat spreading angle as defined in Eq. 5.45 can be plotted as shown in Fig. 5.24. When $K_1 = K_2$, the spreading angle is 45° as indicated by point a on the graph. As the ratio of K_1/K_2 increases, the spreading angle asymptotically approaches 90°.

Example For the configuration shown in Fig. 5.25, the spreading angle in the silicon chip is

$$\alpha = \tan^{-1}\frac{K_1}{K_2} = \tan^{-1}\frac{2.4}{10.2} = 13.2°$$

The spreading angle in the copper is

$$\alpha = \tan^{-1}\frac{K_1}{K_2} = \tan^{-1}\frac{10.2}{0.56} = 86.9°$$

In many applications, one can assume a 45° spreading angle to obtain first-order approximations of temperature rise. This spreading angle approximation is the most

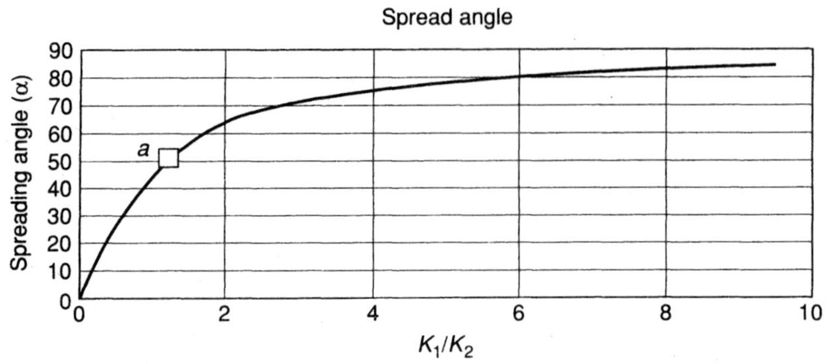

FIGURE 5.24 Spread angle versus the ratio of thermal conductivities of two materials.

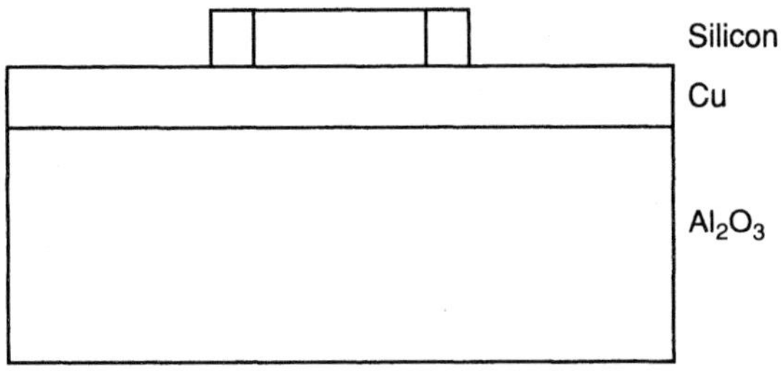

FIGURE 5.25 Heat spreading in copper on alumina.

widely used and holds if there are no severe thermal mismatches and if no thermal boundaries come into play. However, the 45° spreading angle model produces a mathematical case where the thickness t approaches infinity and theta approaches 0. This is incorrect.

For a square heat-dissipating element, we can define the ratio of the heat-dissipating element side L to that of the heat spreader thickness t as the normalized thickness τ_L.

$$\tau_L = \frac{t}{L} \tag{5.46}$$

A more exact single angle model [5] uses 26.6°C. This model is accurate for a normalized thickness τ_L of less than 2.

5.8.2 Restricted Spreading

If thermal boundaries come into play as shown in Fig. 5.26, then the analysis needs to be broken down into two components in series. The first component would have the trapezoidal portion of the heat path with spreading as shown in Fig. 5.27*b*. The second, the rectangular section, depicted in Fig. 5.27*c*, would have no heat spreading. The total thermal resistance θ_{total} is the series combination of the thermal resistances of the trapezoidal and rectangular sections.

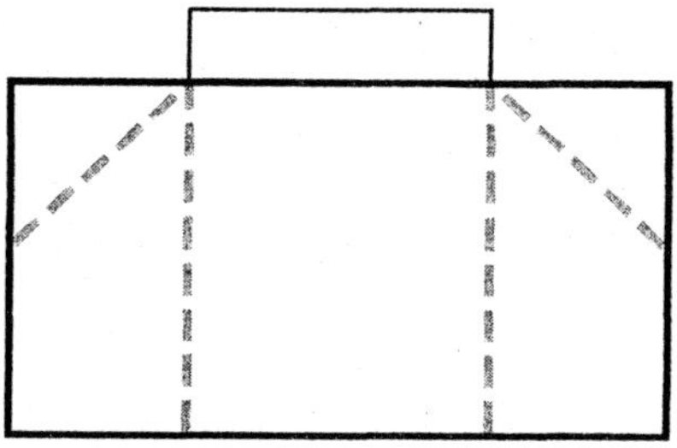

FIGURE 5.26 Example of restricted heat spreading.

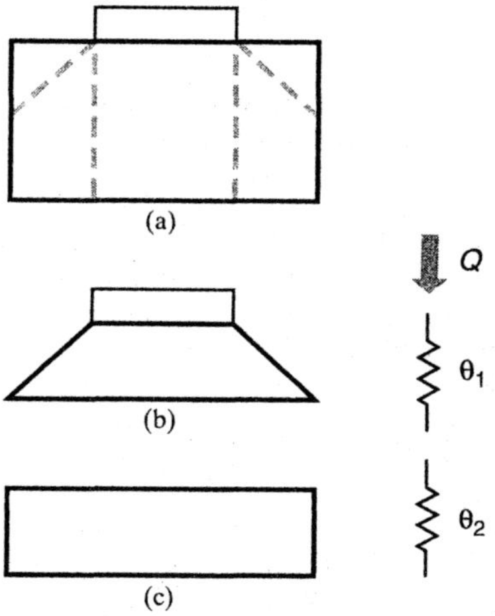

FIGURE 5.27 Thermal resistance for limited spreading.

$$\theta_{total} = \theta_1 + \theta_2 \tag{5.47}$$

where θ_1 = thermal resistance of trapezoidal section
θ_2 = thermal resistance of rectangular section

Heat spreaders are most effective for small heat sources with larger-sized, high thermal conductivity heat spreaders. Materials used for heat spreaders in microelectronic applications include copper, molybdenum, copper-tungsten, copper molybdenum, copper-Invar-copper, beryllium oxide, aluminum nitride, and CVD diamond.

In practical applications, there are bonding and contact thermal resistances between the heat-generating element and the heat spreader. Care must be taken so that these bonding and contact thermal resistances of the heat spreader do not overwhelm the thermal resistance gain afforded by the heat spreader.

REFERENCES

1. J. Seely and R. Chu, *Heat Transfer in Microelectronic Equipment, A Practical Guide,* Marcel Dekker, New York, 1972.
2. Aavid Thermal Management Catalog 1200, 1989, pp. 6, 57.
3. D. J. Dean, *Thermal Design of Electronic Circuit Boards and Packages,* Electrochemical Publications Limited, Tiptree, Essex, UK, 1985.
4. N. B. Nguyen, "Properly Implementing Thermal Spreading Will Cut Cost While Improving Device Reliability," *Proceedings, 1996 International Symposium on Microelectronics,* October 1996, pp. 385–386.
5. R. David, "Computerized Thermal Analysis of Hybrid Circuits," *IEEE Transactions on Parts, Hybrids, and Packaging,* **PHP-13**(3) 283–290, September 1977.

FURTHER READING

Beccue, P., "Micro-Based Thermal Model for Power Hybrids," *PCIM,* pp. 12–18, February 1987.

Bilson, R., M. Hepher, and J. McCarthy, "The Impact of Surface Mounted Chip Carrier Packaging on Thermal Management in Hybrid Microcircuits," *Thermal Management Concepts in Microelectronic Packaging, From Component to System,* R. Howard, ed., ISHM Technical Monograph Series 6984-003, Silver Spring, Maryland, 1984, p. 187.

Designing with Beryllia, Engineering Seminar Notes, National Beryllia Corporation, pp. 27–31.

Ellison, G. N., *Thermal Computation for Electronic Equipment,* Van Nostrand Reinhold Company, New York, 1984.

Fukuoka, Y., and M. Ishizuka, "Transient Temperature Rise for Multi-Chip Packages," *Thermal Management Concepts in Microelectronic Packaging, From Component to System,* R. Howard, ed., ISHM Technical Monograph Series 6984-003, Silver Spring, Maryland, 1984, pp. 313–333.

Kuhlman, M., and H. Schitoglu, "A Thermal Evaluation of Multichip Module (MCM) Materials and Designs," *Eighth IEEE Semi-Therm Symposium,* New York, 1992, pp. 110–116.

Meeks, D., Fundamentals of Heat Transfer in a Multilayer System," *Microwave Journal,* pp. 165–172, January 1992.

Moore, M., *Thermalloy Technical Report,* EIR 84-1012, December 13, 1984.

Motorola Semiconductor Application Note AN1040, "Mounting Considerations for Power Semiconductors," 1988, pp. 1–3.

"Thermal Guide for Reliability Engineers," Hughes Aircraft Company, January 1982.

Wilcox, W., "Heat Transfer in Power Transistors," IEEE Transactions on Electron Devices, New York, September 1963, pp. 308–313.

CHAPTER 6
COMPUTER-BASED THERMAL ANALYSIS

Debabrata Pal
Garron K. Morris*

6.1 INTRODUCTION

As demands on the performance and packaging of electronic assemblies have continually increased, the problem of thermal management has become more challenging. The trend toward miniaturization has lead to an increase in package-level heat fluxes while the temperature limits have remained unchanged. The impact of temperature on reliability was shown in a survey conducted by the United States Air Force which reported that temperature contributed to more than 50 percent of all electronic failures. [1] In industry, thermal analyses are often ignored in the design process or performed too late—when design changes are limited and become too costly. Integration of thermal analyses into the design cycle (from conception to final product) will ultimately lead to more robust and reliable electronic assemblies.

The motivation for this chapter is to present a practical guide for performing thermal analyses using commercially available software packages. It is beyond the scope of this chapter to evaluate or recommend thermal analysis software and an effort has been made to present the information without bias toward a particular software package. This section first introduces the governing equations of fluid flow and heat transfer and is followed by brief discussions of the finite difference and finite element methods. A short discussion of solution methods closes the section.

6.1.1 Governing Equations

The conservation equations for mass, momentum, and energy are written in the form of nonlinear partial differential equations. These partial differential equations have similar forms, which can be used to develop a numerical procedure to solve these equations. For the prediction of fluid flow and heat transfer, the conservation equations for mass, momentum, and energy are described in the following narrative.

Conservation of Mass. The conservation of mass can be written in index notation using an orthogonal set of axes (x_1, x_2, x_3) with respective velocities (u_1, u_2, u_3) where the subscripts i and j can take on the values of 1, 2, or 3:

$$\frac{\partial \rho}{\partial t} + \frac{\partial}{\partial x_i}(\rho u_i) = 0 \tag{6.1}$$

Here, ρ is density, t is time, and u_i is the velocity in the direction of x_i.

* Debabrata Pal and Garron K. Morris, Advanced Thermal Laboratory Cellular Infrastructure Group, Motorola, Inc.

Conservation of Momentum. The conservation of momentum for a Newtonian fluid (for most of the fluids encountered in practice) can be written as

$$\frac{\partial}{\partial t}(\rho u_j) + \frac{\partial}{\partial x_i}(\rho u_i u_j) = \frac{\partial}{\partial x_i}\left(\mu \frac{\partial u_j}{\partial x_i}\right) - \frac{\partial p}{\partial x_j} + S_j \tag{6.2}$$

where μ is viscosity, p is pressure, u is velocity, and S_j indicates the body force term per unit volume. The body force term S_j can be due to gravity, rotational forces, and so on.

Conservation of Energy. The energy equation is expressed as

$$\frac{\partial}{\partial t}(\rho c_p T) + \frac{\partial}{\partial x_i}(\rho u_i T) = \frac{\partial}{\partial x_i}\left(k \frac{\partial T}{\partial x_i}\right) + S_h \tag{6.3}$$

where T is temperature, c_p is specific heat, k is thermal conductivity, and S_h is the energy generation per unit volume. For the energy equation in fluid flow, S_h will be set to 0. For heat conduction through solids, the preceding energy equation reduces to

$$\frac{\partial}{\partial t}(\rho c_p T) = \frac{\partial}{\partial x_i}\left(k \frac{\partial T}{\partial x_i}\right) + S_h \tag{6.4}$$

Here S_h is the energy dissipation per unit volume for the situations where heat dissipation is assumed uniformly from a given volume, such as the silicon die in power transistors. A more in-depth discussion of the derivation of the governing equations of mass, momentum, and energy is available in textbooks such as Kays and Crawford [2] and White [3].

Conservation Equations for Turbulence. Turbulence is commonly encountered in most practical situations. To relate the effects of turbulence to the overall heat transfer and fluid flow, the turbulent diffusion flux and turbulent heat fluxes are used in the fluid flow and energy equations, respectively. There are existing models which can express the turbulent diffusion or heat fluxes as time-averaged properties of the flow. An example of this can be the k-ε model, which introduces turbulent viscosity and turbulent thermal conductivity parameters to account for the mean effects of turbulence on diffusion and heat fluxes, respectively. For this model, two conservation equations are written for the generation of the turbulence kinetic energy (k) and its dissipation (ε):

$$\frac{\partial}{\partial t}(\rho k) + \frac{\partial}{\partial x_i}(\rho u_i k) = \frac{\partial}{\partial x_i}\left(\frac{\mu_t}{\sigma_k}\frac{\partial k}{\partial x_i}\right) + G - \rho\varepsilon \tag{6.5}$$

$$\frac{\partial}{\partial t}(\rho \varepsilon) + \frac{\partial}{\partial x_i}(\rho u_i \varepsilon) = \frac{\partial}{\partial x_i}\left(\frac{\mu_t}{\sigma_\varepsilon}\frac{\partial \varepsilon}{\partial x_i}\right) + (c_1 G - c_2\rho\varepsilon)\left(\frac{\varepsilon}{k}\right) \tag{6.6}$$

where c_1, c_2, s_k, s_e are constants and G is the turbulence energy generation per unit volume. A more detailed discussion of the applicability and selection of various turbulence models is presented in the Sec. 6.2.4.

6.1.2 Finite Volume Method

Thc *finite volume method* is a technique for solving partial differential equations in a domain, using control volume-based discretization. It is very similar to the finite difference method, the only distinction being that it uses the balance of fluxes and forces along control volume boundaries as a physical significance. The general form the conservation equations for a variable ϕ can be written as: [4]

$$\frac{\partial}{\partial t}(\rho\phi) + \frac{\partial}{\partial x_i}(\rho u_i \phi) = \frac{\partial}{\partial x_i}\left(\Gamma \frac{\partial \phi}{\partial x_i}\right) + S_\phi \tag{6.7}$$

where G is the diffusivity and S_ϕ is the source term (such as S_j and S_h) in Eqs. 6.1 and 6.2. Integrating the preceding equation over a control volume surrounding node P, and after simplification, the resulting discretization equation becomes

$$A_P\phi_P = A_E\phi_E + A_W\phi_W + A_N\phi_N + A_S\phi_S + A_T\phi_T + A_B\phi_B + b \tag{6.8}$$

where A and b are values that are calculated from the neighboring control volume faces and the subscripts E, W, N, S, T, and B denote the adjacent east, west, north, south, top, and bottom nodes, respectively. This equation can be solved using an iterative procedure. See Fig. 6.1 for a three-dimensional representation of a control volume.

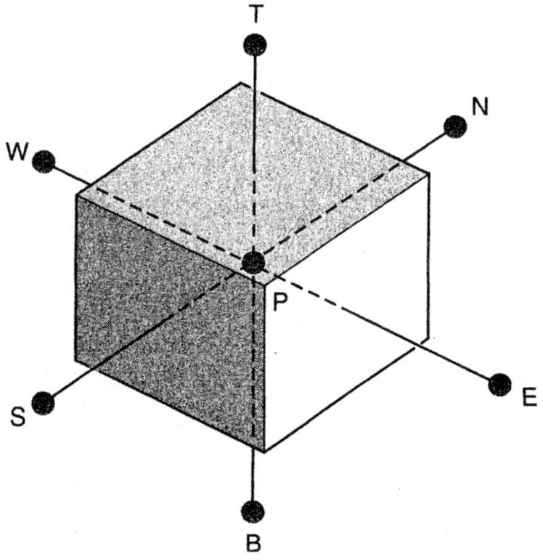

FIGURE 6.1 Three-dimensional control volume surrounding node P.

6.1.3 Finite Element Method

The finite element method uses global error minimization as the solution strategy. The main advantages of the finite element method are the ability to model complex shapes and to use an unstructured grid to discretize the computational domain. In the finite element method, the partial differential equation to be solved can be represented as

$$L(\phi) - f = 0 \tag{6.9}$$

where L is a partial differential operator, ϕ is the variable such as temperature, and f is either a constant or a function. In the Galerkin finite element method, a shape function is defined in such a way that

$$\hat{\phi}(\bar{x}, t) = \sum_{i=1}^{N} W_i(\bar{x})\phi_i(t) \tag{6.10}$$

where W_i and ϕ_i are functions, $\bar{x} = (x, y, z)$ is a coordinate vector, and t is time. For error minimization, the error is defined as

$$\varepsilon(\bar{x}, t) = L(\hat{\phi}(\bar{x}, t)) - f \tag{6.11}$$

and then integrated over a volume to give the overall error minimization

$$\int_V \varepsilon(\bar{x}, t) W_i(\bar{x}, t)\, dV = 0 \tag{6.12}$$

For a two-dimensional incompressible flow, the momentum equations are written as

$$\frac{\partial}{\partial x}(\rho uu) + \frac{\partial}{\partial y}(\rho uv) - \frac{\partial}{\partial x}\left(\mu \frac{\partial u}{\partial x}\right) - \frac{\partial}{\partial y}\left(\mu \frac{\partial u}{\partial y}\right) + \frac{\partial p}{\partial x} = S_x \tag{6.13}$$

$$\frac{\partial}{\partial x}(\rho uv) + \frac{\partial}{\partial y}(\rho vv) - \frac{\partial}{\partial x}\left(\mu \frac{\partial v}{\partial x}\right) - \frac{\partial}{\partial y}\left(\mu \frac{\partial v}{\partial y}\right) + \frac{\partial p}{\partial y} = S_y \tag{6.14}$$

Using the variational formulation, [5] the preceding equations can be expressed in the variational forms as

$$\int_\Omega w_1\left[\frac{\partial}{\partial x}(\rho uu) + \frac{\partial}{\partial y}(\rho uv) - \frac{\partial}{\partial x}\left(\mu \frac{\partial u}{\partial x}\right) - \frac{\partial}{\partial y}\left(\mu \frac{\partial u}{\partial y}\right) + \frac{\partial p}{\partial x} - S_x\right] = 0 \tag{6.15}$$

$$\int_\Omega w_2\left[\frac{\partial}{\partial x}(\rho uv) + \frac{\partial}{\partial y}(\rho vv) - \frac{\partial}{\partial x}\left(\mu \frac{\partial v}{\partial x}\right) - \frac{\partial}{\partial y}\left(\mu \frac{\partial v}{\partial y}\right) + \frac{\partial p}{\partial y} - S_y\right] = 0 \tag{6.16}$$

where w_1 and w_2 are the corresponding test functions for u and v. Following mathematical manipulation, and using the variational form for the continuity equation, the final form of the algebraic equation is given by

$$\begin{bmatrix} [K^{11}] & [K^{12}] & [K^{13}] \\ & [K^{22}] & [K^{23}] \\ & & [K^{33}] \end{bmatrix} \begin{Bmatrix} \{u\} \\ \{v\} \\ \{p\} \end{Bmatrix} = \begin{Bmatrix} \{F^1\} \\ \{F^2\} \\ 0 \end{Bmatrix} \tag{6.17}$$

The preceding equation can be solved in an iterative manner.

6.1.4 Solution of Equations

The governing partial differential equations from the conservation equations usually result in coupled sets of algebraic equations. Solution of such systems of equations requires an iterative strategy. The solution of the velocity and pressure fields poses a problem since the conservation equations of mass and momentum are nonlinear and coupled. Also, for the prediction of turbulent flows and for natural convection problems, the conservation equations for turbulent kinetic energy (k) and dissipation (ε) are coupled. The various issues related to solutions of flow-related equations are described here while a more general overview of solution strategies can be found in Sec. 6.2.7.

In finite volume-based solvers, the solution is usually resolved by using semi-implicit techniques, where the pressure and velocity equations are coupled using the equations for conservation of mass. These methods are briefly discussed in the following.

- *Simple:* In this methodology, the pressure field is guessed. A velocity field is computed using the guessed pressure field. Then a pressure correction equation is solved and the pressure field is corrected. Using velocity correction formulae, the velocity field is then corrected. This is done in an iterative process until the pressure and the velocity fields are converged. [4]
- *Simpler:* This methodology starts the solution with a guessed pressure field. Using the continuity and the momentum equation, a conservation equation is developed for pressure. Following the solution of this equation, a pressure field is obtained. This is then used to correct the velocity field. This is also performed in an iterative process until convergence. [4]

Finite element method–based solvers usually incorporate a method that uses a penalty parameter to couple the pressure and the velocity terms. [5]

6.2 GENERAL METHODOLOGY

Taking time to plan a thermal analysis significantly reduces the bad modeling assumptions that occur when an engineer jumps headfirst into model creation. There are several questions that should be answered prior to creating and running a model of an electronic assembly:

1. Problem definition
   - What are the objectives of the model?
   - How accurate does the model need to be?
2. Physical/computational domain
   - Should the whole system be modeled or a portion of it?
   - Will the geometry be created in the CFD/thermal code or imported from CAD?
   - Will the model be two-dimensional, three-dimensional, or axisymmetric?
   - Can symmetry be used?
3. Computational grid
   - What type of grid/mesh will be used?
   - Are higher-order elements needed?
   - What regions of the model require course/fine meshes?
4. Model selection
   - Is the problem isothermal, conjugate conduction/convection, or conduction only?
   - Is the flow laminar or turbulent, steady or unsteady, compressible or incompressible?
   - What is the appropriate turbulence closure model and near-wall function?
   - Is buoyancy or radiation expected to significantly influence the model?

5. Material selection and properties
   - Are the thermophysical properties affected by temperature?
   - Should an effective thermal conductivity be used for composite materials like printed wiring boards?
   - Are the material properties isotropic or orthotropic?
6. Boundary conditions
   - Can the model be simplified by symmetry, cyclic, or periodic boundary conditions?
   - Are all the boundary conditions known?
   - Can convection boundary conditions be approximated using empirical correlations?
7. Solution strategy
   - What solver/solution algorithm will be used?
   - Is a higher-order interpolation scheme needed?
   - How long will it take for the model to converge?
8. Post-processing
   - Do the results make sense?
   - Does the model need refinement?
   - Are the problem objectives satisfied?

Successful analysts are those that *keep the model simple in the beginning* and remember that finite element and finite volume methods are merely numerical techniques that provide *approximate* solutions to the governing equations. Furthermore, the models are only as good as the perceived boundary conditions and assumptions. Perhaps the most important thing to remember in examining a converged solution is to ask yourself, "Do the results make sense?" Even though CFD/FEA can produce detailed temperature and velocity maps in the solution domain, the results could be unrealistic if proper attention was not paid in developing the model. Only experimental/analytical validation can prove the predicted model results correct.

The following sections present a general methodology for performing thermal analyses of electronic assemblies. Each section gives practical information covering one aspect of model creation or execution. Where applicable, tips and techniques for component-, board-, and system-level analyses are presented at the end of each section.

6.2.1 Problem Definition

The need for thermal analysis was described in the previous section. This section is devoted to the fundamental concepts of heat transfer required for thermal analyses in electronic packaging. Before beginning the modeling process, the objectives and required accuracy of any thermal analysis can be decided by considering the scope of the analysis. There are three main levels of electronic packaging where thermal analyses are performed: at the component, board, and system level.

Component Level. Component-level thermal analyses are performed mainly by solving the energy equation in multimaterial structures using a finite element or a finite volume code. The essential needs for component-level analyses are described below:

1. A detailed temperature field within the component can be obtained. This can be used for performed thermomechanical analyses to estimate the stresses within the structure, enabling the proper design of the component subsystems such as die leadframe, die attach, silicon dice, and various adhesive layers.
2. The overall thermal characteristic of the package can be determined. This is often reported as junction-to-case thermal resistance (θ_{jc}). The choice of the component cooling scheme is dependent on this number. Instead of performing detailed measurements, thermal analyses can be used to perform what-if studies in order to enable the reduction of the thermal resistance of the package.
3. With the increasing power dissipation, component-level thermal design is becoming more and more complex. For the present generation of area-array technologies such as BGA or PGAs, or the high–pin count PQFPs, the heat transfer path from the die to the case is often complex. Proper choices of heat spreaders or other thermal enhancements are needed in order to maintain a low thermal resistance. Thermal analysis can be used to optimize the substrate and spreader sizes.

Board Level. Board-level analyses are usually performed on printed wiring boards (PWBs) populated with components. This is essentially a conduction-type analysis, with prescribed heat transfer coefficients on various surfaces. The essential needs are as follows:

1. A detailed temperature map of the PWB can be obtained. The temperature map of the PWB and the component can be used to determine the solder joint stresses and fatigue calculation. For efficient interconnection design between the component and the board, this provides vital information.
2. For the optimal component placement, an overall thermal map is required, which can be provided by the board-level analysis. The optimal component placement basically aims for minimization of the temperatures for various components. However, it should be mentioned that optimum component placement is often a complex task, and the design objectives and optimal solutions can be very different for thermal, mechanical, and manufacturing cost points of view.
3. Design of the thermal vias and other thermal enhancement for the printed wiring board requires thermal analyses. These may include the choice of substrate and cooling technique for the PWB or module.

System Level. The motivation for system-level thermal analyses is typically to investigate the interactions among the heat-generating devices, heat dissipation equipment such as heat sinks, and the primary airflow path. Unlike component- or board-level analyses, which are based on the solution of temperature, system-level analysis requires the solution of the thermal and flow fields inside the system to obtain complete thermal information. The essential needs of such an analysis are as follows:

1. An overall thermal analysis of a system is required to size air-moving devices such as fans or blowers for forced convection–cooled systems. The flow paths within enclosures are usually complicated which makes sizing fans and blowers difficult. Models of the enclosure can be used to generate a system pressure drop versus flow rate curve that can be used to select an air-moving device. The placement of vents, filters, or EMI shields can also be examined.
2. Regions that may need additional cooling due to obstructions and recirculation zones in the primary flow path can be identified through careful examination of the velocity fields.

3. The placement of heat-generating components and vents for natural convection cooling can be optimized.
4. System-level analyses can be used to verify component- or board-level heat transfer coefficient assumptions. If there is a discrepancy between the assumed and predicted boundary condition values, the component-level thermal models may be refined as needed. The local air temperatures and predicted heat transfer coefficients near a circuit board may be used to calculate the junction temperature of a component using q_{jc} or q_{ja} values from the manufacturer.
5. When modeling an electronic enclosure with intake fans or exhaust vents, the presence of nearby walls, other enclosures, or even room air-conditioning may have a significant effect on the flow path or temperature inside the enclosure. The thermal and flow field around enclosures in rooms can be easily modeled by using the same approaches used in system-level modeling.
6. Different cooling schemes (free convection, forced convection, liquid cooling, etc.) can be investigated.

6.2.2 Physical/Computational Domain

Once the focus of the thermal analysis is defined, the next step is to model features in the physical domain where the analyses will take place. Typically, the geometry is created in the software preprocessor or imported from CAD packages via IGES or DXF file formats. As mentioned earlier, it is important to keep the geometry simple in the early stages of the thermal analyses. Techniques for reducing the complexity of the physical domain are presented for component/board- and system-level analyses.

Component/Board-Level Analyses. For component- or board-level analyses, the choice of the right computational domain can be a critical step. Reduction of the computational domain is primarily dependent on three factors:

- Geometrical symmetry
- Symmetry of the external thermal boundary conditions
- Thermal connection to the next level of packaging

Due to the nature of the packages, the computational domain can be simplified by using symmetry. However, geometrical symmetry refers to the symmetrical internal structure of the component, in terms of location of the die, leadframe, interconnect layers, and other parameters. On the other hand, the application environment of the package also determines if a symmetry can be used. For natural convection cooling, if the component is in vertical orientation as opposed to horizontal orientation in an actual environment, only half of the component can be considered for symmetry because of gravitational effects.

Thermal connection to the next level of packaging refers to the modeling of the printed wiring board along with the package. Due to the solder joints, there is always a thermal connection between the package and the PWB. Hence the proper thermal analysis on a component level should also include the proper size of the PWB underneath. The choice of proper size of the PWB is often dependent on the problem, the structure of the package, and the power levels.

The wide variety of length scales and various materials such as die, leadframe, solder balls, die attach, and so forth, also make the modeling task complicated. Modeling of all these features in a computational model would require a very large mesh size. Mesh size reduction can be achieved through layer or lumped models.

Layer Models. In these models, the interconnect structures are modeled as layers. Examples for this are PQFPs, in which the layer containing the leadframe is modeled as a material with effective thermal properties. This can also be used to model a layer containing solder balls and another material. The effective thermal conductivity k_{eff} then becomes

$$k_{eff} = k_A\phi + k_B(1 - \beta) \tag{6.18}$$

where k_A is the conductivity of the material *A*, k_B is the thermal conductivity of material *B*, and *b* is the volume fraction of material *A*.

Lumped Models. Another way of simplifying interconnect structures such as solder balls or leads is to use lumped models. In this method, several leads or balls are lumped together into one to enable the use of a coarser mesh size. In the example in Sec. 6.3, the actual package studied has a total of 208 leads. Modeling each of these leads could result in a very large mesh size. The modeling is done by lumping every 5.2 leads into one, resulting in the reduction in the total number of leads from 208 to 40.

System-Level Analyses. System-level thermal models typically include the elements that have a significant effect on the flow and thermal fields within an electronics enclosure. Several elements that influence the flow field in enclosures include cabling, connectors, filters, vents, and fans. Large electronic components, printed wiring boards, power supplies, and heat sinks are examples of heat-generating and -dissipating equipment. Various modeling techniques for elements and scenarios encountered in electronic enclosures are discussed here.

The elements that influence the thermal or flow fields are not usually modeled in detail. For example, if the exhaust temperature of an enclosure containing rack-mounted PWBs is needed, it would be computationally wasteful to include a detailed model of a BGA chip mounted on one of the printed wiring boards. The BGA chip can be approximated as a volumetric heat source for the system-level model. The thermal and flow fields in enclosures are typically not affected by the small details (such as leadframe and solder balls) that are included in package- or board-level analyses. As with board- and component-level analyses, symmetry should be used whenever possible to reduce the size of the computational domain.

Fans and Blowers. Air-moving devices such as fans and blowers deliver a volumetric flow rate at a given pressure. Manufacturer catalogs generally plot the performance of a fan or blower in terms of static pressure (p_s) versus volumetric flow rate (G). Although modeling a single fan or blower is achievable, such models require complex rotating or sliding computational meshes, and the level of detail is unacceptable for system-level models. The thermal engineer is usually interested in the flow rate that the fan is capable of delivering in an enclosure.

Many software packages have fan elements or boundary conditions that allow the user to enter the fan or blower performance curve as a series of data points (p_s, G) or polynomial functions $p_s = \Sigma_i A_i G^i$ where A_i is a constant. The use of these elements greatly simplifies the model. If fan elements are not available, a fan can be represented as a planar momentum source and sink with a given velocity normal to the plane. If the fan is located inside the enclosure, it can be modeled as a sink on the upstream side of the fan and a source of equal magnitude on the downstream side. A solid element or wall (with a thickness on the order of the fan depth) should be placed between the sink and source to prevent convergence difficulty.

Care should also be taken to note the air temperature at which the performance curves were generated. Changes in air density ($r_1 \rightarrow r_2$) change the performance of the fan according to fan laws:

$$p_s|_2 = p_s|_1\left(\frac{\rho_2}{\rho_1}\right)^2 \quad \text{for a constant } G\text{, fan diameter, and fan speed} \tag{6.19}$$

The fan or blower performance curves should be adjusted to account for expected changes in air density due to temperature or altitude. If the fan element in the software does not compensate for temperature, then an iterative procedure may be used if the air temperature rise is expected to be large.

Vents, Grills, and EMI Shields. In many air-cooled enclosures, air intake and exhaust vents are encountered in the primary flow path. Grills, EMI shields, and screens are additional examples of planar flow resistance elements. Since it is computationally wasteful to model every hole or opening, these elements are typically approximated in software packages as pressure loss (or resistance) regions. The pressure drop, D_p, across a flow resistance is usually specified in terms of a dimensionless loss coefficient K:

$$K = \frac{\Delta p}{1/2\rho u^2} \tag{6.20}$$

where r is the fluid density and u is the average velocity upstream of the blockage. Loss coefficients may also be functions of flow resistance geometry. An excellent compendium of empirical loss coefficient correlations for various kinds of flow resistances is Fried and Idelcheck [6]. Tables 6.1 to 6.4 summarize loss coefficient equations for screens and grills encountered in system-level CFD analyses.

The loss coefficient for a given flow blockage may be fairly constant over a range of velocities or strongly dependent on velocity and flow regime (laminar, transition, turbulent). Most software packages only allow the user to enter a constant K value, which means that it may become necessary to refine the values of velocity-dependent K coefficients as the model converges.

Cables and Wires. Wires and cables are found in any electronic enclosure and may need to be included in a system-level model if their proximity is close to the primary flow path. Wires or cables that are bundled together may be modeled as a solid element. For widely spaced wires, a planar pressure loss region may be specified similar to those used for vents. Flow resistance elements may be used with the resistance calculated from the pressure loss across a bank or inline or staggered tubes given in Incropera and DeWitt [7] or Fried and Idelcheck [6].

Heat Sinks. Heat sinks can be approximated as equivalent-sized volumes with prescribed flow and thermal resistances. These values can usually be obtained for a specific heat sink from the respective manufacturer's catalog. For custom forced-convection heat sinks, it becomes necessary to specify the thermal resistance q, which can be calculated from

$$\theta = \frac{1}{\eta h A_s} + \frac{1}{\dot{m} c_p} \tag{6.21}$$

where η is the fin efficiency, A_s is the wetted area of the heat sink, $\dot{m}$ is the coolant mass flow rate, c_p is the coolant specific heat, and h is the average convective heat transfer coefficient that may be calculated (via Nusselt number) from correlations and tables presented in Sec. 6.2.6. The flow resistance K_{hs} through the same heat sink may be approximated as

TABLE 6.1 Pressure Loss Coefficient K for Flow through a Circular Metal Wire Screen

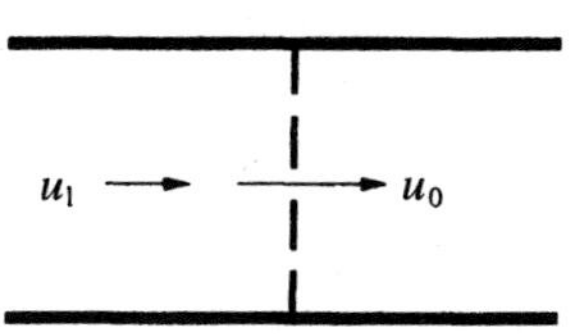

$$K = \frac{\Delta p}{1/2\rho u^2}$$

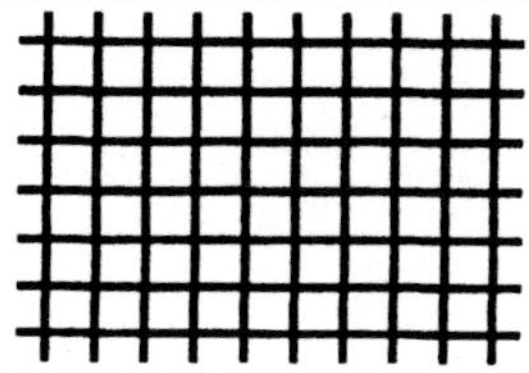

A_0 = area of one orifice
A_1 = area of upstream duct
P_1 = perimeter of upstream duct

$$D_h = \frac{4A_0}{P_1} \qquad \frac{F_0}{F_1} = \frac{\sum A_0}{A_1} \qquad \mathrm{Re}_{Dh} = \frac{u_0 D_h}{\nu}$$

	Re_{Dh}					
F_0/F_1	100	200	300	400	500	>1,000
0.05	450	392	374	367	367	363
0.10	102	88.6	84.5	82.8	82.8	82.0
0.20	21.1	18.4	17.5	17.2	17.2	17.0
0.30	7.69	6.70	6.39	6.26	6.26	6.20
0.40	3.72	3.24	3.09	3.03	3.03	3.00
0.50	2.05	1.78	1.70	1.67	1.67	1.65
0.60	1.20	1.05	1.00	0.98	0.98	0.97
0.70	0.72	0.63	0.60	0.59	0.59	0.58
0.80	0.40	0.35	0.33	0.32	0.32	0.32
0.90	0.17	0.15	0.14	0.14	0.14	0.14

Source: Adapted from Fried and Idelcheck [6].

TABLE 6.2 Pressure Loss Coefficient K for Flow through a Thin-walled Grid ($l/D_h < 0 - 0.015$) with Sharp-edged Orifices of Different Shapes

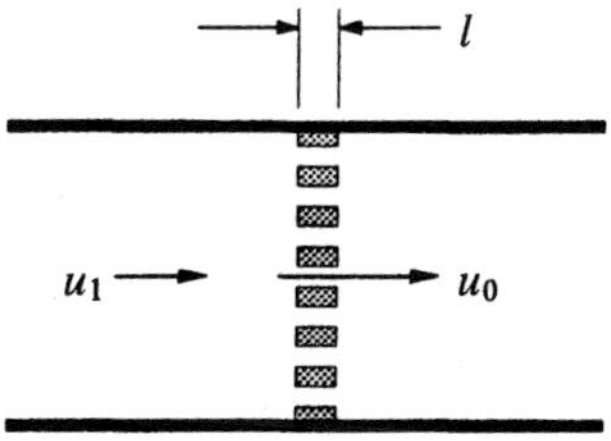

$$K = \frac{\Delta p}{1/2\rho u_1^2}$$

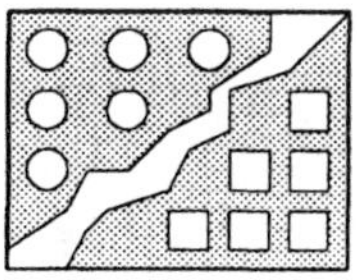

A_0 = area of one orifice
A_1 = area of upstream duct
P_1 = perimeter of upstream duct

$$D_h = \frac{4A_0}{P_1} \qquad \frac{F_0}{F_1} = \frac{\sum A_0}{A_1} \qquad \mathrm{Re}_{Dh} = \frac{u_0 D_h}{\nu}$$

	Re_{Dh}							
F_0/F_1	100	200	400	1,000	2,000	4,000	10,000	20,000
0.20	41.8	38.3	37.8	36.2	36.8	37.8	41.2	43.8
0.30	15.7	14.1	13.3	13.2	13.4	13.8	14.8	15.8
0.40	6.86	5.96	5.54	5.55	5.73	5.89	6.47	6.92
0.50	2.96	2.44	2.32	2.40	2.56	2.68	3.04	3.28
0.60	1.33	1.12	1.08	1.14	1.23	1.30	1.50	1.61
0.70	0.61	0.53	0.50	0.55	0.59	0.62	0.71	0.78
0.80	0.26	0.22	0.22	0.24	0.25	0.27	0.31	0.34
0.90	0.09	0.07	0.06	0.07	0.08	0.08	0.10	0.10
0.95	0.02	0.02	0.02	0.03	0.03	0.03	0.04	0.04

Source: Adapted from Fried and Idelcheck [6].

TABLE 6.3 Pressure Loss Coefficient K for Entrance Flow through a Thin-walled Grid ($l/D_h < 0 - 0.015$) with Sharp-edged Orifices of Different Shapes

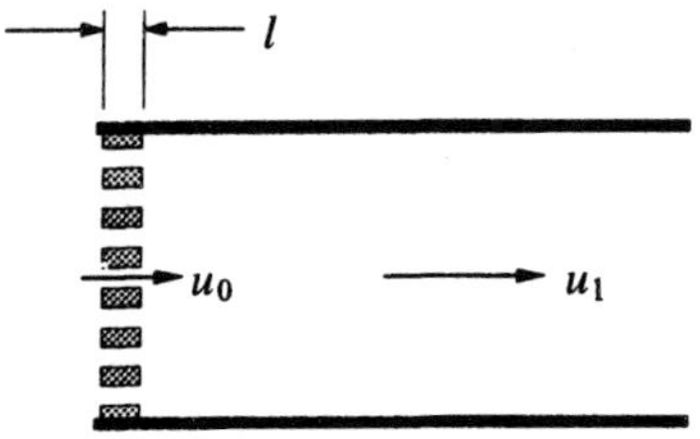

$$K = \frac{\Delta p}{1/2 \rho u_1^2}$$

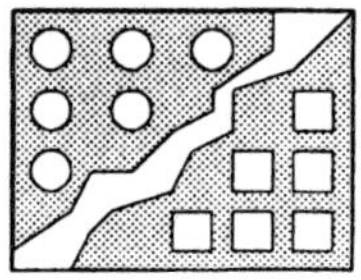

A_0 = area of one orifice
A_1 = area of downstream duct
P_1 = perimeter of downstream duct

$$D_h = \frac{4A_0}{P_1} \qquad \frac{F_0}{F_1} = \frac{\sum A_0}{A_1} \qquad \mathrm{Re}_{Dh} = \frac{u_0 D_h}{\nu}$$

	Re_{Dh}							
F_0/F_1	100	400	1,000	2,000	4,000	10,000	20,000	100,000
0.20	22.2	15.3	10.2	7.84	6.45	4.93	4.09	2.88
0.30	9.13	5.69	4.38	3.61	3.16	2.58	2.38	2.08
0.40	4.25	2.54	2.09	1.88	1.72	1.64	1.63	1.73
0.50	1.94	1.15	1.05	1.06	1.05	1.16	1.22	1.41
0.60	1.02	0.73	0.73	0.78	0.81	0.93	0.99	1.15
0.70	0.63	0.53	0.58	0.62	0.65	0.75	0.82	0.95
0.80	0.42	0.41	0.45	0.49	0.53	0.61	0.67	0.77
0.90	0.30	0.30	0.35	0.38	0.42	0.48	0.53	0.61
0.95	0.23	0.26	0.30	0.34	0.37	0.42	0.46	0.54

Source: Adapted from Fried and Idelcheck [6].

$$K_{hs} = f\frac{L}{D_h} + 1 \tag{6.22}$$

where f is a friction factor, D_h is the hydraulic diameter of one channel (see Eq. 6.45), and L is the heat sink length in the flow direction. An equation for the laminar friction factor in rectangular ducts was given in Shah and London [8] as

$$f = \frac{96}{\mathrm{Re}_{Dh}}[1 - 1.3553\alpha + 1.9467\alpha^2 - 1.7012\alpha^3 + 0.9564\alpha^4 - 0.2537\alpha^5] \tag{6.23}$$

where α is the aspect ratio of one heat sink channel ($0 \le \alpha \le 1$) and Re_{Dh} is given by Eq. 6.26 where D_h is the characteristic length.

6.2.3 Computational Grid

After the physical domain is created, it needs to be discretized into the control volumes (or elements) that were introduced in Sec. 6.1. The complete set of control volumes within the physical domain is called the *computational grid* or *mesh* and it determines the accuracy of the model as well as the numerical stability of the solution. Since grid creation and manipulation capabilities vary from software to software and differ for finite volume and finite element methods, this section is limited to brief descriptions of grid types, grid generation and refinement, and grid inde-

TABLE 6.4 Pressure Loss Coefficient K for Discharge Flow through a Thin-walled Grid ($l/D_h < 0 - 0.015$) with Sharp-edged Orifices of Different Shapes

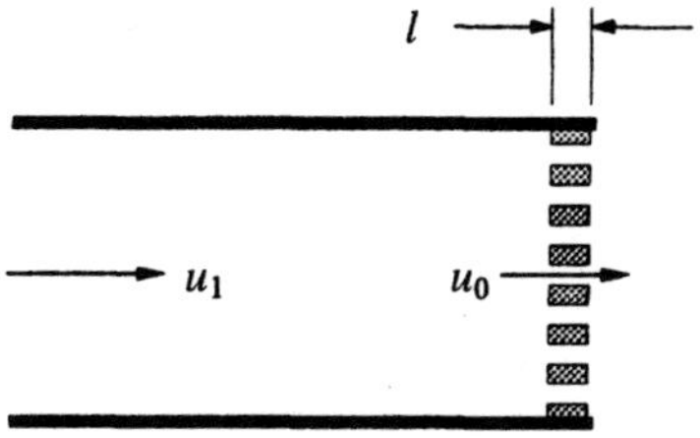

$$K = \frac{\Delta p}{1/2 \rho u_1^2}$$

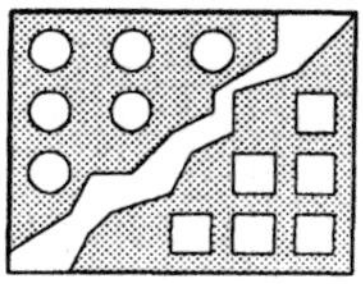

A_0 = area of one orifice
A_1 = area of upstream duct
P_1 = perimeter of upstream duct

$$D_h = \frac{4A_0}{P_1} \qquad \frac{F_0}{F_1} = \frac{\sum A_0}{A_1} \qquad \mathrm{Re}_{Dh} = \frac{u_0 D_h}{\nu}$$

	Re_{Dh}							
F_0/F_1	100	400	1,000	2,000	4,000	10,000	20,000	100,000
0.20	22.3	15.5	10.4	8.07	6.71	5.22	4.41	3.25
0.30	9.35	5.94	4.68	3.94	3.51	2.99	2.83	2.60
0.40	4.52	2.85	2.46	2.29	2.16	2.15	2.19	2.38
0.50	2.26	1.51	1.47	1.53	1.56	1.74	1.86	2.15
0.60	1.37	1.13	1.19	1.29	1.37	1.58	1.70	1.97
0.70	0.99	0.95	1.06	1.16	1.23	1.42	1.56	1.81
0.80	0.79	0.83	0.93	1.02	1.11	1.28	1.40	1.63
0.90	0.64	0.69	0.79	0.88	0.96	1.11	1.21	1.41
0.95	0.54	0.62	0.71	0.79	0.86	0.99	1.09	1.26

Source: Adapted from Fried and Idelcheck [6].

pendence. For an explanation of grid generation theory, see Thompson et al. [9] and Thompson [10].

Grid Types. Discretization of the computational domain into control volumes is achieved by using two types of grids for finite volume-based codes. These are

1. *Structured grids.* This grid type is used for rectangular cartesian grids. Body-fitted grids also fall under this category.
2. *Unstructured grids.* This grid type uses shapes such as tetrahedrons and can be used more efficiently for modeling complex shapes and for local grid refinement.

Finite element–based codes typically use unstructured grids. Some codes have the option of higher-order elements that have additional nodes for increased accuracy.

Grid Generation. The grid generation process starts after the geometrical model of the computational domain with all the required parts and features, such as printed wiring boards, components, and so forth, is established. Typical grid generation steps are as follows:

1. Various surfaces within the solution domain are identified. These surfaces include interfaces among different materials, such as between a PWB and air or between a board and a component mounted on it.

2. The surface mesh is generated on internal and external surfaces. This is done by defining two-dimensional mesh such as a thin shell triangular mesh on the various surfaces.
3. A volume mesh is then generated in the different regions within the computational domain using the appropriate elements (tetrahedral, brick, and so on).

Grid Refinement. The ultimate goal of mesh creation is to create a grid that resolves all the relevant details of the model without significantly impacting computational time or stability. Grid refinement is a modeling technique that adds grids in regions that are of interest while using sparse grids in areas of lesser interest. By creating localized regions of dense grids around relevant elements in the model (components, inlets, and heat sinks, for example) the computational time is decreased while increasing stability and accuracy.

Grid refinement is usually performed during the volume mesh generation process. For situations with conduction analyses (board or component level with prescribed heat transfer boundary), a uniform mesh can be used; however, grid refinement is necessary to efficiently resolve the details of the packaging structure. For conjugate problems involving laminar flows, attention should be paid to grid refinement in the solid-fluid interfaces, in order to capture the boundary layers. A very fine grid spacing should be used near the wall, and the mesh density can be increased gradually away from the boundary layer into the fluid. Figure 6.2 illustrates the influence of the near-wall grid spacing on wall gradients in laminar flow. The recommended distance Y between the wall and the near-wall grid for laminar flows is given in Fluent [11] as:

$$Y \le \sqrt{\frac{x\nu}{u_\infty}} \tag{6.24}$$

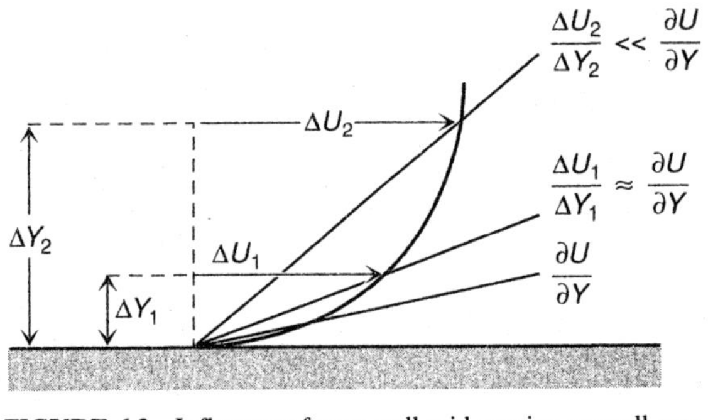

FIGURE 6.2 Influence of near-wall grid spacing on wall gradients in laminar flow. (After Fluent, Ref. 11.)

where x is the distance along the wall from the start of the boundary layer, n is kinematic viscosity, and u_∞ is the free-stream velocity. See Sec. 6.2.4 for a discussion of required near-wall grid spacing for turbulent flows.

Grid Independence. In modeling, there is often a trade-off between solution accuracy and the number of computational grids that is illustrated in Fig. 6.3. As the number of grids increases, so does the solution time; however, there is a point where

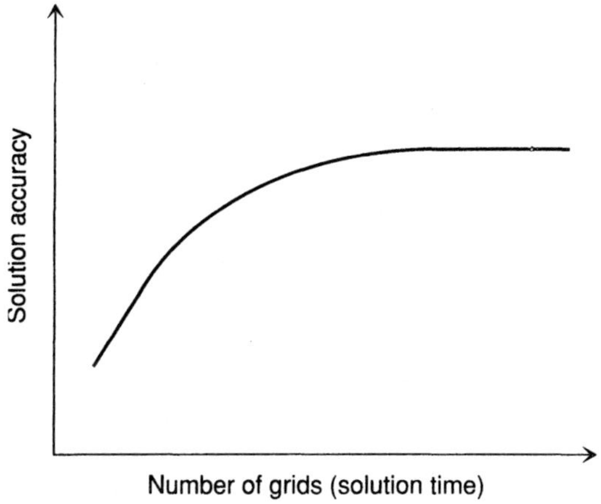

FIGURE 6.3 Influence of the number of grids on the model accuracy.

additional grids have little impact on the solution. At this point, any converged solution is said to be grid-independent. By running several models with a different number of grids, a solution can be declared grid-independent when the difference between relevant solution values (temperature, pressure drop, and so forth) in any two meshes is negligible (less than 5 percent).

6.2.4 Physical Model Selection

Once the computational domain is created and meshed, the type and scope of the thermal analysis is identified by selecting appropriate physical models. These models, in turn, determine the boundary conditions required for solving the problem, which is addressed in the next section. The basic physical models described in this section include steady state and transient problems, conjugate and radiation heat transfer, laminar and turbulent flow, and incompressible and compressible flow.

Although it appears that prior knowledge of the physics within electronic assemblies is required to choose the physical models, assumptions and appropriate rules of thumb can be used in the absence of information or experimental data. The predicted thermal and flow fields can also be refined by rerunning the model with updated information from experiments or the initial solution. For example, if the flow in a heat sink was assumed to be laminar and examination of the velocities in the converged solution showed (via Reynolds number) that the flow was turbulent, the appropriate turbulence closure model could be selected and a new solution can be obtained.

Isothermal, Conduction, and Conjugate Problems. Problems that ignore heat transfer and calculate the flow field are called *isothermal* (only the governing equations for conservation of mass and momentum are solved). Conversely, only the energy equation is solved in *conduction* heat transfer problems. In thermal analysis of electronic assemblies, however, the heat transfer from the component to the surrounding is a combination of conduction, convection, and radiation heat transfer. *Conjugate* heat transfer refers to the effect of multiple modes of heat transfer.

Numerous published studies include all three modes of heat transfer, such as Yu and Joshi [12] or Smith et al. [13].

Transient Models. Time-dependent thermal analyses may be required to accurately predict the temperature of microelectronic devices with transient power dissipations or other time-varying boundary conditions. Transient modeling can also be used to simulate the device temperature from initial power-up to steady state. Steady state problems that tend to be unstable may be solved as transient problems where the iterations are terminated when a steady state solution is achieved. Modeling transient situations requires selecting the time interval *Dt,* imposing any time-dependent boundary conditions, and applying initial conditions ($t = 0$).

Compressible and Incompressible Flow. In electronics cooling, the air velocity and pressure variations are usually low enough to assume that the flow is subsonic and therefore incompressible (e.g., the density only varies with temperature). The effects of compressibility may be considered negligible when the Mach number M is less than 1 ($M < 0.3$). The Mach number is defined as

$$M = \frac{u}{\sqrt{\gamma RT}} \tag{6.25}$$

where u is the mean gas velocity, R is the universal gas constant, T is the gas temperature, and g is the ratio of specific heats (c_p/c_v). If the Mach number is greater than one, the flow is supersonic and compressible (the gas density becomes a function of temperature and pressure). Some thermal analysis codes incorporate the special equations and solution methods that are needed to solve compressible flow fields. These equations and techniques are beyond the scope of this chapter, but may be found in White [3].

Laminar and Turbulent Flow. In addition to being subsonic and incompressible, the relatively low velocity in electronic assemblies ensures that the flow is in the laminar regime where the flow is governed by the equations of motion presented in Sec. 6.1.1. Occasionally, the velocity may be high enough to cause the flow to become turbulent, which is characterized by a random fluid motion identified by fluctuations, eddies, random variations in fluid properties, self-sustaining motion, and mixing. [3] Before choosing physical models, it is important to know whether the flow field is laminar or turbulent.

In electronic assemblies the flow is usually wall-bounded; thus, a good indication of the flow regime is through the Reynolds number:

$$\mathrm{Re}_{Lc} = \frac{\rho u L_c}{\mu} \tag{6.26}$$

where u is an average velocity, ρ is density, μ is dynamic viscosity, and L_c is a characteristic length. Given the geometric complexities and varying scales in electronic systems, however, the choice of characteristic length is often confusing. Several characteristic length values typically encountered in electronic packaging are shown in Table 6.5. In a circular pipe, the flow regimes are laminar for $\mathrm{Re}_{Lc} > 2300$, transition for $2300 < \mathrm{Re}_{Lc} < 4000$, and turbulent for $\mathrm{Re}_{Lc} > 4000$. These ranges, along with the characteristic lengths in Table 6.5, can be used to determine whether the flow is laminar or turbulent in the region of interest. If the Reynolds number is in the tran-

TABLE 6.5 Velocities (u) and Characteristic Length Scales (L_c) for Reynolds Number Given by Eq. 6.26

Configuration	Velocity (u)	Characteristic length (L_c)
Rack-mounted PWBs	Average velocity between boards	Distance between boards
Circular ducts	Average velocity in duct	Diameter
Noncircular ducts	Average velocity in duct	Hydraulic diameter
Inline array of components	Average duct velocity upstream of components	Streamwise component length

sition regime, solutions should be obtained using both laminar and turbulence settings and carefully examined.

Turbulence Closure Models. Since turbulent flow was first recognized by Osbourne Reynolds in 1883, the interest in predicting velocity and temperature profiles along with wall friction and heat transfer has grown considerably. The basis for many predictive models for turbulence is the use of Reynolds averaging, where the random fluid motion is broken into steady and fluctuating components. A complete discussion of this may be found in White [3].

Selecting a Turbulence Model. After examining five benchmark flows using five commercial CFD codes, Freitas [14] concluded that the appropriate turbulence model must be used along with a balance between grid resolution and discretization accuracy. Even if a grid-independent solution is achieved, the flow field may not be in agreement with experimental results due to the limits or assumptions in the turbulence closure model. [14] It is important to know the strengths and weaknesses of the turbulence models in a CFD code. The goal of the section is not to go into detailed derivations of these turbulence models, but to present a general overview of common turbulence models and their merits and limitations. Comprehensive reviews of the turbulence closure models can be found in papers by Launder [15,16]. A review of advanced turbulence models is available in Hanjalic [17].

Turbulence models are typically classified as zero-equation, one-equation, two-equation, and Reynolds stress models. While many analytical and empirical models have been proposed for predicting the effects of turbulence, application-oriented problems with turbulent flow are solved using a handful of robust turbulence models. The mixing-length model, k-equation model, k-ε model, Reynolds stress, algebraic stress, and RNG k-ε models are examples of time-proven models that will be briefly examined here and are summarized in Table 6.6.

Mixing-Length Model. The mixing-length model (or zero-equation model) proposed by Prandtl [18] was one of the first turbulence models used for turbulent flow calculations and related turbulence fluctuations to a length scale and velocity gradient:

$$\mu_t \approx \rho l^2 \left| \frac{\partial u}{\partial y} \right| \tag{6.27}$$

where μ_t is the effective turbulent viscosity and l is the mixing length. The mixing-length model has been proven successful in predicting simple shear flows (jets and wakes) and thin shear layers near walls, provided that an appropriate mixing length (l) is chosen.

TABLE 6.6 Comparison of Turbulence Closure Models

Model	Applications	Strengths	Deficiencies	References
Mixing-length (zero-equation)	Thin shear layers near walls, free shear (unbounded) flows, jets, plumes, wakes	Simple, allows realistic predictions of velocity and shear-stress distributions, and general boundary-layer behavior	Poor at predicting recirculating flows, ignores convection and diffusion of turbulence, stress vanishes as mean velocity gradient becomes zero, length scale model required	Prandtl [18] White [3]
One-equation	Shear layers, jets, wakes, pipe flows, simple developing boundary layer flows	One additional equation needs to be solved, more realistic representation of turbulence energy than mixing-length model	Length scale model is needed, mixing-layer model will give similar results for boundary-layer flows in less time	Kolomogorov [19] Prandtl [20] White [3]
k-ε model	Proven engineering accuracy for a wide range of turbulent flows (ducts, jets, etc.)	Merits and demerits well understood, best suited for application-oriented problems	Isotropic m_t (velocity and length scales same in all directions), semiempirical model, requires use of wall functions	Fluent [21] Jones and Launder [22] Launder and Spalding [23]
Reynolds stress	Swirling flows, separated and reattached flows, complex recirculation	Most sophisticated turbulence model, nonisotropic, improved evaluation of momentum diffusion rates	Computationally intensive, possible convergence difficulties, 50–60 percent more computational time/iteration over k-ε model	Fluent [21] Rodi [24] Launder et al. [25]
Algebraic stress	Swirling flows, separated and reattached flows, complex recirculation	Less computationally intensive than Reynolds stress model, nonisotropic	Ignores convective and diffusive transport or Reynolds stresses, possible convergence difficulties	Hanjalic and Launder [26] Hutchings and Iannuzzelli [27]
RNG k-ε	Separated and recirculating flows, time-dependent vortex shedding, liquid metal heat transfer, low–Reynolds number flows, transition, swirling flows, relaminarization	Constants in model derived analytically, additional term in ε-equation, accounts for effect of swirl on turbulence, analytically derived m_t formula accounts for low-Re effects	Computationally intensive, relatively new model, small database of test cases, isotropic conditions are assumed, 10–15 percent increase in computational time over k-ε model	Fluent [21] Kirtley [28] Yakhot and Orszag [29]

One-Equation Model. Kolmogorov [19] and Prandtl [20] addressed the shortcomings of the mixing-length model (see Table 6.6) by deriving a differential equation based on the transport of turbulence kinetic energy, *k:*

$$\rho\left(u\,\frac{\partial k}{\partial x}+v\,\frac{\partial k}{\partial y}\right)=\frac{\partial}{\partial y}\left(\frac{\mu_t}{\sigma_k}\,\frac{\partial k}{\partial y}\right)+\mu_t\left(\frac{\partial u}{\partial y}\right)^2-C_D\,\frac{\rho k^{3/2}}{l} \tag{6.28}$$

where C_D and σ_k must be prescribed as well as the length-scale distribution l. The k-equation provides a better representation of turbulence energy than the mixing-length model in such cases as pipe flow where the velocity gradient at the centerline is zero, but turbulence is not. This model is called the *one-equation model* because the k-equation must be solved along with the momentum, continuity, and energy equations.

Two-Equation Model. The two-equation turbulence closure model was introduced in Sec. 6.1.1 as the k-ε model. Over the years, there have been many variations on the two-equation model (k-w, k-w^2, k-l), but the k-ε model is perhaps the most robust (and popular) turbulence model for solving application-oriented problems with proven engineering accuracy. The k-ε model, however, is not without shortcomings. The empirical constants in the k-ε model are not universal and must be modified for non-boundary-layer flows such as jet and wakes. The k-ε model assumes isotropic conditions (uniform velocity and length scales in all directions) and does not model viscous and heat transfer effects in the region near walls. These effects must be included using wall functions (which are also discussed in this section). Since Jones and Launder [22] popularized the k-ε model in their paper, it can be found in virtually all computational fluid dynamics (CFD) software.

Reynolds Stress Model. The primary shortcoming of the k-ε model is that the magnitudes of the Reynolds stresses are assumed to be equal or isotropic in all directions. The second-order Reynolds stress model employs six transport equations for the individual stresses (*uv, uw, vw, uu, vv, ww*) and allows for improved predictions of complex nonisotropic flows with high rates of strain. The stress equations must be solved simultaneously along with the continuity, momentum, and dissipation equations and require a large amount of computing time and storage. Overall, the Reynolds stress model is considered to be one of the most sophisticated (and computationally intensive) turbulence closure models. Reviews of this model can be found in Rodi [24] and Launder et al. [25].

Algebraic Stress Model. To simplify the Reynolds stress model, Hanjalic and Launder [26] and others proposed that the six stress transport equations be replaced with approximate algebraic relations (see Ref. 3). Hutchings and Iannuzzelli [27] reported an improvement in the algebraic stress model over the k-ε model in predicting velocity distributions in highly swirling flow. Despite the simplifications (and reduction in computational time), the algebraic stress model does not take the convective and diffusive transport of Reynolds stresses into consideration and should not be omitted in high–Reynolds number flows.

Renormalization-Group-Theory–Based Model. The relatively new RNG k-ε model is a modified k-ε model based on renormalization group (RNG) theory. The RNG method can describe small-scale phenomena (such as eddies) that lack externally imposed length and time scales. Once the small scales (degrees of freedom) are quantitatively accounted for, they are removed and the equations of motion are modified to include a characterization of these degrees of freedom. The RNG method gives a theoretical basis for obtaining the various constants and function in the k-ε model otherwise obtained empirically [28,29].

Near-Wall Models. Wall functions (or near-wall models) account for the shortcomings of the high–Reynolds number k-ε models in the region near walls where molecular viscosity affects the generation, destruction, and transport of turbulence energy and heat transfer is dominated by molecular conductivity. Low–Reynolds number k-ε models and the recent RNG k-ε model address these viscous effects; however, they typically require a large number of computational grids in the region near the wall. The advantage of near-wall functions is that viscous effects are related to the grids located adjacent to the wall in the fully turbulent region (where viscous effects are negligible), reducing the mesh size and CPU time. Although flow field predictions away from the wall are somewhat insensitive to the near-wall model used, local heat transfer predictions at the wall are highly dependent on the choice of near-wall model. Knowing the limitations of wall functions is important for accurate prediction of heat transfer.

Many thermal analysis codes incorporate wall functions that are used with the standard high–Reynolds number k-ε turbulence model. Popular wall functions include equilibrium, two-layer, and three-layer models that differ either in the velocity scale or division of the near-wall region into three distinct regions (see Table 6.7). White [3] labels these regions the *viscous sublayer, buffer layer,* and *overlap regions* that are located in terms of a dimensionless wall coordinate:

$$y^+ = \frac{\sqrt{\tau_w/\rho}\,y}{\nu} \qquad \text{for equilibrium near-wall models} \tag{6.29}$$

$$y^+ = \frac{C_\mu^{1/4} k^{1/2} y}{\nu} \qquad \text{for two- and three-layer models} \tag{6.30}$$

where y^+ is the distance from the wall, τ_w is the wall shear stress, ρ is density, ν is kinematic viscosity, k is turbulence energy in the wall-adjacent cell, and C_μ is an empirical constant ($\approx$0.09). For a review and evaluation of various near-wall models, see Djilali et al. [34].

Since wall functions relate the near-wall viscous effects to the fully turbulent region, it is important to ensure that all the wall-adjacent grid cells are sufficiently far from the wall for the viscous effects to be negligible ($30 < y^+ < \approx 200$). The value of y^+ is likely to vary as it is dependent on the flow field; thus, it may become necessary to adjust the location of the near-wall grid as the solution converges by periodically examining the distribution of y^+ along the walls (especially at heat-transfer surfaces).

TABLE 6.7 Comparison of Near-wall Turbulence Models

Near-wall model	Equations	Layers	y^+ Range	Reference
Equilibrium	k equation e equation	Equilibrium layer	$20 < y^+ < 60$	Launder and Spalding [23]
Two-layer	k equation	Viscous sublayer Overlap layer	$0 < y^+ < 11$ $11 < y^+ < 400$	Chieng and Launder [30] Johnson and Launder [31]
Two-layer	e equation	Viscous sublayer Overlap layer	$0 < y^+ < 11$ $11 < y^+ < 400$	Amano [32]
Three-layer	k equation e equation	Viscous sublayer Buffer layer Overlap layer	$0 < y^+ < 5$ $5 < y^+ < 30$ $30 < y^+ < 400$	Amano [33]

Always consult software documentation when using near-wall functions as there may be additional limitations or requirements not addressed here. See Table 6.8.

Radiation Models. Including radiation effects in the thermal analysis complicates the model, but results in a more accurate prediction of the physical phenomena. Radiation effects should be included when radiation heat transfer Q_{rad} is expected to be large compared to convective or conduction heat transfer. Radiation heat transfer can be estimated by

$$Q_{\text{rad}} = \sigma(T_{\text{max}}^4 - T_{\text{min}}^4) \tag{6.31}$$

where T_{max} and T_{min} are the maximum and minimum temperature in the model, and σ is the Stephan-Boltzmann constant [5.671×10^{-8} W/(m-K^4)].

Modeling of radiation heat transfer for systems containing multiple cards and electrical assemblies requires the radiant exchange between various surfaces. Computation of view factors is required for such calculations. This is complicated by the three-dimensional nature of the problem. Smith et al. [13] used irradiation-radiosity formulation to solve conjugated conduction-convection-radiation in an enclosure containing multiple heat-generating cards. This technique was extended by Yu and Joshi [12] in three dimensions. The net radiation heat flux at a surface is given by

$$q_i = A_i(J_i - G_i) \tag{6.32}$$

where irradiation J_i and G_i are given by

$$J_i = \varepsilon_i \sigma T_i^4 + (1 - \varepsilon_i) \sum_{j=1}^{N} F_{ij} J_j \tag{6.33}$$

$$G_i = \sum_{j=1}^{N} F_{ij} J_j \tag{6.34}$$

where σ is the Stefan-Boltzmann constant, ε_i is the emissivity, F_{ij} is the view factor, and A_i is the area of zone i. To model radiation using this technique the following steps are as follows:

1. Identify the various surfaces participating in radiation. For this method all the participating surfaces are assumed to be isothermal. To ensure this, one surface may have to be discretized into several isothermal zones.
2. Compute the view factor matrix. For N isothermal surfaces, the view factor matrix F_{ij} can be computed. Equations for computing radiation exchange are compiled by Siegel and Howell [39].
3. The irradiation J_i can be computed from Eq. 6.27, and then the radiosity and the net heat flux can be computed from Eqs. 6.26 and 6.28. The radiation heat flux then can be used to update the temperature field.
4. An iterative solution between conduction-convection and radiation is then performed to obtain a converged solution.

This technique can be applied to a component/board-level conduction solution or a system-level conjugate conduction-convection-radiation solution. There are sev-

TABLE 6.8 Correlations for Modeling the Thermal Contact Resistance Coefficient h_c

Correlation	Nomenclature	Reference
where $h_c = 0.55m(k_s/\sigma)(p/H)^{0.85}$ $k_s = k\frac{2k_1k_2}{k_1 + k_2}$	k_1: thermal conductivity of material 1 k_2: thermal conductivity of material 2 σ: rms value of surface roughness m: rms value of the slope of the contacting asperities p: the contact pressure H: the hardness of the softer of the two materials in contact	Tien [35]
$h_c = \frac{2k_s na}{\left(1 - \sqrt{\frac{A_r}{A_a}}\right)^{1.5}}$	k_s: the harmonic mean thermal conductivity n: the contact spot density (m^{-2}) a: the mean radius of circular contact (m^2) A_a: the apparent contact area (m^2) A_r: the real contact area (m^2)	Cooper et al. [36]
where $h_c = 3.81 \frac{k_s}{\sigma}\left(\frac{p}{E'}\right)^{0.93}\left(\frac{\sigma}{\beta}\right)^{0.035}$ $E' = \left[\frac{(1-\nu_1^2)}{E_1} + \frac{(1-\nu_2^2)}{E_2}\right]^{-1}$	p: applied pressure σ: rms value of surface roughness E': equivalent elastic modulus, MPa β: mean asperity peak radius n_1: Poisson's ratio of material 1 n_2: Poisson's ratio of material 2	Blahey et al. [37]
$h_{ce} = k_s \frac{m}{\sigma}(1.18 + 0.161 \ln \sigma)\left(\frac{\sqrt{2}p}{E'm}\right)^{0.922\alpha^{1/205.54}}$	p: applied pressure σ: rms value of surface roughness E': equivalent elastic modulus, MPa	Sridhar [38]

eral higher-order techniques to incorporate radiation. These are discrete ordinate method [39], *P-n* method [40], and exchange factor method [41].

6.2.5 Material Selection and Properties

Following the definition of the physical models, the various materials and their properties within the solution domain are to be set. Most of the available software packages have a built-in library of commonly used materials; however, the property values in the library should be checked with reference books first. These materials are typically either for fluids or solids. The material properties needed depend on the type of analysis that is to be performed, as shown in Table 6.9. If radiation is considered in the model, then the emissivity and absorptivity of the material may also be required. Care should be taken to ensure that the values of the properties are entered in the units specified by the software.

TABLE 6.9 Required Thermophysical Properties for a Given Analysis

Analysis type	Required properties	Temperature dependence
Flow only	Density	No
	Viscosity	
Steady state thermal only	Thermal conductivity	Yes
Transient thermal only	Density	Yes
	Heat capacity	
	Thermal conductivity	
Flow and thermal	Density	Yes
	Viscosity	
	Heat capacity	
	Thermal conductivity	

Note: Adapted from Fluent [11].

Temperature-Dependent Properties. Although material properties can usually be considered constant, the use of temperature-dependent properties is important for accurate modeling of heat transfer and fluid flow. There are several cases where temperature-dependent material properties are required. For example, in problems with natural or mixed convection, the density must be defined as a function of temperature. In advanced thermal management systems that use liquids, viscosity is strongly affected by temperature and should be included in the model. Some software packages allow the user to enter temperature-dependent thermophysical properties as piecewise linear variations, polynomial functions, or use the ideal gas law with kinetic theory. The thermophysical properties of dry air are presented as polynomial functions of temperature in Table 6.10.

Effective Thermal Conductivity in Printed Wiring Boards. Multilayer PW boards can be modeled as a single-layer board by imposing an effective (board-averaged) thermal conductivity that takes into account the higher thermal conductivity of

TABLE 6.10 Polynomial Curve-Fits for Temperature-Dependent Thermophysical Properties of Dry Air

			$\text{Property} = \sum_n A_n(T)^n$ where T is in Kelvin				
Property	Units	Range	A_4	A_3	A_2	A_1	A_0
c_p	cal/(g-K)	260–610K	0	-6.4613×10^{-11}	1.6919×10^{-7}	-7.5518×10^{-5}	0.249679
c_p	cal/(g-K)	610–1500K	0	-6.4073×10^{-11}	1.3612×10^{-7}	-3.89538×10^{-5}	0.239496
m	kg/(m-s)	200–1500K	0	8.1504×10^{-15}	-3.1320×10^{-11}	6.25989×10^{-8}	2.2880×10^{-6}
k	W/(m-K)	200–1500K	-6.5998×10^{-15}	2.3172×10^{-11}	-4.4425×10^{-8}	9.3677×10^{-5}	1.3003×10^{-3}
r	kg/m^3	200–1500K			$\rho = 352.989/T$		

Source: Adapted from General Electric [42].

copper sheets between layers of glass epoxy. Graebner [43] proposed the following correlations for the anisotropic effective thermal conductivities in directions perpendicular and parallel to the plane of the PWB:

$$k_{\perp} = \left[1.69\left(1 - \frac{Z_{\mathrm{Cu}}}{Z}\right) + 0.0026\left(\frac{Z_{\mathrm{Cu}}}{Z}\right)\right]^{-1} \quad \text{in W/(m-K)} \tag{6.35}$$

$$k_{\parallel} = 0.8 + 350\left(\frac{Z_{\mathrm{Cu}}}{Z}\right) \quad \text{in W/(m-K)} \tag{6.36}$$

In the preceding equations, Z_{Cu} is the total thickness of the copper layers in a PWB of thickness Z. Graebner [43] reported agreement with experimental measurements to within 10 percent using Eqs. 6.35 and 6.36.

6.2.6 Boundary Conditions

Before solving the model, the known conditions on external and internal boundaries of the computational domain must be specified. Boundary conditions impose known or assumed vector and scalar quantities that will be used in the software to predict the thermal and/or flow fields. In this section, geometric boundary conditions such as symmetry are discussed in addition to flow and thermal boundary conditions.

Geometry Boundary Conditions. Although not scalar or vector quantities, symmetry, cyclic, and periodic boundary conditions need to be specified on the computational domain to enable the solver to take these effects into account.

Symmetry. As discussed in Sec. 6.2.2, symmetry boundary conditions can be used to simplify the physical domain of a model. These boundary conditions can also be applied when the geometry, flow, and thermal fields have mirror symmetry. At symmetry boundaries, the normal gradients of all variables and the normal velocity are equal to zero. These conditions result in all variables having zero flux across the symmetry plane. For problems with axisymmetric geometry, the symmetry boundary conditions can be applied at the centerline of the geometry. Software packages may specify that the centerline of the problem must be in a designated axis direction as well as limit the number of axisymmetric boundary conditions that are problems (usually one). It is important to remember when applying symmetric or axisymmetric boundary conditions to a model with geometric symmetry that the accompanying thermal and flow fields must also obey symmetry conditions.

Cyclic and Periodic. When the problem geometry, flow, and thermal fields have a cyclical or periodic repeating nature, cyclic and periodic boundary conditions may be imposed, respectively. Flow over a bank of offset-strip fins and flow over an array of heated components are examples of such flow conditions. These boundary conditions are imposed in pairs and the computational domain is usually limited to only one pair. Some software packages may specify in what planes the boundaries must be located.

Flow Boundary Conditions. For isothermal or conjugate heat transfer problems, there are two main boundary conditions: velocity and pressure. For turbulent flow, additional boundary conditions such as turbulence energy and dissipation, characteristic length, or turbulence intensity may be required.

Velocity. Depending on the problem geometry or grid definition, flow velocities can be applied in terms of cartesian, tangential, cylindrical-polar, or angular components. The velocity components may be imposed at flow inlets, exits, and walls. Since velocities are vector quantities, the sign of the velocity magnitude dictates whether the vector points in the positive or negative coordinate direction.

Pressure. Fluid pressures at inlets and outlets can be imposed using pressure boundary conditions. These can be used when the flow rate or velocity is unknown, but the pressure drop across the model is known. Pressure boundary conditions may be used to model situations where the flow exits the domain at more than one location, as found in electronic enclosures with more than one vent. In free convection cases, where the direction and magnitude of the flow field are unknown, pressure boundary conditions can be imposed at the model boundaries as the flow is allowed to enter and exit the computational domain. Pressure boundaries can also model mixed conditions where the flow enters and exits a domain at a given location (e.g., recirculation at a vent). For inflows and outflows, pressure boundary conditions are given in terms of total and static pressures, respectively. The total pressure p_t of an incompressible fluid is given by

$$p_t = p_s + \tfrac{1}{2}\rho u^2 \tag{6.37}$$

where p_s is the static pressure, ρ is density, and u is the velocity normal to the pressure boundary.

For electronics cooling problems, the pressure boundary condition is a better choice than the so-called outlet boundary condition that assumes a one-dimensional exit flow with zero diffusion flux for flow variables. The outlet boundary condition forces all the flow that enters the computational domain to exit through the designated outlet region. Pressure boundary conditions allow for multiple outlets with exit flow recirculation, while outlet boundary conditions will lead to unconverged solutions if these situations are encountered. System-level analyses can use velocity boundary conditions at inlets (modeling fans) and pressure boundary conditions at outlets (to model multiple vents or flow exits). If the velocity is unknown or flow is expected to enter and exit the inlet région, a pressure boundary condition at the inlet can be imposed using an arbitrary value. As the solution converges, this value can be adjusted as needed by comparing the predicted mass flow rate at a location in the computational domain with an expected or known mass flow rate.

Thermal Boundary Conditions. There are basically three types of boundary conditions for conjugate and conduction thermal analyses: temperature, heat flux, and convection. Approximate radiation boundary conditions can be imposed using an effective heat transfer coefficient that includes radiation effects.

Constant Temperature. If the temperature of a surface is known and is constant, a constant temperature boundary condition can be applied, although this condition is highly unlikely in practical electronics cooling situations. Constant temperature boundary conditions are more often applied at flow inlets. However, for various types of parametric studies to determine package or system characteristics, such a boundary condition may be necessary on surfaces within the computational domain.

Constant Heat Flux. If any surface is dissipating a constant heat flux, this boundary condition can be used. This is also less frequently encountered in practical situations. A constant heat flux value of zero can be used to model adiabatic (insulating) walls.

Convection Boundary Condition. Most electronics cooling problems deal with this type of boundary condition. The convection boundary condition in thermal modeling is often difficult to quantify since the temperature of the heat-dissipating surface, as well as the local air temperature must be known. Convective heat transfer is strongly dependent on the fluid motion; thus, prior knowledge of the air velocity is also required. Empirical and analytical natural and forced convection correlations can be used to approximate the convection heat transfer from a surface with a high degree of accuracy. If the velocity of the air near the convecting surface is unknown, an isothermal model can be run to obtain approximate local velocity magnitudes.

Radiation. Heat transfer analysis from electronic assemblies often ignores the effects of radiation heat transfer. While this can provide a good approximation for situations such as forced convection, it is not a good approximation for natural convection situations. Radiation heat transfer from a surface is given by

$$Q_{\text{rad}} = \sigma A(T_w^4 - T_\infty^4) \tag{6.38}$$

where T_w is the temperature of a surface in a surrounding at T_∞.

Radiation for component- and board-level simulation can be handled by following a relatively simple approach that utilizes a view factor of one and is useful only when there is a high heat-flux component in a relatively big enclosure. The radiation heat transfer coefficient h_r can be defined as

$$h_r = \frac{q_r}{A\Delta T} = \sigma(T_w^2 + T_\infty^2)(T_w + T_\infty) \tag{6.39}$$

The obtained radiation heat transfer coefficient h_r is added to the convection heat transfer coefficient to account for the effect of radiation. The effective heat transfer coefficient h_T is given by

$$h_T = h + h_r \tag{6.40}$$

where h is the convection heat transfer coefficient and is applied to a convection-radiation surface as a convection boundary condition. This approximation cannot be used for densely packaged systems containing multiple assemblies and radiation effects should be included using more robust models described in Sec. 6.2.4.

Correlations for Natural Convection. Natural convection is the mechanism of the heat transfer caused by the buoyancy-induced flow of fluid. This is also referred to as *free convection* as opposed to *forced convection,* where an external air mover such as a blower or fan is necessary. Many heat transfer situations in electronics packaging require cooling by natural convection. Amount of heat flux removed by natural convection from a given surface depends on the fluid thermophysical properties such as the density, specific heat, thermal conductivity, viscosity, and coefficient of thermal expansion. The characteristic number to represent this is called the Rayleigh number (Ra).

Uniform Temperature Wall. A generic natural convection situation is shown in Table 6.11. This figure represents the fluid flow past a heated vertical plate. The plate can be the heat sink surface, kept at a constant temperature of T_w in an ambient temperature of T_∞. For such a heat transfer situation, the Rayleigh number is given by

$$\text{Ra}_{Lc} = \frac{g\beta(T_w - T_\infty)L_c^3}{\nu\alpha} \tag{6.41}$$

where L_c is the characteristic length, g is the acceleration due to gravity, β is the thermal expansion coefficient, ν is kinematic viscosity, and α is thermal diffusivity of the

TABLE 6.11 Correlations for Natural Convection Heat Transfer from Vertical Plates

Configuration	Illustration	Correlation and limits	Reference
Isothermal plate	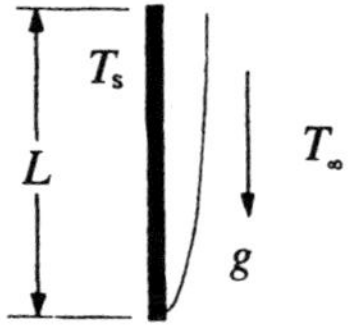	$\mathrm{Nu}_L = \left[0.825 + \frac{0.387\mathrm{Ra}_L^{1/6}}{[1 + (0.492/\mathrm{Pr})^{9/16}]^{8/27}}\right]^2$	Churchill and Chu [44]
Isothermal plates separated by a spacing *b*	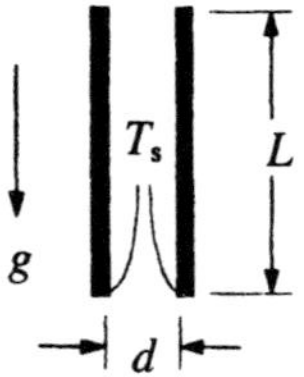	*Isothermal and symmetrically heated plates* $\mathrm{Nu}_d = \frac{1}{24}\mathrm{Ra}_d\left(\frac{d}{L}\right)\left\{1 - \exp\left[\frac{-35}{\mathrm{Ra}_d\left(\frac{d}{L}\right)}\right]\right\}^{0.75}$	Elenbaas [45]

fluid. For this instance, the characteristic length is the height of the heat-dissipating wall. Correct specification of this characteristic length is necessary, which is often derived from the geometry of the heat transfer situation. For this instance, the characteristic length is the height of the heat-dissipating wall. Also, for many practical heat transfer situations, it is not possible to have a uniform temperature boundary condition, since the temperature field is unknown. For such circumstances, two generic situations can occur, which are presented in the following sections.

Uniform Heat Flux Wall. In this situation, the heat flux is assumed constant from a given surface. In practical situation, such a description of heat transfer can be assumed from the heat transfer side of a densely populated PCB. The Rayleigh number in this situation is defined as

$$\mathrm{Ra}_{Lc} = \frac{g\beta q'' D L_c^3}{k\nu\alpha} \tag{6.42}$$

where q'' is the uniform heat flux, and D is another characteristic length associated with the heat transfer situation. Ideally, this should be equal to the average thermal boundary layer thickness.

Uniform Heat Generation. In this situation, the heat generation is assumed uniform from a given volume. In practical situations, this can occur from a single component, such as a PQFP. The Rayleigh number for this situation is defined as

$$\mathrm{Ra}_{Lc} = \frac{g\beta \dot{q}''' V D L_c^3}{k\nu\alpha A} \tag{6.43}$$

where q''' is the rate of volumetric heat generation, V is the volume generating the heat, and A is the wetted area or the heat transfer area of the package.

Heat Transfer Correlations. The heat transfer correlations for the natural convection situation are often represented by the generic equation

$$\mathrm{Nu}_{Lc} = \frac{hL_c}{k} = C\,\mathrm{Ra}^m\,\mathrm{Pr}^n \tag{6.44}$$

where k is the fluid thermal conductivity, Nu is the Nusselt number, Pr is the Prandtl number, and C, m, and n are constants. Depending on heat transfer situations and the range of the Rayleigh number, various correlations are available. The choice of a natural convection correlation should be done with caution, and appropriate choice of length scale and the Rayleigh numbers are the two most important deciding factors. Several important correlations are presented in Tables 6.11 and 6.12. Teertstra et al. [48] compiled a review of natural and mixed convection correlations for circuit board arrays.

Correlations for Forced Convection. The forced convection heat transfer correlations that exist in the heat transfer literature are classified based on the flow configuration. External flow correlations are available for flows over heated flat plates, cylinders, or spheres, where the boundary layers on these objects are free to grow and the fluid is unconstrained. Flows where the boundary-layer growth or fluid is constrained are called *internal flows.* Examples of internal flows are flows in heated ducts or between PW boards.

External Flows. Correlations for the average convective heat transfer coefficient h for a flat plate and cylinder are given in Table 6.13 as functions of Reynolds and Nusselt numbers which are defined by Eqs. 6.26 and 6.44, respectively. The characteristic lengths L_c are the plate length L and the cylinder diameter D. The average

TABLE 6.12 Correlations for Natural Convection Heat Transfer from Horizontal Plates

Configuration	Illustration	Correlation and limits	Reference
Isothermal plate	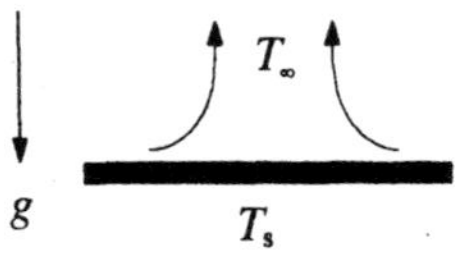	*Isothermal boundary condition* $Nu_L = 0.54Ra^{0.25}$ for $10^4 < Ra < 10^7$ $Nu_L = 0.15Ra^{0.333}$ for $10^7 < Ra < 10^{11}$ where the Nusselt and Rayleigh numbers are given by $Nu_L = \frac{hL}{k}$ and $Ra = \frac{g\beta(T_s - T_\infty)L^3}{\nu\alpha}$ The length scale $L = A_s/P$, where A_s and P are the surface area and the perimeter, respectively.	Incropera and DeWitt [7]
Isothermal and isoflux plates in an enclosure separated by a spacing d	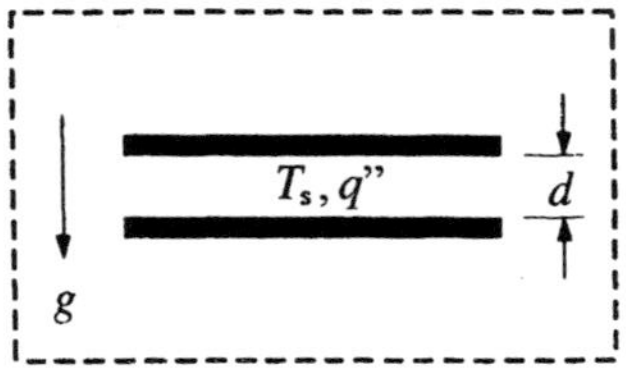	*Isothermal boundary condition* $Nu_d = 0.069Ra_d^{0.3333} Pr^{0.074}$ $3 \times 10^3 < Ra_d < 7 \times 10^9$ where $Ra_d = \frac{g\beta\Delta T d^3}{\nu\alpha}$	Globe and Dropkin [46]
		Constant heat flux boundary condition Plate-averaged Nusselt number Nu_d $Nu_d = 1$ for $Gr_d < 2000$ $Nu_d = 0.18Gr_d^{0.25} (L/d)^{-0.111}$ for $Gr_d > 2000$	Jakob [47]

TABLE 6.13 Correlations for External Forced Convection over a Flat Plate and Cylinder

Configuration	Average Nusselt number	Conditions
Flat plate	$Nu_{L,T} = 0.664 \cdot Re_L^{1/2} \cdot Pr^{1/3}$ $Nu_{L,H} = 0.906 \cdot Re_L^{1/2} \cdot Pr^{1/3}$	$Re_L < 5 \times 10^5$ $Pr > 0.6$
Flat plate	$Nu_{L,T} = 0.0370 Re_L^{4/5} Pr^{1/3}$ $Nu_{L,H} = 0.0390 Re_L^{4/5} Pr^{1/3}$	$Re_L > 5 \times 10^5$ $0.6 < Pr < 60$
Cylinder	$Nu_{D,T} = 0.3 + \frac{0.62\, Re_D^{1/2} Pr^{1/3}}{[1 + (0.4/Pr)^{2/3}]^{1/4}} \left[1 + \left(\frac{Re_D}{282{,}000}\right)^{5/8}\right]^{4/5}$	$Re_D \, Pr > 0.2$

T = constant temperature
H = constant heat flux
Source: Adapted from Incropera and DeWitt [7].

convective heat transfer coefficient can be calculated from the Nusselt number correlations in Table 6.13 and imposed as a boundary condition in thermal models.

Internal Flow—Rectangular Ducts. Heat transfer textbooks normally present flow and heat transfer correlations for circular ducts or pipes. Most ducts encountered in electronic cooling are rectangular ducts as in the channels formed between rack-mounted PC boards. As with external flows, the heat transfer is a function of Nusselt and Reynolds numbers, and instead of the characteristic length in Eqs. 6.20 and 6.37, there is the hydraulic diameter D_h which is defined as

$$D_h = \frac{4 \cdot \text{Cross-sectional area}}{\text{Perimeter}} = \frac{2ab}{a + b} \tag{6.45}$$

for a rectangular duct with sides a and b. In electronics cooling, the duct length may be short such that the laminar flow and thermal fields are not fully developed at the exit. Tables 6.14 and 6.15 list Nusselt number values for hydrodynamically and thermally (simultaneously) developing laminar flow as a function of the dimensionless inverse Graetz number x^+, which is defined as

$$x^+ = \frac{L/D_h}{Re_{Dh}\, Pr} \tag{6.46}$$

where L is the streamwise duct length and Pr is the fluid Prandtl number. The flow may be considered to be fully developed when $x^+ > 0.1$. The bottom rows of Tables 6.16 and 6.17 give fully developed Nusselt number values for laminar flow in rectangular ducts with constant temperature or heat flux walls. In turbulent duct flow ($Re_{Dh} > 4000$), the influence of duct aspect ratio is negligible on heat transfer; thus, the heat transfer coefficient may be calculated from standard circular tube-flow correlations published in standard heat transfer textbooks (e.g., Ref. 7). Additional values are listed in Tables 6.16 and 6.17 for ducts with one or more insulated sides.

Internal Flow—Inline Arrays of Packages. A common configuration encountered in electronics cooling is the thermal management of board-mounted packages. These boards are often rack-mounted and the coolant (typically air) passes between the boards, effectively cooling the packages and PW boards [49]. Although PW boards are diversely populated with capacitors, packages, microprocessors, cabling, and so forth, the simplest configuration would be an inline array of uniform-size packages. Wirtz [49] and Morris and Garimella [50] proposed empirical correlations for fully developed package-averaged heat transfer coefficient for air and liquid coolants, respectively, listed in Table 6.18. Garimella and Eibeck [51] established that

TABLE 6.14 Average Nusselt Numbers for Simultaneously Developing Flow in Ducts with Constant Surface Temperatures for Pr ≈ 0.7

		$Nu_{Dh,T}$				
x^+		1	2	3	4	6
	○	□	▭	▭	▭	▭
0.0040	16.7					
0.0045		9.70	10.00	10.30	10.58	10.90
0.0050		9.30	9.60	9.91	10.18	10.51
0.0055		8.91	9.20	9.50	9.77	10.12
0.0060	14.1					
0.0063		8.50	8.80	9.10	9.36	9.72
0.0072		8.06	8.37	8.66	8.93	9.28
0.0083		7.61	7.91	8.18	8.48	8.85
0.010	11.3	7.10	7.42	7.70	7.98	8.38
0.013		6.57	6.88	7.17	7.47	7.90
0.017		5.95	6.27	6.60	6.90	7.35
0.020	8.7	5.63	5.95	6.28	6.61	7.07
0.025		5.27	5.61	5.96	6.27	6.78
0.033		4.88	5.23	5.60	5.93	6.47
0.050	6.1	4.39	4.79	5.17	5.56	6.13
0.10		3.75	4.20	4.67	5.11	5.72
∞	3.66	2.98	3.39	3.96	4.44	5.14

Source: Adapted from Shah and London [8] and Kays and Crawford [2].

TABLE 6.15 Average Nusselt Numbers for Simultaneously Developing Flow in Ducts with Constant Heat Flux Walls for Pr ≈ 0.7

		$Nu_{Dh,H}$			
x^+		1	2	3	4
	○	□	▭	▭	▭
0.0020	17.8				
0.0045		15.03	15.36	15.83	16.02
0.0050		14.55	14.88	15.21	15.49
0.0055		14.05	14.35	14.70	14.95
0.0063		13.50	13.79	14.10	14.48
0.0072		12.87	13.15	13.47	13.73
0.0083		12.19	12.48	12.78	13.03
0.010	9.12	11.43	11.70	12.00	12.23
0.013		10.53	10.83	11.13	11.35
0.017		9.49	9.77	10.07	10.32
0.020	7.14	8.90	9.17	9.48	9.70
0.025		8.25	8.54	8.85	9.07
0.033		7.52	7.83	8.13	8.37
0.050		6.60	6.94	7.31	7.58
0.10	4.72	5.43	5.77	6.27	6.65
0.20		4.60	5.00	5.57	6.06
∞	4.36	3.61	4.12	4.99	

Source: Adapted from Shah and London [8] and Kays and Crawford [2].

TABLE 6.16 Fully Developed Nusselt Number of Rectangular Ducts of Varying Aspect Ratio with One or More Constant-Temperature Walls for Pr ≈ 0.7

	$Nu_{Dh,T}$				
b/*a*	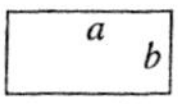				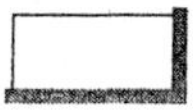
0	7.541	7.541	7.541	0	4.861
½	3.383	3.842	4.619	1.647	2.613
1	2.970	3.018	3.703	2.437	2.375
2	3.383	2.602	2.657	3.185	2.613
2½	3.670	2.603	2.333	3.390	2.768
3⅓	4.114	2.703	1.946	3.626	2.996
5	4.803	2.982	1.467	3.909	3.330
10	5.858	3.590	0.834	4.270	3.823
∞	7.541	4.861	0	4.861	4.861

Note: Shaded lines represent adiabatic surfaces.
Source: Adapted from Shah and London [8].

TABLE 6.17 Fully Developed Nusselt Number of Rectangular Ducts of Varying Aspect Ratio with One or More Constant–Heat Flux Walls for Pr ≈ 0.7

	$Nu_{Dh,H}$				
b/*a*	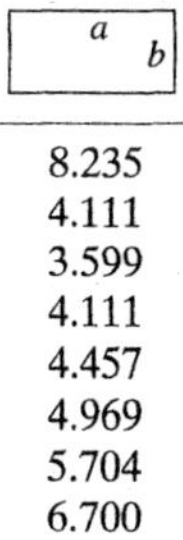	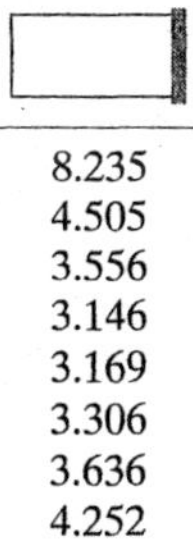	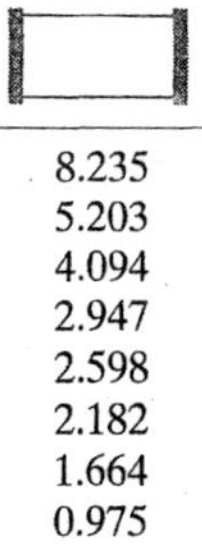	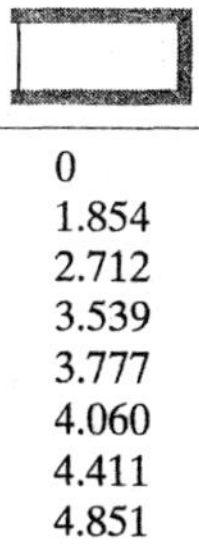	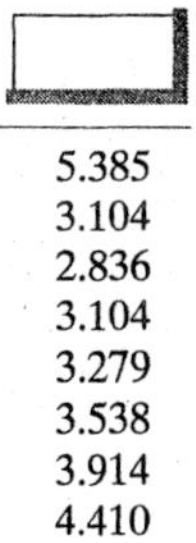
0	8.235	8.235	8.235	0	5.385
½	4.111	4.505	5.203	1.854	3.104
1	3.599	3.556	4.094	2.712	2.836
2	4.111	3.146	2.947	3.539	3.104
2½	4.457	3.169	2.598	3.777	3.279
3⅓	4.969	3.306	2.182	4.060	3.538
5	5.704	3.636	1.664	4.411	3.914
10	6.700	4.252	0.975	4.851	4.410
∞	8.235	5.385	0	5.385	5.385

Note: Shaded lines represent adiabatic surfaces.
Source: Adapted from Shah and London [8].

the flow becomes hydrodynamically and thermally fully developed by the fourth row of an array of packages; thus, the convective heat transfer coefficients obtained from Table 6.18 should be applied for arrays with five or more rows. The local convective heat transfer coefficients for the first four array rows may be calculated from thermal wake functions found in Anderson and Moffat [52], Kang [53], and Wirtz [49].

Modeling Contact Resistance. In electronic assemblies, interfaces between materials are frequently encountered. These include situations such as the placement of a circuit card on a heat sink or attachment of a heat sink on an electronic package. There exists a thermal contact resistance at such an interface which is a function of the following parameters [54]:

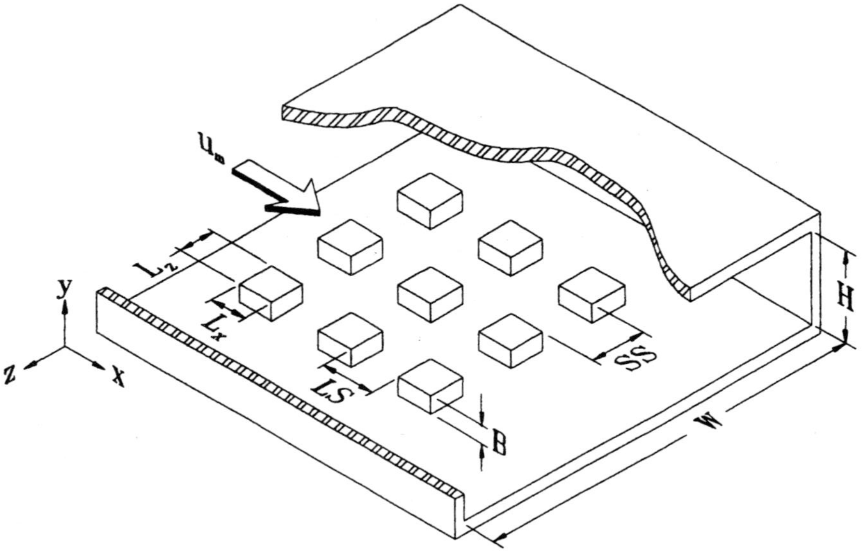

TABLE 6.18 Correlations for In-Line Arrays of Heated Three-Dimensional Components

Coolant	Average Nusselt number	Conditions	Reference
Air	$Nu_{Lx} = 0.6Re_{Lx}^{0.5} Pr^{0.33}$ $\quad Re_{Lx} \le 5000$	$2.67 < L_x/B < 8.75$	Wirtz [49]
		$1.5 < H/B < 10.0$	
	$Nu_{Lx} = 0.082Re_{Lx}^{0.72}$ $\quad Re_{Lx} > 5000$	$0.25 < \dfrac{L_x L_z}{LS\ SS} < 0.69$	
Liquids	$Nu_{Lx} = 0.158Re_{Lx}^{0.655} Pr^{0.5} \left(\dfrac{\mu_i}{\mu_h}\right)^{0.13} \left(\dfrac{LS}{H}\right)^{0.125} \left(\dfrac{H-B}{L_x}\right)^{0.05}$	$610 < Re_{Lx} < 68{,}850$	Morris and Garimella [50]
		$6 < Pr < 25.2$	
	m_i = evaluated at inlet temperature	$0.025 < LS/H < 3.29$	
	m_h = evaluated at component temperature	$0.06 < [(H-B)/L_x] < 3.87$	

- Surface roughness
- Surface waviness
- Thermal conductivity
- Poisson's ratio
- Yield stress
- Hardness
- Applied load
- Geometry

The effect of contact is often expressed as an effective heat transfer coefficient. The existing data on the contact resistances are presented in Table 6.8.

For thermal simulations involving contact resistances at any dissimilar interface, an effective heat transfer coefficient h_c is prescribed. Such a scheme is shown in Fig. 6.4 where the heat flow Q through the surface A is expressed as

$$Q = h_c A(T_1 - T_2) \tag{6.47}$$

where T_1 and T_2 are the temperatures at the interface of materials 1 and 2, respectively.

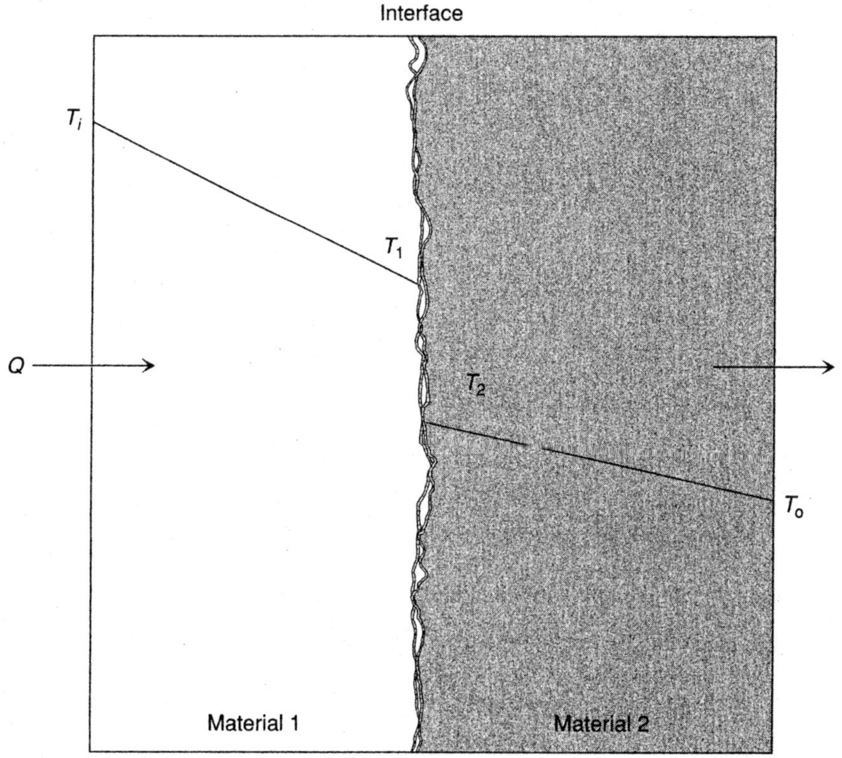

FIGURE 6.4 Imperfectly bonded surfaces producing a contact resistance in the interface.

6.2.7 Solution Strategy

For finite volume and finite element methods, various approaches are used to solve the algebraic equations. These equations are coupled and require an iterative strategy. This section discusses various solvers, interpolation schemes, and solver parameters for various types of flow and heat transfer problems.

Solvers

Point-by-Point Gauss-Siedel Solver. This is a simple iterative solver in which the variables at various grid points are solved in a specific order. For an example, a discretized equation written as [4]:

$$A_P T_P = \sum A_{nb} T_{nb} + b \tag{6.48}$$

where nb represents a neighbor point of the point P. The solution for T_P can be written as

$$T_P = \frac{\sum A_{nb} T^*_{nb} + b}{A_P} \tag{6.49}$$

where T^*_{nb} denotes the value at the previous iteration, and T_P represents the current iteration.

Line-by-Line Gauss-Siedel Solver. The line-by-line Gauss-Siedel solver [21] uses the solution scheme for a line of grid points of the entire computational domain. The direction of the line is called *sweep direction.* Once the solution for one line is obtained, the solver moves to the next row of cells to perform computation for that line. The direction in which the solution progresses from one line to the next is called *marching direction.*

Tridiagonal Matrix Solver (TDMA). The TDMA algorithm is useful for solving banded matrices. This method also requires the solution to be done in a group of cells in a line called sweeping direction, and then marching forward in the marching direction. Details about this can be found in Patankar [4].

Guidelines for Using Iterative Solvers. The choice of marching and sweeping direction is important for the particular problem. For a problem where the flow direction is known (non-recirculating flows such as forced convection flow over a set of blocks or over heat sink fins), the marching direction is usually chosen as the direction of the flow, and sweeping direction is chosen as the perpendicular to the flow. This ensures the uniform flow of boundary condition information in the flow direction. For a problem involving no particular direction of flow (natural convection in an enclosed box such as a laptop computer), it is difficult to choose the marching and sweep directions. In such situations, usually alternate sweep directions are used. An alternate sweep is done by sweeping the lines in the I direction and then in the J direction.

Number of sweeps is also another important parameter and usually is chosen by trial and error. The number of sweeps for an equation refers to the number of times the equation is solved. Usually the momentum and scalar equations can be solved once each (one sweep), and the pressure correction equation can be solved with five sweeps, in one iteration. However, for problems with conjugate conduction, the energy equation may need multiple sweeps (5–20) in order to assure convergence. Choices of optimum number of sweeps depend on the particular heat transfer problem.

Interpolation Schemes. During solution of the discretized equations, interpolation is performed to get the values at the control volume boundaries. Several interpolation schemes such as central difference, power law, exponential, and so on, have been summarized by Patankar [4]. The most commonly applicable one is power law scheme. In this scheme, the exact solution for a convection-diffusion equation in one direction is approximated using a power law equation. For convergence acceleration for flows with curvature, second-order methods such as QUICK [55] are used. The QUICK scheme involves the interpolation using a quadratic function involving the values at three adjacent nodal values. QUICK can reduce errors related to numerical diffusion.

Convergence Acceleration Schemes. In order to accelerate convergence, the following steps or their combinations can be utilized.

Underrelaxation. The choice of underrelaxation parameters depends on the nature of the problem and have to be employed using trial and error. Usually the underrelaxation factors for the velocities can be 0.2 to 0.5, for pressure and pressure corrections, 0.4 to 0.8, and for temperature, 0.9 to 1.0. The underrelaxation parameters can be increased to increase convergence. On the other hand, if a solution is found to diverge, the underrelaxation factors need to be reduced in order to assure convergence.

Addition of Complexity. Improved solver performance can be achieved by slowly imposing the required complexities in the problem. For example, for achieving faster convergence in a conjugate heat transfer problem, the heat loads can be slowly increased from a low value to the required value. For forced convection heat and mass transfer problems, the fluid flow solution can be obtained first, before switching on the heat sources. For problems involving radiation, a converged conduction and convection solution can be obtained before the radiation model is turned on.

Multigrid Acceleration. For line-by-line iterative solvers, the communication of information from one point to another is limited by the sweep and marching. For problems involving larger grid sizes, and for problems with anistropic coefficients in the flow field (for conjugate problems with various materials such as aluminum, FR-4, air, and so forth), the convergence can be slow. In order to improve the acceleration, a multigrid algorithm [56] can be used which uses a coarse and a fine grid to perform solution together. This results in the global reduction of errors.

Convergence Criteria. The convergence criterion is defined as the condition when the solution is said to be converged. One commonly used criterion is to check the sum of the normalized residuals. At the same time, an overall energy balance residual should also be calculated to confirm the validity of the solution.

6.2.8 Postprocessing

Once the solution has converged, the predicted thermal and flow fields can be analyzed graphically. While the capabilities for graphically examining the solution varies, most commercial software has the ability to plot velocity vector and contour plots on planes within the computational domain. Velocity vectors show the direction and magnitude of the flow. The velocity magnitude may be presented by the length of the vector, by color-coding, or both. Contour plots show lines or filled regions of constant magnitude for a chosen variable (temperature, pressure, and so forth). For software that lacks graphical postprocessing abilities, it is usually possible to extract solution data as a series of datum points in the form (x_1, x_2, x_3, f) where x_1, x_2, x_3 are orthogonal coordinates and f is a variable (e.g., temperature). The data set can then be inspected or plotted as needed.

6.3 ILLUSTRATIVE EXAMPLE

6.3.1 Example of a Conjugate Analysis: A Plastic Quad Flat Package

The following example illustrates a conjugate thermal analysis of a plastic quad flat package in natural convection. Figure 6.5*a* and *b* show the internal structure of a plastic quad flat package (PQFP). It consists of the silicon die mounted on the pad-

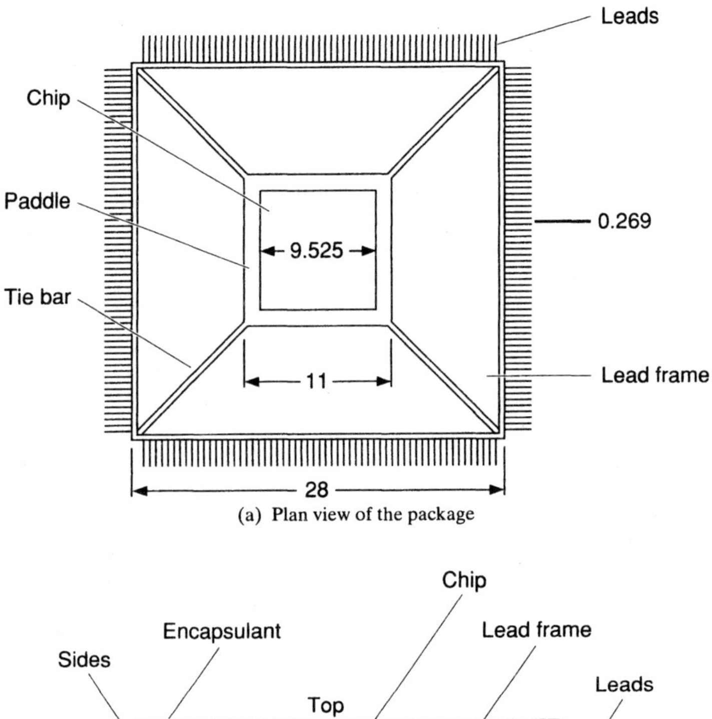

(a) Plan view of the package

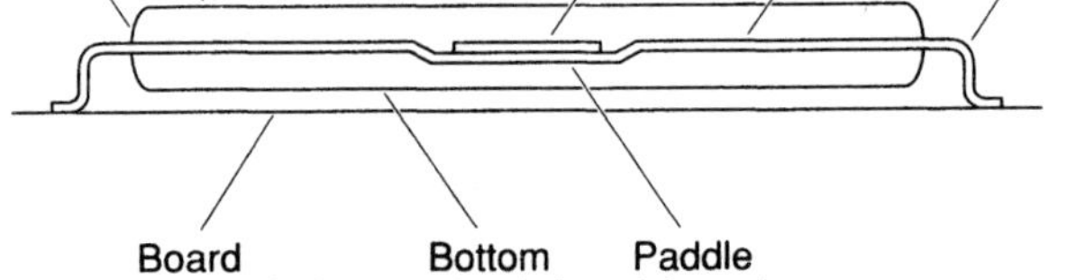

(b) Cross-sectional view of the package

FIGURE 6.5 Details of the plastic quad flat package (all dimensions are in mm): (*a*) plan view of the package; (*b*) cross-sectional view of the package.

dle, with the paddle being attached to the leadframe. The entire package is encapsulated in plastic for protection from the environment. For this particular example, a 208 lead package was considered [57].

Problem Definition. The objective of this study was to evaluate the applicability of phase-change material (PCM) for thermal control purposes. A layer of PCM was placed under the printed wiring board (PWB). The dissipated heat from the component was absorbed in the PCM as it changed phase from solid to liquid. The purpose for performing a system-level conjugate analysis was to determine the major heat transfer paths from the silicon die to the surrounding and the fluid flow and temperature patterns within the air and the phase change material (PCM).

Physical/Computational Domain. The computational domain is shown in Fig. 6.6. The package is mounted on a 1-mm-thick PWB. A layer of 10 mm organic PCM called *n-Eicosene* is placed below the PWB. The assembly is then placed inside a 50- × 50- × 50-mm enclosure. The package-level details were modeled to determine the dominant heat transfer path from the package to the surrounding, as well as for accurate determination of the die temperature. For such an analysis, the following simplifications were made:

- Every 5.2 leads were lumped into one lead, resulting in 10 effective leads per side. This was done in order to use a reasonable size mesh.

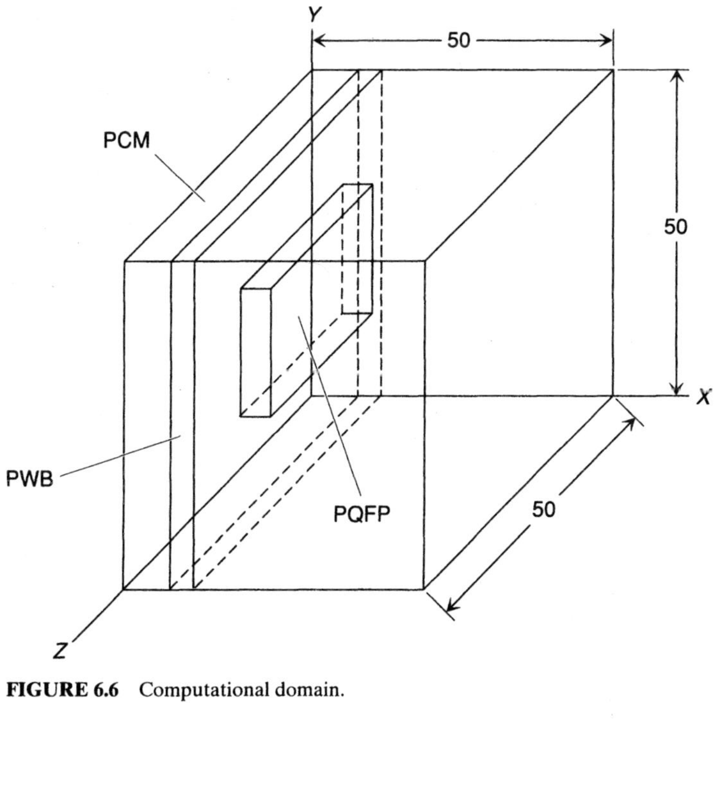

FIGURE 6.6 Computational domain.

- Several minor de-featurings were done to simplify the model. This required approximation of the package and lead outlines as rectangular.

Computational Grid. The computational mesh size for this problem was 39 × 39 × 34, which was reasonable from a transient analysis point of view. This model used a structured nonuniform mesh. Local grid refinement was incorporated near all the walls, to capture the boundary layers. To test the grid independence, computations were also performed on a finer grid of 47 × 42 × 47. The results showed a difference of 2 percent in the maximum temperature after one hour of operation from an initial temperature of 25°C.

Physical Model Selection. This analysis is a typical example of conjugate heat transfer modeling, where the conduction from the die to the exterior of the package, solid-liquid phase change, and the convection in the air were modeled simultaneously. Convection and phase change in the PCM were modeled using Enthalpy-Porosity technique. [58] The convection in air was assumed to be laminar. The Boussinesq approximation was used to model the buoyancy terms of the governing momentum equation.

Material Selection and Properties. Constant thermophysical properties were used for the PCM, air, and various solids, except the density of air in the body force term of the y-momentum equation. A total of seven different materials were used. The leadframe layer in the plastic in the package was treated as a composite material with effective thermophysical properties as mentioned in Sec. 6.2.2. The material properties used are presented in Table 6.19.

Boundary Conditions. For the thermal analysis, all the walls except the one at x = 50 mm were assumed adiabatic. The wall at x = 50 mm was assumed isothermal at an external ambient temperature of 25°C. Heat dissipation from the die was assumed to be volumetrically uniform. Heat-dissipation levels from the die were assumed to be 1 W.

Solution Strategy. The computational domain was discretized into control volumes, using a staggered grid. Using this approach, each of the independent variables is defined at different grid points. To resolve the velocity-pressure coupling in the

TABLE 6.19 Material Properties Used for Computation

Materials	Thermal conductivity, W/(m-K)	Density, kg/m³	Specific heat, J/(kg-K)	Dynamic viscosity, kg/(m-s)
Air	0.0261	1.177	1,005	1.85×10^{-5}
FR-4	0.35	1,938	1,600	—
Ceramic (C-786)	18.0	3,875	840	—
Leads and paddle (copper)	385	8,933	385	—
Encapsulant	0.31	1,070	100	—
Lead frame encapsulant mixture	154.17	4,215.2	214	—
Chip (silicon)	$154.86\left(\frac{300}{T(K)}\right)^{4/3}$	691	2,330	—

momentum equations, SIMPLER methodology was used, which is described in Patankar [4]. The discretized equations were solved using a Tridiagonal Matrix Algorithm (TDMA). The underrelaxation factors for u, v, and w were varied from 0.1 to 0.5, and from 0.4 to 0.8 for the pressure and pressure correction equations, and 0.9 to 1.0 for the energy conservation equation.

Two separate criteria were used for convergence. The first criterion was satisfied when the maximum change in the primitive variables, such as u, v, w, and T fell below their corresponding termination residual set value. For all computations, this was set as 1×10^{-6} for T and 1×10^{-4} for u, v, and w, respectively. The second criterion satisfied energy balance. At every time step, the energy residual from the system was set as 1×10^{-4}.

The conjugate heat transfer by convection and conduction in various materials was handled in a straightforward manner. The fluid flow and energy equations were solved for the entire domain. Velocities in the solid region (except the solid PCM) were artificially set to zero, by using a very high coefficient of viscosity. Harmonic mean formulations for diffusivities were used to assure continuity of flux at the interfaces of dissimilar materials. A fully implicit time stepping was used to march the solution in time. A control of time step was necessary to ensure convergence, during different phases of computation.

Postprocessing. Figure 6.7 shows the velocity vectors of a vertical plane along the computational domain. The velocity vectors show a plume generated above the package. This hot air moves up, impinges on the upper surface, and flows down along

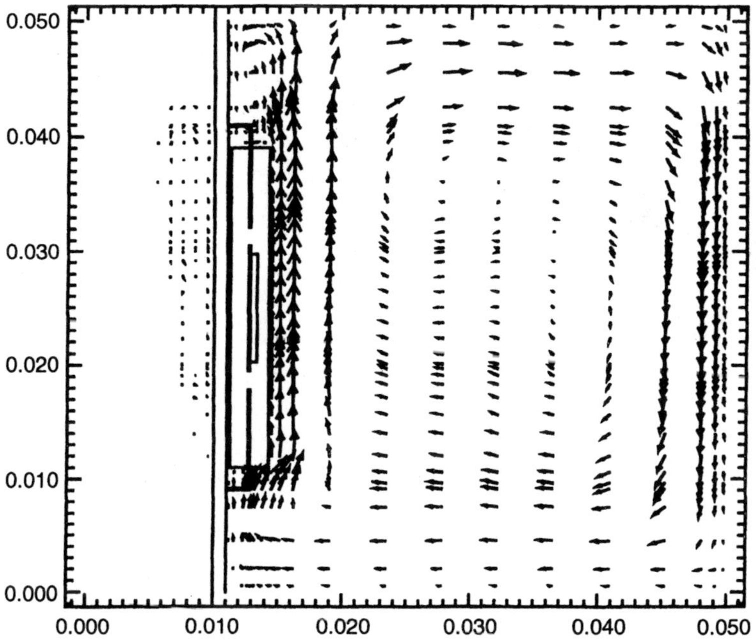

FIGURE 6.7 Velocity vectors for 1 W of power and FR-4 as the PWB material.

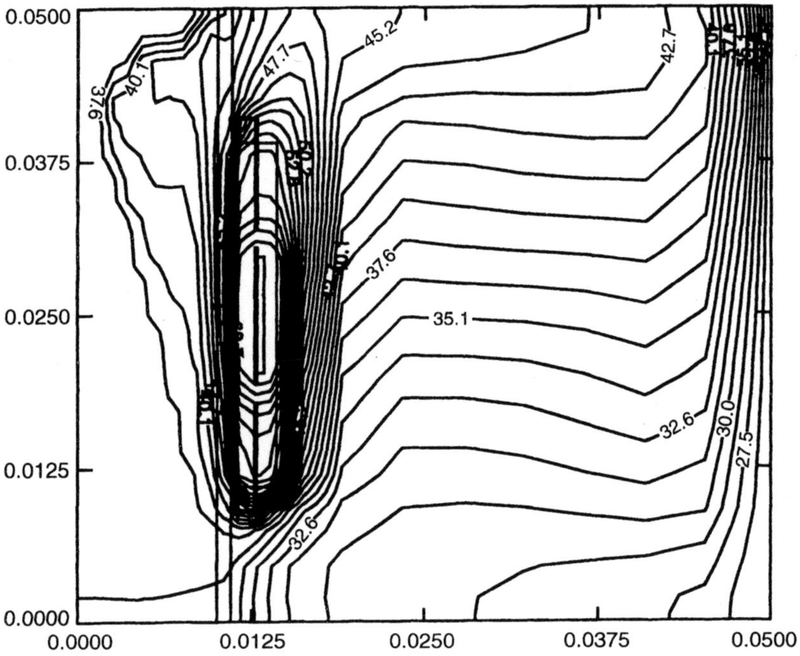

FIGURE 6.8 Temperature contours.

the cold wall at $x = 50$ mm, as it gets colder and denser. This sets up a single cell pattern of airflow inside the enclosure. Figure 6.8 shows the corresponding thermal contours in the plane.

The temperature contours show strong boundary layers at the package and the cooler wall surface, and natural convection throughout the enclosure. Table 6.20 shows some data extracted from one thermal analysis showing the relative heat transfer budget in the analysis.

TABLE 6.20 Results from PQFP Analysis Showing the Thermal Budget

Package-level thermal budget	
Heat transfer path	Relative percentage, %
Convection from top of the package	8
Convection from bottom of the package	24
Conduction through the leads	68
System-level thermal budget	
Heat transfer path	Relative percentage, %
Heat transfer to the PCM	82
Heat transfer to the right wall ($x = 50$ mm)	18

REFERENCES

1. L. T. Yeh, "Review of Heat Transfer Technologies in Electronic Equipment," *Journal of Electronic Packaging,* **117**:333–339, 1995.
2. W. M. Kays and M. E. Crawford, *Convective Heat and Mass Transfer,* McGraw-Hill, New York, 1993.
3. F. M. White, *Viscous Fluid Flow,* 2d ed., McGraw-Hill, New York, 1991.
4. S. Patankar, *Numerical Heat Transfer and Fluid Flow,* Hemisphere Publishing Company, Washington D.C., 1980.
5. J. N. Reddy, *An Introduction to Finite Element Method,* McGraw-Hill, New York, 1984.
6. E. Fried and I. E. Idelcheck, *Flow Resistance—A Design Guide for Engineers,* Hemisphere Publishing Company, New York, 1989.
7. F. P. Incropera and D. P. DeWitt, *Introduction to Heat Transfer,* John Wiley & Sons, New York, 1990.
8. R. K. Shah and A. L. London, *Laminar Flow Forced Convection in Ducts,* Academic Press, New York, 1978.
9. J. F. Thompson, Z. U. A. Warsi, and C. W. Mastin, *Numerical Grid Generation: Foundations and Applications,* North Holland, Amsterdam, 1985.
10. J. F. Thompson, "Grid Generation," *Handbook of Numerical Heat Transfer,* eds. Minkowycz et al., Wiley Interscience, New York, 1988.
11. Fluent, FLUENT User's manual, v4.23, 1991.
12. E. Yu and Y. K. Joshi, "Air Cooling of a Vented Enclosure by Combined Conduction, Natural Convection, and Radiation," *Proceedings of the ASME Heat Transfer Division,* HTD 333:323–330, 1996.
13. T. F. Smith, C. Beckermann, and S. W. Weber, "Combined Conduction, Natural Convection, and Radiation Heat Transfer in an Electronic Chassis," *Journal of Electronic Packaging,* **113**:382–391, 1991.
14. C. J. Freitas, "Perspective: Selected Benchmarks from Commercial CFD Codes," *Journal of Fluids Engineering,* **117**:208–218, 1995.
15. B. E. Launder, "On the Computation of Convective Heat Transfer in Complex Turbulent Flows," *Journal of Heat Transfer,* **110**:1112–1127, 1988.
16. B. E. Launder, "Current Capabilities for Modeling Turbulence in Industrial Flows," *Applied Scientific Research,* **48**:247–269, 1991.
17. K. Hanjalıc, "Advanced Turbulence Closure Models: A View of Current Status and Future Prospects," *International Journal of Heat and Fluid Flow,* **15**(3):178–203, 1994.
18. L. Prandtl, "Über die ausgebildete turbulenz," *Z. Angew. Math. Mech.,* **5**:136–139, 1925.
19. A. N. Kolmogorov, "Equations of Turbulent Motion of an Incompressible Fluid," *Izvestiya AN SSSR,* **6**(1–2):56–58, 1942.
20. L. Prandtl, "Über ein neues formelsystem für die ausgebildete turbulenz," *Nachr. Akad. Wiss. Göttingen Math.-Phys. Kl.,* **2**:6, 1945.
21. Fluent, FLUENT/UNS User's manuals, v4.4, 1996.
22. W. P. Jones and B. E. Launder, "Prediction of Laminarization with a Two-Equation Model of Turbulence," *International Journal of Heat and Mass Transfer,* **15**:301–314, 1972.
23. B. E. Launder and D. B. Spalding, "The Numerical Computation of Turbulent Flow," *Computational Methods in Applied Mechanical Engineering,* **3**:263–289, 1974.
24. W. Rodi, *Turbulence Models and Their Application in Hydraulics,* Brookfield Publishing, Brookfield, Vermont, 1984.

25. B. E. Launder, B. J. Reece, and W. Rodi, "Progress in the Development of a Reynolds Stress Turbulence Closure," *Journal of Fluid Mechanics,* **144**:79–101, 1975.

26. K. Hanjalic and B. E. Launder, "A Reynolds Stress Model of Turbulence and Its Application to Thin Shear Flows," *Journal of Fluid Mechanics,* **52**:609–638, 1972.

27. B. Hutchings and R. Iannuzzelli, "Taking the Measure of Fluid Dynamic Software," *Mechanical Engineering,* **109**(5):72–76, 1987.

28. K. R. Kirtley, "Renormalization Group Based Algebraic Turbulence Model for Three-dimensional Turbomachinery Flows," *AIAA Journal* **30**(6):1500–1506, 1992.

29. V. Yakot and S. A. Orszag, "Renormalization Group Analysis of Turbulence: Basic Theory," *Journal of Scientific Computing,* **1**:1–51, 1986.

30. C. C. Chieng and B. E. Launder, "On the Calculation of Turbulent Heat Transport Downstream from an Abrupt Pipe Expansion," *Numerical Heat Transfer,* **3**:189–207, 1980.

31. R. W. Johnson and B. E. Launder, Discussion of "On the calculation of turbulent heat transport from an abrupt pipe expansion," *Numerical Heat Transfer,* **5**:493–496, 1982.

32. R. S. Amano, "On the Calculation of Turbulent Heat and Mass Transport Downstream from an Abrupt Pipe Expansion," *AIAA Paper 82-1269, AIAA/SAE/ASME 18th Joint Propulsion Conference,* June 1982.

33. R. S. Amano, "Development of a Turbulence Near-Wall Model and Its Application to Separated and Reattached Flows," *Numerical Heat Transfer,* **7**:59–75, 1984.

34. N. Djilali, I. Gartshore, and M. Salcudean, "Calculation of Convective Heat Transfer in Recirculating Turbulent Flow Using Various Near-wall Turbulence Models," *Numerical Heat Transfer Part A,* **16**:189–212, 1989.

35. C. L. Tien, "A Correlation for Thermal Contact Conductance of Nominally Flat Surfaces in a Vacuum," *Special Publication 302,* National Bureau of Standards, 1968.

36. M. G. Cooper, B. B. Mikic, and M. M. Yovanovich, "Thermal Contact Conductance," *International Journal of Heat and Mass Transfer,* 279–300, 1969.

37. A. Blahey, J. L. Tevaarwerk, and M. M. Yovanovich, "Contact Conductance Correlations of Elastically Deforming Flat Rough Surfaces," *AIAA Paper No. 80-1470,* presented at the *AIAA 5th Thermophysics Conference,* Snowmass, Colorado, 1980.

38. M. R. Sridhar and M. M. Yovanovich, "Review of Elastic and Plastic Conductance Models: Comparison with Experiment," *Journal of Thermophysics and Heat Transfer,* **2**(1):43–47, 1994.

39. R. Siegel and J. R. Howell, *Thermal Radiation Heat Transfer,* Hemisphere Publishing Company, Washington D.C., 1992.

40. W. A. Fiveland, "Discrete-Ordinates Solutions of the Radiative Transport Equation for Rectangular Enclosures," *Journal of Heat Transfer,* **106**(4):699–706, 1984.

41. M. H. N. Naraghi and M. Kassemi, "Analysis of Radiative Transfer in Rectangular Enclosures Using a Discrete Exchange Factor Method," *Journal of Heat Transfer,* **109**(2):470–477, 1987.

42. General Electric Fluid Flow Division, "Properties of Gases—Air," **410**(2):6, 1982.

43. J. E. Graebner, "Technical Brief: Thermal Conductivity of Printed Wiring Boards," *Electronics Cooling,* **1**(2):27, 1995.

44. S. W. Churchill and H. H. S. Chu, "Correlating Equations for Laminar and Turbulent Free Convection from a Vertical Flat Plate," *International Journal of Heat and Mass Transfer,* **18**:1323, 1975.

45. W. Elenbass, "Heat Dissipation of Parallel Plates by Free Convection," *Physica,* **9**:1–28, 1942.

46. S. Globe and D. Dropkin, "Natural Convection Heat Transfer in Liquids Confined between Two Horizontal Plates," *Journal of Heat Transfer,* 81C, 1959.

47. M. Jakob, "Free Convection Through Enclosed Plane Layers," *ASME* **68**:489, 1946.

48. R. Teertstra, J. R. Culham, and M. M. Yovanovich, "Comprehensive Review of Natural and Mixed Convection Heat Transfer Models for Circuit Board Arrays," presented at *IEPS Advanced Technology for Electronic Packaging Conference,* September 29–October 2, Austin, Texas, 1996.

49. R. A. Wirtz, "Forced Air Cooling of Low-Profile Package Arrays," in *Air Cooling Technology for Electronic Equipment,* eds. S. J. Kim and S. W. Lee, CRC Press, New York, 1996, pp. 81–101.

50. G. K. Morris and S. V. Garimella, "Composite Correlations for Convective Heat Transfer from Arrays of Three-dimensional Obstacles," *International Journal of Heat and Mass Transfer,* **40**(2):493–498, 1997.

51. S. V. Garimella and P. A. Eibeck, "Heat Transfer Characteristics from an Array of Protruding Elements in Single Phase Forced Convection," *International Journal of Heat and Mass Transfer,* **33**:2659–2669, 1990.

52. A. M. Anderson and R. J. Moffat, "The Adiabatic Heat Transfer Coefficient and the Superposition Kernel Function: Part I—Data for Arrays of Flatpacks at Different Flow Conditions," *Journal of Electronic Packaging,* **12**:14–21, 1992.

53. S. S. Kang, "The Thermal Wake Function for Rectangular Modules," *Journal of Electronic Packaging,* **116**:55–59, 1994.

54. Lee, et al., 1995.

55. B. P. Leonard, "A Stable and Accurate Convective Modeling Procedure Based on Quadratic Upstream Interpolation," *Computer Methods in Applied Mechanical Engineering,* **19**:59–98, 1979.

56. B. R. Hutchinson and G. D. Raithby, "A Multigrid Method Based on the Additive Correction Strategy," *Numerical Heat Transfer,* **9**:511–537, 1986.

57. D. Pal and Y. K. Joshi, "Application of Phase Change Materials for Passive Thermal Control of Plastic Quad Flat Packages: A Computational Study," *Numerical Heat Transfer, Part A,* **30**:19–34, 1996.

58. A. D. Brent, V. R. Voller, and K. J. Reid, "Enthalpy-porosity Technique for Modeling Convection-Diffusion Phase Change: Application to the Melting of a Pure Metal," *Numerical Heat Transfer,* **13**:297–318, 1988.

CHAPTER 7
THERMAL MANAGEMENT

Jerry Sergent

7.1 INTRODUCTION

The goal of the packaging engineer in performing a thermal design is to remove sufficient heat from the circuit such that the temperature remains below the maximum operating temperature as established by the circuit specification. This parameter is commonly expressed in two ways: the maximum ambient temperature in which the circuit must operate and the maximum junction temperature of the semiconductor devices. Often, particularly in power devices, the specification is given as a combination of both, as in "a maximum junction temperature of 150°C in an ambient of 125°C." The thermal design must meet this requirement while also keeping the size, weight, and cost to a minimum. Thermal design is not a well-understood technology, except for those who practice it on a regular basis, and there is a tendency to overdesign, just to be on the safe side, which can render a product just as impractical as underdesigning.

Fundamentally, the thermal designer must create a path with low thermal impedance from the device level to a point where it may be dissipated safely without damage to the circuit or to the environment. The task begins with the electrical circuit analysis, which determines the amount of power dissipated in the individual components and elements of the circuit, and the layout, which determines the physical location of the generated heat. From this data, the heat flux, in watts/unit area, may be calculated, providing a basis for selecting a thermal design approach.

Thermal design is an iterative process. The temperature of the various points in the circuit depends on both the location of the heat-generating elements and the thermal path. A process flow chart for thermal design is shown in Fig. 7.1. The layout is generated and the heat flux at each point of the topography is calculated. From this data, a preliminary cooling approach is selected and a thermal analysis is performed on the overall combination. The layout and/or the thermal path is adjusted until the thermal analysis indicates that the temperature is below the desired level. Several approaches may be compared for cost, weight, and size, and the thermal design may be optimized for all these criteria in the shortest time.

The task is somewhat complicated by the fact that thermal analysis is not an exact science. There are simply too many variables and the mathematical models are not in closed form in most cases. However, the more sophisticated numerical methods allow for an arbitrary degree of precision, and sufficient accuracy can be obtained to make a valid engineering decision.

The thermal design affects most of the decisions made during the packaging phase and, in many cases, is the ultimate criteria in selecting an approach. For example, the power dissipation in a FET may determine whether the FET is mounted on a printed circuit board in packaged form or is mounted on a ceramic substrate in die form. Beginning with the manner in which the components are packaged and continuing to the point where external cooling is or is not required, the thermal design affects every aspect of the packaging design process.

This chapter considers the options available to the packaging engineer in selecting a thermal approach, including component mounting, substrate selection, and heat exchanger selection. The term *heat exchanger* in this context refers to the means of transferring heat from the electronic circuit to the environment and includes heat sinks, thermoelectric coolers, and heat pipes. For simplicity, the approximate methods of thermal analysis discussed elsewhere in this book are utilized. While not as accurate as numerical methods, they suffice as a basis to compare the various methods of cooling.

7.2 THERMAL STRESS ANALYSIS

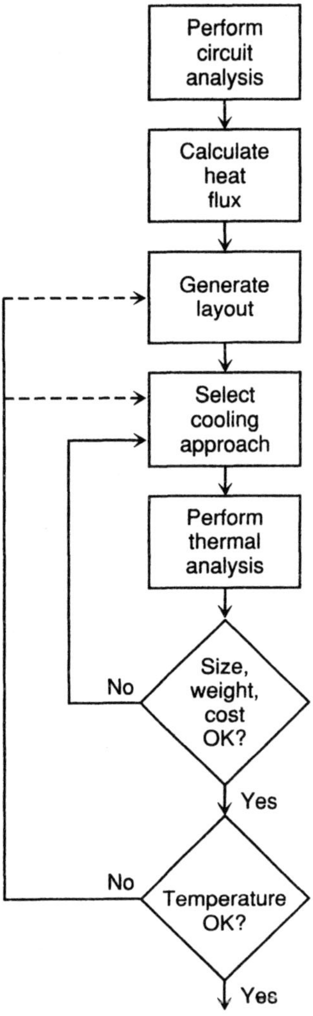

FIGURE 7.1 Flowchart for thermal design.

The temperature coefficient of expansion (TCE) phenomenon described in Chap. 3 has serious implications in thermal design. When a sample of material has one end fixed, which may be considered to be a result of bonding to another material that has a much smaller TCE, the net elongation of the hotter end per unit length, or *strain* (E), of the material is calculated by

$$E = \text{TCE} \times \Delta T \tag{7.1}$$

where E = strain in length/length
ΔT = temperature differential across the sample

Elongation develops a stress (S) per unit length in the sample as given by Hooke's law:

$$S = EY \tag{7.2}$$

where S = stress in psi/in ($N/m^2/m$)
Y = modulus of elasticity in lb/in^2 (N/m^2)

When the total stress, as calculated by multiplying the stress/unit length by the maximum dimension of the sample, exceeds the tensile strength of the material, mechanical cracks will form in the sample which may even propagate to the point of separation. This analysis is somewhat simplistic in nature, but serves as a basic understanding of the mechanical considerations in joining different materials together, a common practice in the assembly of electronic circuits.

Since the volume of a solid is constant, there is also a displacement in the tangential direction when a force is applied in the normal direction. The ration of the tangential force to the normal force is called *Poisson's ratio* and is given by

$$\nu = \frac{e_T}{e_N} \tag{7.3}$$

where e_T = strain in tangential direction
e_N = strain in normal direction

Poisson's ratio is a constant for a given material for elastic strains and can be used to determine the strain in other directions. This is of particular importance in thermal management, since many of the configurations create stresses in all directions. A quantity called the *shear modulus* can be defined which is related to the modulus of elasticity by

$$G = \frac{E}{2(1+\nu)} \tag{7.4}$$

Material parameters of interest are shown in Tables 3.5 to 3.8. In using these tables, it is important to note that there is a wide variation of certain of these parameters in the literature, and values for some of the alloys and composite materials may not be available at all. Material parameters also vary widely with temperature, and more detailed information should be utilized where applicable.

There are two geometries of interest: a component mounted flat to a substrate, such as a semiconductor die, and a component mounted on both ends, such as a chip capacitor.

When a component, such as a semiconductor die, is mounted to a substrate, a three-layer structure composed of the substrate, the die attach material, and the active die is formed as shown in Fig. 7.2. For a structure of this type, excessive stress can occur when either one of two conditions is present [1]:

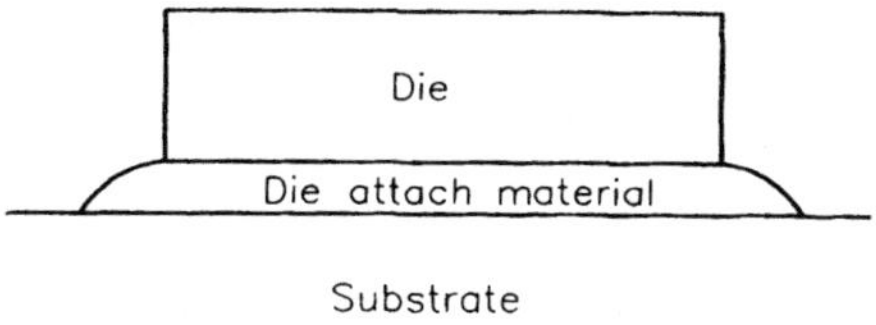

FIGURE 7.2 Die-die attach—substrate structure.

1. When the TCE of the die is sufficiently lower than that of the substrate such that the resultant stress is greater than that of the tensile strength of the die material when the structure is in thermal equilibrium. This condition may occur when no power is applied to the die and the structure is subjected to temperature excursions, or *temperature cycling.*
2. When a nonequilibrium condition exists such that the temperature differential between the die and the substrate is sufficiently large to create a stress condition greater than the tensile strength of the die. This condition may be brought about when the thermal resistance of the die bond is high due to improper selection of materials, when voiding occurs at the die bond interface during transient power conditions, or during power cycling, when the die is operating.

In many practical situations, both conditions exist to a greater or lesser degree. In this case, the strain is given by

$$E = (\mathrm{TCE}_D - \mathrm{TCE}_S)(T_D - T_A) + \mathrm{TCE}_S(T_D - T_S) \tag{7.5}$$

where TCE_D = temperature coefficient of expansion of the die
TCE_S = temperature coefficient of expansion of the substrate
T_D = temperature of the die
T_S = temperature of the substrate
T_A = ambient temperature with the power off

Two situations are of further interest. [2]

1. $T_D = T_S$. In this case, the power dissipated in the die is negligible or nonexistent, and Eq. 7.3 reduces to

$$E = (T_D - T_S)(T_D - T_A) \tag{7.6}$$

This is the situation described by temperature cycling.

2. $TCE_D = TCE_S$. In this case, the TCEs of the die and the substrate have been made equal by proper material selection during the design process. Eq. 7.3 reduces to

$$E = \mathrm{TCE}_S(T_D - T_S) \tag{7.7}$$

This is the condition described by power cycling or high thermal resistance, which creates an excessive temperature gradient from the die to the substrate.

The analysis presented here may be considered a best-case analysis, since most experimental results will result in lower figures than calculated. There is a wide variation in experimental data due to the large number of variables present in the component attach process. There are many other considerations that will determine whether or not a failure occurs, such as a fault in the material, which concentrates the stress at a given point, or failures that occur as a result of excessive mechanical fatigue when a structure is subjected to a number of temperature excursions.

Frequently, the maximum stress is observed when the structure is cooled down from the die attach process, since the process temperature is greater than the operating temperature. The stress is maximum at the corners and is given by [3]

$$S_M = \frac{(\mathrm{TCE}_S - \mathrm{TCE}_D) \times (T_P - T_A) \times L \times G \times \tanh(\beta)}{\beta \times t_B} \tag{7.8}$$

$$\beta = \sqrt{\frac{G}{t_s}\left(\frac{1}{E_D t_D} + \frac{1}{E_S t_S}\right)} \tag{7.9}$$

where S_M = maximum stress at the die corners in Pa or psi
E_D = die modulus of elasticity in Pa or psi
E_S = substrate modulus of elasticity in Pa or psi
G = shear modulus of bonding material
L = maximum die dimension in units of length
t_B = adhesive bond thickness in units of length
t_D = die thickness
t_S = substrate thickness
T_P = processing temperature in °C
= glass transition temperature for epoxies
= solidification temperature for solders
T_A = ambient temperature or temperature of interest

Note:

$$\text{lb/in}^2 = 6.90 \times 10^{-3} \times \text{n/mm}^2$$
$$\text{N/mm}^2 = 1.45 \times 10^2 \times \text{lb/in}^2$$
$$\text{N/mm}^{3/2} = 28.77 \times \text{lb/mil}^{3/2}$$
$$\text{lb/mil}^{3/2} = 3.48 \times 10^{-2}\ \text{N/mm}^{3/2}$$

Data obtained by Bolger and Mooney [4] and Van Kessel et al. [5] agrees with Eq. 7.8 within the accuracy of the measurements.

It should also be noted that many of the parameters in Eq. 7.8 and Eq. 7.9 are functions of temperature. Where the variation is extreme over the temperature range of interest, this should be considered when making direct comparisons between materials.

The brittle nature of certain of the materials used in electronic circuits plays a role in the packaging of these components. In practice, the force required to fracture a semiconductor die or ceramic substrate is much lower than predicted by theory. The discrepancy is due to small flaws or cracks residing within these materials as a result of processing. For example, when a semiconductor die is sawed, small edge cracks may be created. Similarly, when a ceramic substrate is fired, trapped organic material may outgas during firing, leaving a microscopic void in the bulk. The result is an amplification of the applied stress in the vicinity of the void which may exceed the tensile strength of the material and create a fracture. If the microcrack is assumed to be elliptical with the major axis perpendicular to the applied stress, the maximum stress at the tip of the crack may be approximated by [6]

$$S_M = 2S_O \leq \left(\frac{a}{\rho_t}\right)^{1/2} \tag{7.10}$$

where S_M = maximum stress at the tip of the crack
S_O = nominal applied stress
a = length of the crack as defined in Fig. 7.3
ρ_t = radius of the crack tip

The ratio of the maximum stress to the applied stress may be defined as

$$K_t = \frac{S_M}{S_O} = 2\left(\frac{a}{\rho_t}\right)^{1/2} \tag{7.11}$$

where K_t = stress concentration factor

For certain geometries, such as a long crack with a small tip radius, K_t may be much larger than one, and the force at the tip may be substantially larger than the applied force.

Based on this analysis, a material parameter called the *plain strain fracture toughness,* which is a measure of the ability of the material to resist fracture, can be defined as

$$K_{IC} = ZS_C\sqrt{\pi a} \tag{7.12}$$

where K_{IC} = plain strain fracture toughness in psi-in$^{1/2}$ or MPa-m$^{1/2}$
Z = dimensionless constant, typically 1.2 [5]
S_C = critical force required to cause breakage

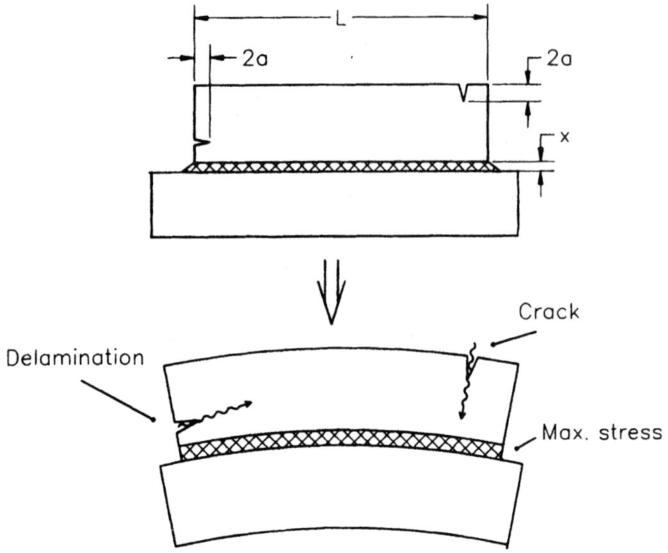

FIGURE 7.3 Cracks and chip-outs in semiconductor devices. *(From Ref. 5.)*

From Eq. 7.12, the expression for the critical force can be defined as

$$S_C = Z \frac{K_{IC}}{\sqrt{\pi a}} \tag{7.13}$$

When the applied force on the die due to TCE or thermal differences exceeds this figure, fracture is likely. The plain strain fracture toughness for selected materials is presented in Table 7.1. It should be noted that Eq. 7.13 is a function of die thickness up to a point, but is approximately constant for the area to thickness ratio normally found in semiconductor die.

Van Kessel et al. [5] presents data obtained by mounting strain gauges with epoxy and by eutectic means. This data indicates that such factors as bond thickness and bond voids may reduce the applied stress by more than an order of magnitude from that calculated by Hooke's law.

Figure 7.4 plots the calculated stress on a die 0.250 × 0.250 × 0.015 inches mounted on an alumina substrate 0.025 in thick as a function of bond thickness for three different materials: conductive epoxy, Sn10/Pb90 solder, and Sn96.5/Ag3.5 solder. Note that the processing temperature plays an important role. Solders with similar moduli of elasticity but with drastically different melting points will create higher stress in the

TABLE 7.1 Fracture Toughness for Selected Materials

Material	Fracture toughness, MPa-m$^{1/2}$
Silicon	0.8
Alumina (96%)	4.6
Molding compound	2.0

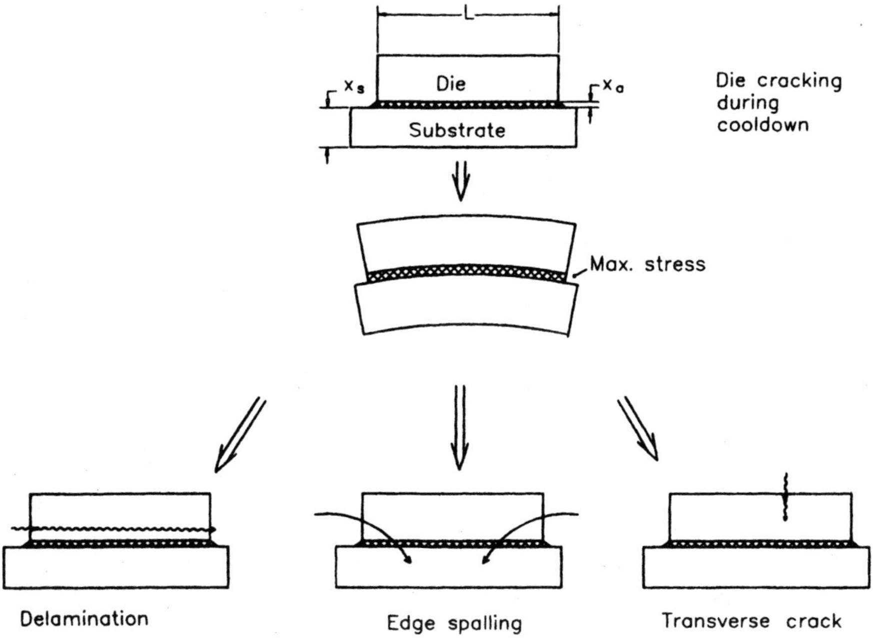

FIGURE 7.7 Types of cracks in a die as a result of thermal stress. *(From Ref. 4.)*

Calculate the stress and determine if a potential for cracking exists if the die is mounted to

A. A copper header 0.050 in thick

B. An alumina substrate 0.050 in thick

Solution The first step is to perform an approximate thermal analysis using the method developed in Chap. 5.

The thermal resistance of a square die is calculated from Eq. 5.17

$$R_{TH} = \frac{L}{kA} \tag{5.17}$$

where L = die thickness in inches
k = thermal conductivity in W/°C-in
A = area of die in in^2

For this case,

$$R_{TH\ \mathrm{Si}} = \frac{0.012}{3.81 \times 0.250^2} = 0.05\ \frac{°\mathrm{C}}{\mathrm{W}}$$

Spreading may be neglected through the thin solder layer as it has virtually the same lateral dimensions as the silicon die. The thermal resistance of the solder layer is

$$R_{TH\ \mathrm{So}} = \frac{0.002}{0.914 \times 0.250^2} = 0.035\ \frac{°\mathrm{C}}{\mathrm{W}}$$

which is surprisingly close to that of the die.

A. Copper As the heat reaches the copper, it begins to spread in accordance with Eq. 5.45. The spreading angle α is

$$\alpha = \tan^{-1} \frac{k_1}{k_2} \qquad (5.45)$$

where k_1 = thermal conductivity of current layer
k_2 = thermal conductivity of the underlying layer

The spreading angle into the copper is

$$\alpha = \tan^{-1} \frac{0.914}{9.804} = 5.33 \text{ deg} = 0.093 \text{ rad}$$

Where spreading is occurring in a square area, the thermal impedance can be determined by calculating the effective width of the spread as shown in Fig. 7.8. The extent of the spread is calculated from the spreading angle, and the effective width is the geometric mean of the widths before and after spreading. The width after spreading is given by

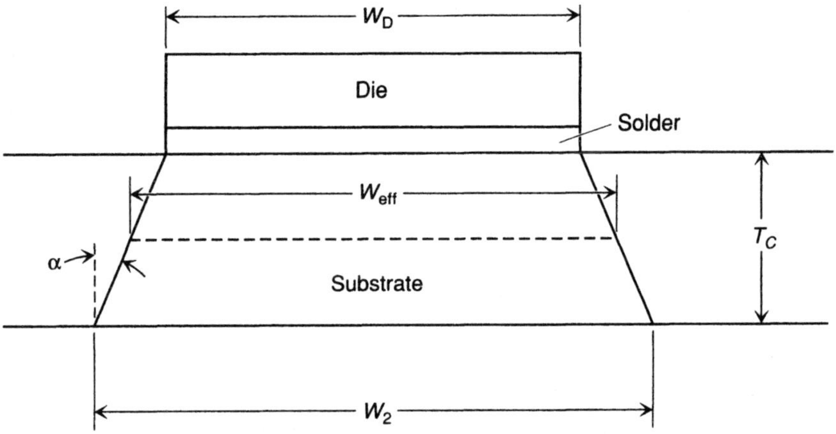

FIGURE 7.8 Spreading angle and effective width.

$$W_2 = W_D + 2T_C \sin \alpha \qquad (7.14)$$

where W_2 = width after spreading
W_D = width of die
T_C = thickness of copper

For this case,

$$W_2 = 0.250 + 2 \times 0.05 \times \sin(0.093) = 0.259 \text{ in}$$

The effective width of the copper is given by

$$W_{\text{eff}} = \sqrt{W_D W_2} \qquad (7.15)$$

and the thermal resistance of the copper is

$$\theta_{Cu} = \frac{0.05}{9.8 \times 0.254^2} = 0.079 \frac{°C}{W}$$

The net thermal resistance is the sum of the three elements.

$$\theta_T = \theta_{Si} + \theta_{So} + \theta_{Cu} = 0.05 + 0.035 + 0.079 = 0.164 \frac{°C}{W}$$

and the temperature of the die is

$$T_D = \theta_T P_D + T_A \tag{7.16}$$

where T_D = temperature of the die
P_D = power dissipated in die
T_A = ambient temperature

$$T_D = 0.164 \times 12 + 25 = 2 + 25 = 27°C$$

The stress on the die is calculated from Eq. 7.8.

$$\beta = \sqrt{\frac{1 \times 10^6}{0.05}\left(\frac{1}{18.7 \times 10^6 \times 0.015} + \frac{1}{16.96 \times 10^6 \times 0.05}\right)} = 9.74$$

$$S_M = \frac{(16.5 \times 10^{-6} - 2.8 \times 10^{-6}) \times (290 - 27) \times 1 \times 10^6 \times 0.250 \times 1.414 \times \tanh(9.74)}{9.74 \times 0.001}$$

$$= 1.31 \times 10^5 \text{ psi}$$

The critical stress is calculated from Eq. 7.13.

$$S_C = 1.2 \frac{0.8 \times 10^6}{\sqrt{\pi 1.0 \times 10^{-6}}} = 4.51 \times 10^8 \text{ Pa} = 6.54 \times 10^4 \text{ psi}$$

The maximum stress is much greater than the critical stress, indicating that cracking of the die is likely. It is significant to note that the dissipated power contributes very little to the stress in the die; it is primarily due to the combination of materials and the processing parameters.

B. Alumina The thermal resistance of the die and the solder are the same for this case. It is assumed that the effects of the metallization on the alumina may be neglected. The spreading angle in the alumina is

$$\alpha = \tan^{-1}\left(\frac{0.914}{0.533}\right) = 1.043 \text{ rad} = 59.75 \text{ deg}$$

The width after spreading is

$$W_2 = 0.25 + 2 \times 0.05 \times \sin(1.043) = 0.336 \text{ in}$$

and the effective width is

$$W_{Eff} = \sqrt{0.25 \times 0.336} = 0.290 \text{ in}$$

The thermal resistance of the alumina is

$$\theta_{Al} = \frac{0.05}{0.533 \times 0.290^2} = 1.115 \frac{°C}{W}$$

and the total thermal resistance is

$$\theta_T = 0.05 + 0.035 + 1.115 = 1.2\ \frac{°\mathrm{C}}{\mathrm{W}}$$

The die temperature is

$$T_D = 12 \times 1.21 + 25 = 39.4°\mathrm{C}$$

The maximum stress on the die is

$$\beta = \sqrt{\frac{1 \times 10^6}{0.05}\left(\frac{1}{18.7 \times 10^6 \times 0.015} + \frac{1}{49.9 \times 10^6 \times 0.05}\right)} = 8.91$$

$$S_M = \frac{(6.5 \times 10^{-6} - 2.8 \times 10^{-6}) \times (290 - 39.4) \times 0.250 \times 1.414 \times 1 \times 10^6 \times \tanh(8.91)}{8.91 \times 0.002}$$

$$= 3.68 \times 10^4 \text{ psi}$$

This figure is well below the critical stress level and should not result in fracture of the die. Both results presented here are consistent with experimental results. Large die are not amenable to mounting directly on most metals unless more compliant bonding materials are utilized.

The results presented here can be extrapolated to extend to three layers, a common configuration when a die is mounted on a substrate which in turn is mounted in a package. The stress created by the substrate-package interface is, in turn, added to the stress created by the die-substrate interface.

Example 7.2 The alumina substrate in Example 7.1 is soldered to a copper header which is 0.100 in thick with Au80/Sn20 solder (MP = 280°C) which is 0.005 in thick. Determine the stress on the die.

The spreading angle from the alumina to the Au80/Sn20 solder is assumed to be zero. From Example 7.1, $W_2 = 0.336$ in at the bottom of the alumina substrate. The thermal resistance of the Au80/Sn20 solder is therefore

$$\theta_{\mathrm{AuSn}} = \frac{0.005}{1.48 \times 0.336^2} = 0.031\ \frac{°\mathrm{C}}{\mathrm{W}}$$

The spreading angle from the solder to the copper is

$$\alpha = \tan^{-1}\frac{1.48}{9.8} = 8.4 \text{ deg} = 0.147 \text{ rad}$$

The amount of spreading for the copper is

$$_{2\mathrm{Cu}} = 0.336 + 2 \times 0.01 \times \sin 8.4° = 0.339 \text{ in}$$

and the effective width of the copper is

$$W_{\mathrm{EffCu}} = \sqrt{0.336 \times 0.339} = 0.3375$$

and the thermal resistance is

$$\theta_{\mathrm{Cu}} = \frac{0.01}{9.8 \times 0.3375^2} = 0.009\ \frac{°\mathrm{C}}{\mathrm{W}}$$

The total thermal resistance is

$$\theta_T = 1.2 + 0.031 + 0.009 = 1.24 \frac{°C}{W}$$

and the die temperature is

$$T_D = 1.24 \times 12 + 25 = 40°C$$

which is virtually the same as the alumina substrate.

The applicable maximum dimension is the diagonal of the area defined by W_2 at the bottom of the alumina substrate.

$$D_{max} = 0.336 \times 1.414 = 0.475 \text{ in}$$

The stress on the alumina due to the copper header is

$$\beta = \sqrt{\frac{3.06 \times 10^6}{0.05}\left(\frac{1}{16.96 \times 10^6 \times 0.05} + \frac{1}{49.9 \times 10^6 \times 0.05}\right)} = 9.8$$

$$S_M = \frac{(16.5 \times 10^{-6} - 6.5 \times 10^{-6}) \times (280 - 40) \times 0.250 \times 1.414 \times 1 \times 10^6 \times \tanh (9.83)}{9.83 \times 0.005}$$

$$= 7.10 \times 10^4 \text{ psi}$$

This amount of stress must be added to that created by the die-alumina interface.

$$S_{MT} = 3.68 \times 10^4 + 7.10 \times 10^4 = 10.78 \times 10^4 \text{ psi}$$

The total is considerably greater than the critical stress, indicating almost certain die failure. An interesting sidelight is that the stress created on the substrate by the copper is much greater than the tensile strength of the alumina, indicating almost certain fracture of the substrate as well.

Example 7.3 The substrate is mounted to the copper header with conductive epoxy, 0.005 in thick, cured at 150°C. Calculate the stress on the die.

Assuming that there is no spreading, the thermal resistance of the epoxy is

$$\theta_{ep} = \frac{0.005}{0.077 \times 0.336^2} = 0.575 \frac{°C}{W}$$

The spreading angle from the epoxy into the copper is

$$\alpha_{Cu} = \tan^{-1} \frac{0.077}{9.8} = 0.45 \text{ deg} = 0.008 \text{ rad}$$

The amount of spread in the copper is therefore negligible and the thermal resistance of the copper is

$$\theta_{Cu} = \frac{0.01}{9.8 \times 0.336^2} = 0.009 \frac{°C}{W}$$

The total thermal resistance is

$$\theta_T = 1.21 + 0.575 + 0.009 = 1.784 \frac{°C}{W}$$

and the die temperature is

$$T_D = 12 \times 1.784 + 25 = 46.4°C$$

Due to the relatively high thermal resistance of the conductive epoxy, the substrate temperature must be determined, as it is the differential between this element and the die above and the copper below that will determine the maximum stress. The temperature profile is shown in Fig. 7.9.

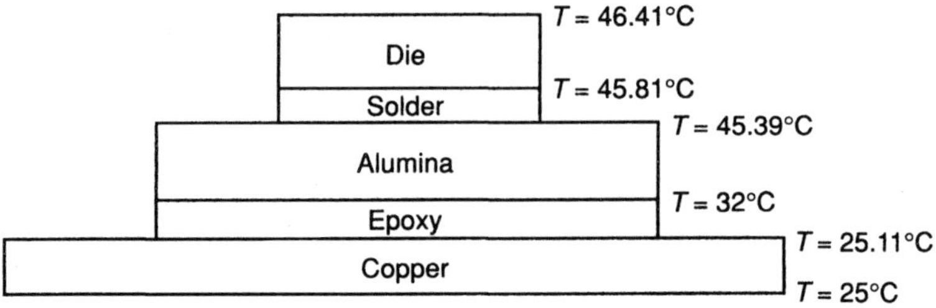

FIGURE 7.9 Temperature profile for Example 7.3.

The stress on the alumina substrate due to the copper is

$$\beta = \sqrt{\frac{0.24 \times 10^6}{0.05}\left(\frac{1}{16.96 \times 10^6 \times 0.05} + \frac{1}{49.9 \times 10^6 \times 0.05}\right)} = 2.75$$

$$S_M = \frac{(16.5 \times 10^{-6} - 6.5 \times 10^{-6}) \times (150 - 46.4) \times 0.336 \times 1.414 \times 1 \times 10^6 \times \tanh(2.75)}{2.75 \times 0.05}$$

$$= 8.51 \times 10^3 \text{ psi}$$

The total stress on the die is

$$S_{MT} = 3.68 \times 10^4 + 8.51 \times 10^3 = 4.53 \times 10^4 \text{ Pa}$$

which is still below the critical stress level, indicating that both the die and the substrate should be safe from fracture.

A second configuration of interest is the case where a component is fixed at both ends, such as a chip capacitor soldered to a printed circuit board or ceramic substrate, as shown in Fig. 7.10. During the soldering process, the ends of the capacitor become fixed as the temperature drops below the melting point. Depending on the relative magnitudes of the TCEs of the component and the substrate, the force may be compressive or tensile in nature. The problem may be complicated further when the substrate is also rigidly mounted to a package, as is sometimes the case when a ceramic substrate is soldered to a metal header to minimize thermal resistance. If the TCE of the header is greater than both the component and substrate, additional force on the component may be generated. On the other hand, if the TCE of the header approaches that of the component, the force may be largely negated. The net elongation of the component is given by

$$\Delta L_C = |D_C(T_P - T_A)(\text{TCE}_S - \text{TCE}_C)| \tag{7.17}$$

where ΔL_C = net elongation of the component
D_C = longest dimension of the component, usually the diagonal
TCE_S = TCE of the substrate

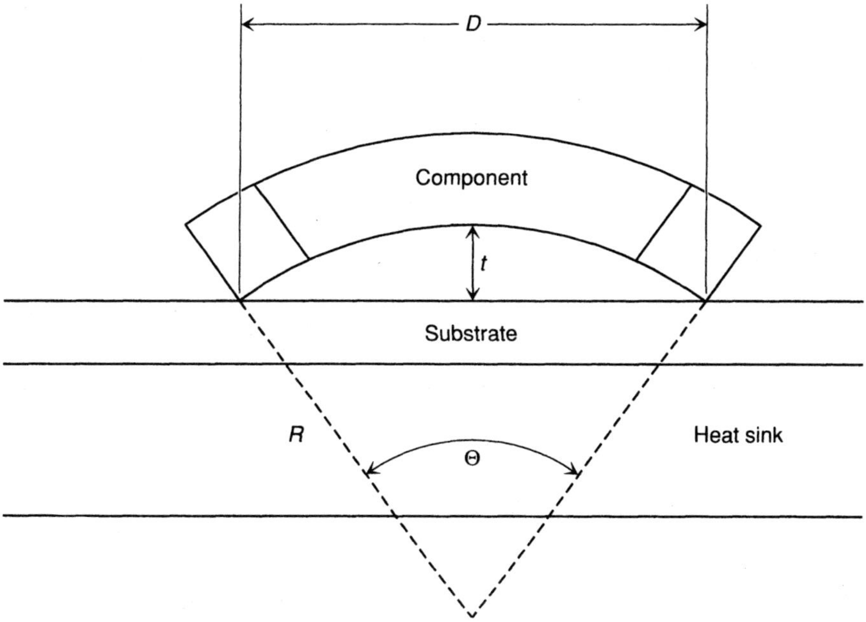

FIGURE 7.10 Bowing in a component due to soldering.

TCE_C = TCE of the component
T_P = processing temperature
T_A = ambient temperature

If Eq. 7.17 results in a negative number, the force on the component will be compressive, and, conversely, if it is a positive number, the force will be tensile.

With the ends fixed, the elongation will cause the component to bow, as shown in Fig. 7.10. The amount of bowing can be calculated by considering the extended length of the component as an arc and the original length as the chord of the arc.

$$L_A = L_C + \Delta L_C \tag{7.18}$$

where L_A = diagonal length of the component after cooling
L_C = original diagonal length of the component before attach

The radius of the arc can be approximated by

$$R = \sqrt{\frac{D_C L_A^2}{12 \Delta L_C}} \tag{7.19}$$

The amount of displacement t can be calculated by

$$t = \frac{D_c}{2} \tan\left(\frac{L_A}{2R}\right) \tag{7.20}$$

The strain is equal to the net elongation divided by the amount of displacement.

$$e = \frac{\Delta L_C}{t} \frac{\text{in}}{\text{in}} \tag{7.21}$$

The stress is given by Hooke's law, Eq. 7.2.

If the stress as calculated from Eq. 7.2 exceeds the tensile strength of the component, catastrophic failure is likely to occur.

The tensile strengths of alumina, beryllia, and barium titanate, a common material used to manufacture chip capacitors, are given in Table 7.3.

Example 7.4 A ceramic capacitor 0.220 × 0.200 in is soldered to an alumina substrate (which may be considered rigid) with Sn96.5/Ag3.5 solder (MP = 221°C). Calculate the stress on the capacitor and determine if breakage is likely.

$$D_C = \sqrt{0.220^2 + 0.200^2} = 0.297 \text{ in}$$

$$\Delta L_C = |0.297(221 - 25)(6.5 \times 10^{-6} - 8.1 \times 10^{-6})| = 9.32 \times 10^{-5} \text{ in}$$

Note that the algebraic sign of ΔL_C is negative, indicating compressive stress on the capacitor.

$$L_A = D_C + \Delta L_C = 0.297 + 9.32 \times 10^{-5} = 0.297415 \text{ in}$$

$$R = \sqrt{\frac{D_C L_A^2}{12\Delta L_C}} = \sqrt{\frac{0.297 \times 0.297415^2}{12 \times 9.32 \times 10^{-5}}} = 16.79 \text{ in}$$

$$t = \frac{D_c}{2} \tan\left(\frac{L_A}{2R}\right) = \frac{0.297}{2} \tan\left(\frac{0.297415}{2 \times 16.79}\right) = 0.00132 \text{ in}$$

The strain is therefore

$$E = \frac{\Delta L_C}{t} = \frac{9.32 \times 10^{-5}}{0.00132} = 0.0708 \frac{\text{in}}{\text{in}}$$

and the stress is

$$S = EY = 0.0708 \times 1.28 \times 10^9 = 90.84 \text{ MPa}$$

which is slightly less than the tensile strength for a compressive force as given in Table 7.3. This is the largest capacitor which can be soldered to an alumina substrate in this fashion.

7.2.1 Factors Which Minimize Stress

Hooke's law, as previously demonstrated, typically results in values of stress that are much greater than those observed experimentally. A variety of factors contribute to the reduction in stress.

TABLE 7.3 Tensile Strength of Selected Ceramic Materials

Material	Tensile strength, MPa
Aluminum oxide (alumina)	172
Beryllium oxide (beryllia)	138
Barium titanate	103 (compressive) 69 (tensile)

TCE Match. The amount of stress generated in a component is directly proportional to the difference in TCE between the component and the substrate. The TCE of clad materials, such as Cu-Mb-Cu, can be tailored to provide a specific TCE by controlling the relative thickness of the layers.

Bond Thickness. An increase in the bond thickness contributes to a reduction in stress on the die by having a greater ability to flex when a force is applied. This principle is commonly employed by increasing the thickness of solder joints when chip carriers, chip capacitors, and other ceramic components are soldered to printed circuit boards.

Bonding Voids. It has been experimentally observed that small voids in the bond distributed over the area of the die reduce the stress. However, voids in the bond area increase the thermal resistance and, consequently, the temperature of the die, which counters the positive effect somewhat. Large voids tend to concentrate the stress at the point of bonding and increase the probability of cracking.

Compliant Bonding Materials. The use of a compliant bonding material, such as a soft solder or epoxy, enables the bond to absorb much of the stress, minimizing the stress on the die. Examples of soft solders include PbIn solders and high tin content solders, such as Sn96.5/Ag3.5. It must be noted, however, that even filled epoxies are relatively poor thermal conductors and tend to increase the junction temperature of the die.

Intermediate Layers. The use of an intermediate layer of material with a TCE between that of the die and the substrate is a design approach commonly used to reduce stress. One very common material used for this purpose is molybdenum, which has a TCE of 5.4×10^{-6} in/in/°C as compared to 2.8×10^{-6} in/in/°C for silicon and 6.5×10^{-6} in/in/°C for alumina. Typically, the Mb tab, or *molytab,* is plated with Au and the die is eutectically mounted to it. The molytab with the die attached may be mounted to the substrate with solder or epoxy. The molytab also has the effect of being a heat spreader, increasing the effective area of the bond and reducing the junction temperature of the die.

Processing Temperatures. Selecting materials for minimum processing temperature has a dramatic effect on stress reduction, as the stress is initially applied at the time the bonding material is solidified or cured.

7.3 DIE ATTACH MATERIALS

The die attach material is the first step on the heat removal path and, consequently, is one of the most important. If not properly selected and implemented, the die attach material can dominate the other elements in the path and render futile any further attempts at improving the thermal resistance. The criteria for selecting a die attach material includes thermal resistance, mechanical compliance, compatibility with other materials, and processability. There are three basic categories of die attach materials:

1. *Organic,* including epoxies, thermoset materials, and polyimides
2. *Metallurgical,* including solders and eutectic attach
3. *Silver-filled glass,* a glass matrix with a silver filler

7.3.1 Organic Die Attach Materials

Organic die attach materials include epoxies, polyimides, and thermoplastic adhesives. Although organic materials have a lower thermal conductivity than solder or silver-filled glass, there are numerous applications in power circuits where their use is desirable.

The term *epoxy* encompasses a wide range of materials, all of which contain an epoxide ring as shown in Fig. 7.11. To form a solid structure, the basic resin must be cross-linked (cured) by adding a catalyst to form a thermosetting structure. For die-mounting purposes, it is necessary to use a resin/catalyst combination requiring heat to accelerate the cure, preferably a temperature equal to or greater than 150°C for process compatibility. As it cures, epoxy forms a mechanical bond by flowing into small irregularities in the die and substrate surfaces. No chemical changes occur between the epoxy and the bonding surfaces. The strength of the bond is enhanced by the formation of a fillet around the perimeter of the die.

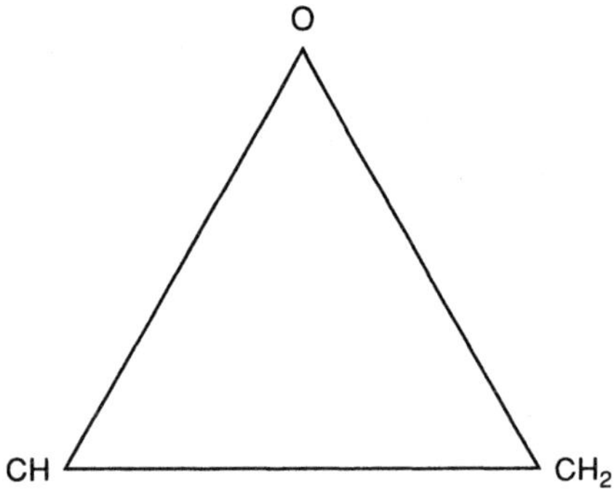

FIGURE 7.11 Basic epoxide ring.

Epoxies are available in a variety of forms, including conductive and nonconductive. Most die utilize the epoxy for electrical contact as well as for mechanical bonding, requiring that the epoxy be filled with a conductive material, usually silver or gold. Silver is preferred because the conductivity is higher than that of gold, and less is required for a given resistance. The amount of silver added is a compromise between conductivity and strength and is typically around 70 percent by weight. Table 7.4 summarizes the composition of solvent-based and non-solvent-based epoxies. The solvent-based epoxy has higher thermal and electrical conductivities due to the higher concentration of silver as a result of solvent evaporation during cure.

TABLE 7.4 Approximate Composition of Solvent and Non-Solvent-Based Epoxies

Composition		Solvent-based, %	Non-solvent-based, %
By weight	Silver	70	70
	Resin/catalyst	20	30
	Solvent	10	N/A
By volume	Silver		
	Before cure	18	20
	After cure	34	20

The selection of an epoxy for mounting a power die requires many compromises. The higher electrical and thermal conductivity of a solvent-based epoxy is balanced by the potential for voids under the die as a result of solvent evaporation, and the curing process must be more closely monitored and controlled. A second potential problem is also alleviated by the higher conductivity. With lower concentrations of silver, there is a tendency for the silver to settle away from the interface over time at high temperature, ultimately resulting in an increase of electrical and thermal resistance.

Epoxies are also more compliant than other bonding materials, with a modulus of elasticity substantially lower than for solder. This feature, along with the lower processing temperature, generates lower stress on the die and increases the range of available substrate materials.

Epoxy may be dispensed by screen printing, by die transfer, and pneumatically through a nozzle. The screen-printing process may be performed with a stainless steel screen or a stencil. The pattern on the screen is fabricated by coating it with a photosensitive material and exposing it to UV light through 1:1 artwork, which hardens the area directly under the openings. The unexposed material is removed by rinsing with water. A stencil may be prepared by electroforming [7], by laser cutting, or by chemical etching. The electroformed stencil is the most precise, allowing dimensions as low as 0.003 in to be printed. The die transfer process requires that a die be dipped in epoxy passed under a doctor blade to form a film of uniform thickness. The epoxy adheres to the die which is then brought into contact with the mounting pad on the substrate. The die is deliberately made smooth, usually with stainless steel or hard rubber, so that most of the material will tend to adhere to the pad. The pneumatic process uses compressed air to force the epoxy through a nozzle, dispensing a dot of material. This process can be made more precise by using a positive placement pump and adding an x-y table. In this manner, a pad can be covered with epoxy of uniform thickness, albeit slower than screen printing. Of the three methods described here, screen printing, particularly with a stencil, provides the most uniform coating. However, it does require a flat surface for printing, and, if other components have been mounted, is not a viable approach. Frequently, the process sequence will dictate the application method. The epoxy is cured by exposing it to an elevated temperature, usually 150°C or above. This may be performed as a batch process in a fixed oven or as an inline process in a conveyer oven.

The characteristics of a typical conductive epoxy used for die attach are summarized in Table 7.5.

TABLE 7.5 Typical Parameters of Thermoset MD-100 Electrically Conductive Die Attach Adhesive

Parameter	Units	Value
Glass transition temperature	°C	80
TCE $<T_g$	ppm/°C	56
$>T_g$		225
Thermal conductivity	W/m-°C	3.02
Die shear strength	psi	5,600
Tensile shear strength	psi	1,600
Volume resistivity at 25°C	Ω-cm	1×10^{-4}
Weight loss at 300°C	%	1.2

Source: Thermoset, Incorporated.

Certain epoxies have the capability of being *B-staged,* or partially cured, in the form of a thin flexible sheet, dry to the touch, which is applicable to bonding large areas. In Example 7.3, it was pointed out that the use of epoxy to mount an alumina substrate to a metal package could eliminate fracture of the substrate and/or the die. B-staged epoxy is ideal for this application, as it is difficult and time-consuming to apply a thin layer of epoxy uniformly to a large surface. By the time the heat reaches the substrate-epoxy interface, it has spread to the point where the larger thermal impedance of the epoxy layer is less significant.

Polyimides are based on aromatic dianhydrides and diamines. Polyimide resins are considered to be thermoplastic materials and do not cross-link as do epoxies. Polyimide adhesives are extremely pure, with chlorine, sodium, and potassium concentrations below 10 ppm. They are also extremely stable at temperatures ranging from cryogenic to well above 300°C. Their primary use is in applications subjected to elevated temperatures in subsequent processing, as in CERDIP sealing.

Most polyimides used for die attach are of the preimidized type. These generally are one-part and have a solvent content even higher than that of the solvent-based epoxies. A wide range of fillers can be used. Silver, while it has the highest electrical and thermal conductivity, has been observed to degrade the thermal performance of polyimides [2], and aluminum is preferred where electrical conductivity is not required. Polyimides require a two-step curing process: at approximately 70°C to evaporate the solvent and at an elevated temperature for the actual cure.

When compared to epoxies, polyimide adhesives have two disadvantages when considered for power applications.

1. The modulus of elasticity is somewhat higher, resulting in greater mechanical stress on the die.
2. The solvent evaporation process may lead to excessive voiding under the die if not properly processed.

Thermoplastic adhesives are similar to epoxies in form and in use. They cure at approximately the same temperature and have similar mechanical properties. Their primary advantage is that repair is substantially facilitated. By nature, the T_g is not radically different from the cure temperature and, when they are heated, become highly fluid, allowing die to be easily removed and replaced.

The properties of organic adhesives are summarized in Table 7.6.

7.3.2 Metallurgical Die Attach Materials

Solders are typically alloys of two or more metals used to join two metal surfaces. When two or more metals are alloyed, the lowest melting point of the alloy can be

TABLE 7.6 Comparison of Organic Adhesives

Property	Epoxy	Polyimide	Thermoplastic
Cure temperature	150°C	>200°C	125°C
Modulus of elasticity	5 GPa	15 GPa	5 Gpa
High temperature stability	<250°C	300–450°C	<250°C
Cure steps	1	2	1

considerably less than the melting points of either of the individual metals, a phenomenon which makes the soldering process possible. The melting point of most pure metals is simply too high to prevent damage to components or PCB laminates. At a certain composition, referred to as the *eutectic composition,* the alloy will have the lowest possible melting point, called the *eutectic temperature.* These two parameters are defined on the phase diagram of lead (Pb) and tin (Sn) in Fig. 7.12. The *eutectic point* on this graph is the point at which the liquidus and solidus states exist in equilibrium. The area marked *liquidus* defines the temperature above which a given composition will be in the liquid state, while the area marked *solidus* defines the temperature below which the composition will be in the solid state. In between is a state which is a mixture of liquid and solid referred to as the *plastic flow* state. Thus, there are two temperatures defining the state of a solder: the liquidus temperature and the solidus temperature. For the eutectic composition, the two temperatures are the same. Table 7.7 lists several common solder compositions and their liquidus and solidus temperatures. Note from Fig. 7.12 that the solidus temperature between lead concentrations of 2.5 to 81 percent by weight is the eutectic temperature. Above or below those levels, the solidus temperature is higher and is defined by the line between the plastic and Pb areas.

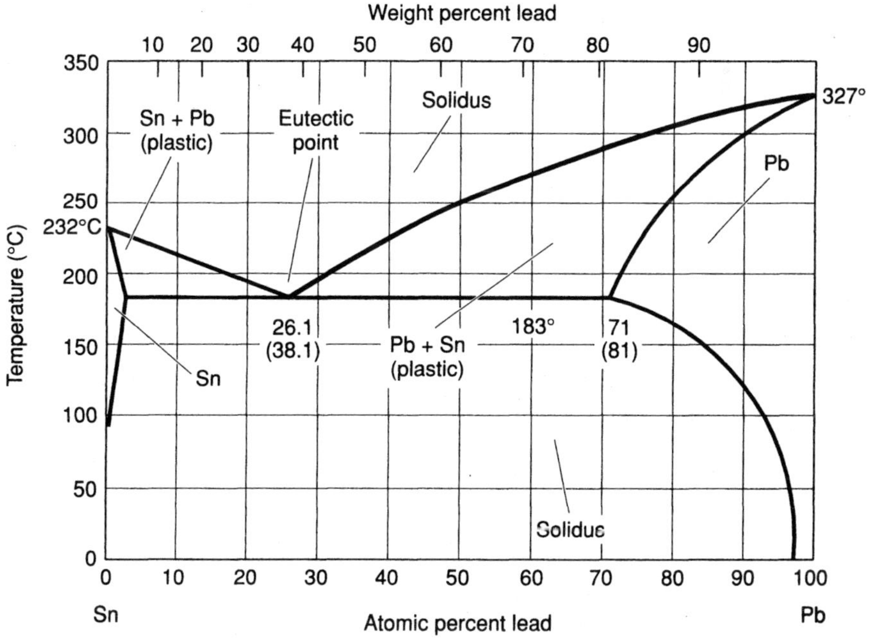

FIGURE 7.12 Tin-lead phase diagram.

There are two fundamental properties that a solder must possess in order to be used in this application:

1. The metal(s) comprising the surfaces must be soluble in the molten solder, that is, the solid solubility coefficient of the metal in the solder must be finite and greater than zero.

TABLE 7.7 Solidus and Liquidus Temperatures of Selected Solder Compositions

Material	Liquidus temperature, °C	Solidus temperature, °C	Plastic range, °C
Sn63/Pb37	183	183	Eutectic
Sn62/Pb36/Ag2	179	179	Eutectic
Sn10/Pb90	302	268	34
Pb50/In50	209	180	29
Sn96.5/Ag3.5	221	221	Eutectic
Au80/Sn20	280	280	Eutectic
Sn10/Pb88/Ag2	290	268	22

2. The solder must wet the surface. When a drop of liquid is placed into contact with a solid surface, the degree to which the liquid will spread, or *wet,* the surface depends on the relative surface energies of the two materials. In particular, the surface energy of the surface must be greater than that of the liquid. Metals have a very high surface energy while that of organic materials is somewhat lower.

When the molten solder is placed in contact with a compatible metal surface meeting these two criteria, the solder spreads and wets the surface. The molten solder dissolves a portion of the metal at the interface, creating an alloy of the solder and the metal. This alloy, which comprises the *solder joint,* solidifies when the solder is cooled below the melting point, forming the bond as shown in Fig. 7.13.

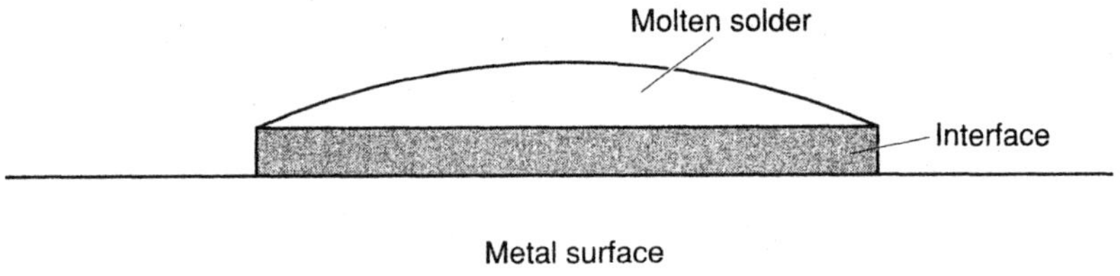

FIGURE 7.13 Solder-metal surface interface. (Not drawn to scale.)

In order for the solder joint to form, both the surface and the solder must be clean and free from oxides. Cleaning may be accomplished with one of several solvents, such as isopropyl alcohol, to remove oils and loose dirt. Oxides must be attacked with an organic acid, or *flux,* which may be contained within the solder or applied externally. The flux activates when heated and reduces the oxides on the surface and on the solder. There are three basic types of flux that may be used:

1. *RMA (resin, mildly activated) flux,* which may be removed only with solvents.
2. *Water-soluble flux,* which may be removed with hot water with a saponifier possibly required.
3. *No-clean flux,* which does not require cleaning and may be left on the substrate after soldering.

While all three types may be used for passive components, depending on the application, the no-clean flux is not suitable for soldering semiconductor die, as the flux will interfere with wire bonding. For many surfaces, the water-soluble flux will suffice, but, for surfaces such as nickel, an RMA flux is necessary. RMA fluxes present extreme challenges for cleaning. This is particularly true if high-temperature solders such as Sn10/Pb90 are used. The high temperatures may cause the flux to polymerize, adding to the difficulty. Many of the solvents which readily remove RMA flux are hazardous, but there are several commercially available compounds which do an excellent job of removing flux and which are safe to use if handled properly. Flux removal is facilitated by cleaning immediately after soldering. For extreme cases, plasma cleaning may be required. Exposure to an argon plasma for 5 to 10 minutes after solvent cleaning is usually sufficient to provide a good wire-bonding surface. Oxygen should not be necessary if proper cleaning procedures are used beforehand.

The solder may be applied as a preform or as solder paste. Preforms are preferred for several reasons:

1. *Thickness control.* The amount of solder is more rigidly controlled. This is important in order to minimize the thermal resistance and mechanical stress.
2. *Vehicle removal.* Solder paste contains an organic vehicle to improve dispensing or screening which is more volatile than the flux. To minimize voiding, the solder paste must be dried to remove this element before soldering.
3. *Uniform coverage.* Voids are minimized by insuring that the pad is completely covered with solder.

The solder composition depends on a number of factors.

1. *Compatibility with the die and substrate metallizations.* The solder must wet both surfaces, and both must react with the solder to form an alloy at the interface. Further, there must be no intermetallic compounds formed which might have a detrimental effect on reliability. Semiconductor devices intended for power applications have a backside metallization solderable by any of the common solder compositions, either gold or an alloy of titanium, nickel, and silver. The Ti/Ni/Ag combination is preferred, as gold forms intermetallic compounds with the tin-based solders known to cause failures.

   The various substrate metallizations, including thick film, plated copper, direct bond copper, and laminated copper (printed circuit boards) all present solderable surfaces, although some solders may be preferable. Table 7.8 summarizes the compatibility issues between metal films and solders.
2. *Temperature hierarchy.* The melting point of the solder must be compatible with other processes and with the maximum temperatures of the components.
3. *Environmental considerations.* The solder must not exert excessive force on the die during temperature excursions and must be able to withstand multiple temperature cycles. Table 7.9 presents the fatigue resistance of selected solder compositions.

7.3.3 Silver-Filled Glass

Silver-filled glass may also be used as a die attach material, particularly for applications seeing high postprocessing temperatures, as in hermetic package sealing with solder or glass. One such material is JM5160 manufactured by Johnson Matthey. The metal-to-glass ratio is about 4:1, resulting in a composition with a thermal conduc-

TABLE 7.8 Compatibility Issues Between Solders and Substrate Metallizations

Metallization		Compatible solders	Comments
Thick film	PdAg	Pb/Sn, Pb/Sn/Ag, Pb/In, Sn/Ag	Ag helps to reduce leaching.
	PtPdAg	Pb/Sn, Pb/Sn/Ag, Pb/In, Sn/Ag	Ag helps to reduce leaching.
	Au	Pb/In, Au/Sn	Pb/Sn solders leach gold.
Direct bond copper	Unplated	Pb/Sn, Pb/Sn/Ag, Pb/In, Sn/Ag	Uncoated Cu may tarnish during soldering.
	Ni Plate	Pb/Sn, Pb/Sn/Ag, Pb/In, Sn/Ag	Ni may require stronger flux.
Plated copper	Unplated	Pb/Sn, Pb/Sn/Ag, Pb/In, Sn/Ag	Uncoated Cu may tarnish during soldering.
	Ni Plate	Pb/Sn, Pb/Sn/Ag, Pb/In, Sn/Ag	Ni may require stronger flux.
Thin film			Not generally soldered due to low film thickness.
Laminated copper (PC board)	Solder plated	Pb/Sn, Pb/Sn/Ag	Higher-temperature solders may damage PC board laminate.

tivity of 78.3 W/m-K and a resistivity ≤20 μΩ-cm. JM5160 is based on a lead borate glass system and contains 89 percent solids by weight with the remainder being the organic vehicle. The presence of an organic vehicle requires a burnout step early in the firing process. The paste must be held above 410°C for 8 to 11 minutes in an oxidizing atmosphere depending on the size of the die.

TABLE 7.9 Comparative Fatigue Characteristics of Selected Solder Compositions*

Performance	Composition
Worst	Sn63/Pb37
	Sn60/Pb40
Poor	Sn62/36Pb/Ag2
	Sn65/In35
	Sn42/Bi58
	Sn50/In50
	Sn50/Pb50
Fair	In99/Cu1
	Sn90/Pb10
	Sn99.25/Cu0.75
Good	Sn99/Sb1
	Pb50/In50
	Sn100
Excellent	Sn96.5/Ag3.5
	Sn95/Sb5

* From Sergent and Harper, Ref. 1.

7.4 SUBSTRATE MATERIALS

Three basic types of materials are used for substrates: metals, ceramics, and plastics. These materials are frequently used in conjunction, and a critical aspect of hybrid design is to insure that the variations in properties of these materials are properly considered. The most important properties are the thermal conductivity, the heat capacity, and the temperature coefficient of expansion (see Table 7.10). Substrates for power applications can be divided into three categories: insulators (including diamond and ceramics), insulated metal substrates, and composite materials (metal matrix composites). In order to properly compare these materials, it is informative to study the thermal conductivity property in depth.

The thermal conductivity of a material is defined as

$$q = -k \frac{dT}{dx} \tag{7.22}$$

where k = thermal conductivity in W/m-°C
q = heat flux in W/cm^2
$\frac{dT}{dx}$ = temperature gradient in °C/m in steady state

The negative sign denotes that heat flows from areas of higher temperature to areas of lower temperature.

There are two mechanisms contributing to thermal conductivity: the movement of free electrons and lattice vibrations, or phonons. When a material is locally heated, the kinetic energy of the free electrons in the vicinity of the heat source increases, causing the electrons to migrate to cooler areas. These electrons undergo collisions with other atoms, losing their kinetic energy in the process. The net result is that heat is drawn away from the source toward cooler areas. In a similar fashion, the increased lattice vibrations as a result of a higher temperature transmit phonons, which also carry energy away from the source. The thermal conductivity of a material is the sum of the contributions of these two mechanisms.

$$k = k_p + k_e \tag{7.23}$$

where k_p = contribution due to phonons
k_e = contribution due to electrons

TABLE 7.10 Properties of Die Attach Materials

Material	Thermal conductivity, W/m-K	Thermal coefficient of expansion, ppm/°C	Modulus of elasticity, GPa
Conductive epoxy*	3	56 ($<T_g$) 225 ($>T_g$)	5
Sn63/Pb37 solder	51	24	18–25
Sn10/Pb90 solder	36	28	19
Sn96.5/Ag3.5	64	30	5.48
Sn62/Pb36/Ag2	52	26	23.2
Silver-filled glass	78	11	3

* Typical.

For metals, with a high concentration of free electrons, $k_e >> k_p$ and the thermal conduction is primarily due to electron movement. The electrical and thermal conductivities are related by the Wiedemann-Franz law

$$L = \frac{k}{\sigma T} \tag{7.24}$$

where L = Lorentz number in Ω-W/(K)^2
k = thermal conductivity in W/mm-K
σ = conductivity in siemens/mm
T = temperature in K

The Lorentz number is independent of temperature if it is assumed that the heat energy is transported entirely by electrons. Theoretically, the Lorentz number under this condition is 2.44×10^{-8} Ω-W/(K)^2, but this number will vary slightly from material to material. Table 7.11 gives the Lorentz number for some of the more common metals used in the fabrication of electronic circuits.

In ceramics, the heat flow is primarily due to phonon generation, and the thermal conductivity is generally lower than that of metals. Crystalline structures, such as alumina and beryllia, are more efficient heat conductors than amorphous structures such as glass. Organic materials used to fabricate printed circuit boards or epoxy attachment materials are electrical insulators and highly amorphous and tend to be very poor thermal conductors.

Impurities or other structural defects in metals and ceramics tend to lower the thermal conductivity by causing the electrons to undergo more collisions, which lowers the mobility and lessens the ability of the electrons to transport heat away from the source. For example, alloys of metals have a lower thermal conductivity than either of the pure metals. By the same token, as the ambient temperature increases, the number of collisions increases, and the thermal conductivity of most materials drops. In most practical cases only the temperature dependence of ceramics is critical, as these materials are frequently the limiting factor in the amount of heat flow.

Ceramic materials are highly desirable for use as substrates for at least two reasons: the electrical resistivity is extremely high and the thermal stability is excellent. The thermal conductivity of ceramics varies widely. Aluminum oxide (Al_2O_3, alumina) is relatively low, while that of beryllium oxide (BeO, beryllia) is higher than that of pure aluminum metal. The nitride ceramics, aluminum nitride (AIN) and boron nitride (BN), have much higher thermal conductivities than alumina. Aluminum nitride in

TABLE 7.11 Lorentz Number for Selected Metals*

Metal	Lorentz number at 0°C (10^{-8} W-Ω/K^2)	Lorentz number at 100°C (10^{-8} W-Ω/K^2)
Silver	2.31	2.37
Aluminum	2.24	2.28
Gold	2.35	2.40
Copper	2.23	2.33
Molybdenum	2.61	2.79
Lead	2.47	2.56
Platinum	2.51	2.60
Tin	2.52	2.49
Tungsten	3.04	3.20

* From Sergent and Harper, Ref. 1.

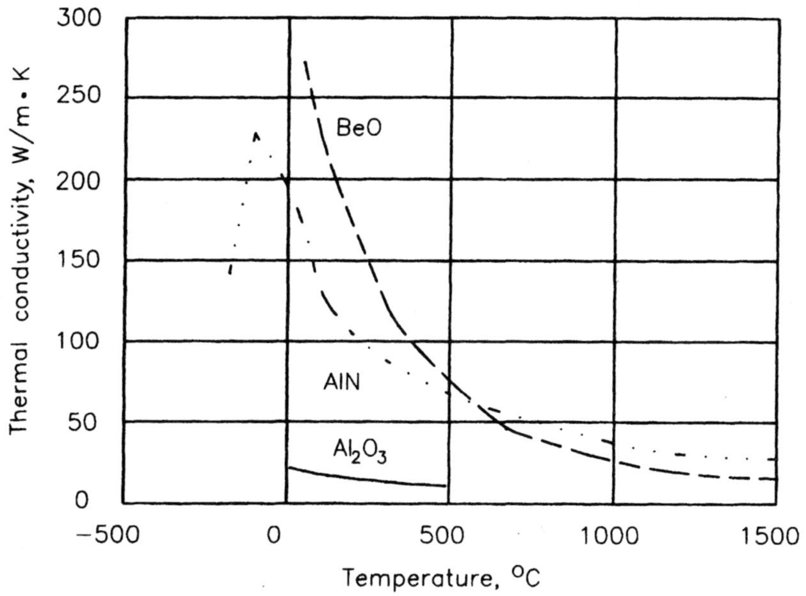

FIGURE 7.14 Thermal conductivity versus temperature for ceramic substrate materials.

particular has been used as a substrate and package material. The thermal conductivity of ceramics is highly dependent on temperature as shown in Fig. 7.14.

Alumina and beryllia can be directly metallized by a variety of methods, including thick film, thin film, direct bond copper (DBC), active metal braze (AMB), and plated copper. Aluminum nitride needs surface preparation by oxidation beforehand to be compatible with many of the metallization processes. Certain of the AMB materials can be used to bond copper directly to AIN, and thick film pastes are available which are directly useable without the need for oxidation. [8] Boron nitride cannot be metallized with any of the standard processes. [9]

The DBC process allows a thick layer of copper, up to 0.050 in, depending on the thickness of the substrate, to be bonded to alumina, beryllia, and aluminum nitride ceramics, with conductor patterns formed by photoetching the copper. This process can dramatically lower the thermal resistance by providing heat spreading directly under the die.

Of the available ceramics, boron nitride has the best CTE match to silicon with aluminum nitride a close second. AIN is widely used to manufacture single-chip packages for power die, but the use of BN is limited simply to interfaces.

Diamond has an extremely high thermal conductivity, five times greater than that of beryllia and four times that of pure copper. Diamond substrates fabricated with the Chemical Vapor Deposition (CVD) method have been successfully metallized and patterned with thin films using three different material combinations, Ti/Pt/Au, WTi/Au, and Nb/Au, and with copper thick film for use as substrates for MCMs. [10] The cost of diamond is still quite high, but it is presently used in large quantities as a substrate for laser diodes. As the process is refined, diamond is expected to see wider use as a substrate material.

Metal matrix composites (MMCs) can be fabricated to have both a low TCE and a high thermal conductivity. One of the more common of these is aluminum/silicon carbide (Al/SiC) which consists of SiC particles in an aluminum matrix. The CTE of this combination can be varied from 6 to 15 ppm by varying the SiC content from 30 to 75 percent by volume. A range of 70 to 73 percent by volume provides the best combination of properties and economics. [11] The resulting structure has a TCE of approximately 7 ppm and a thermal conductivity of about 218 W/m-K. The TCE increases slightly with temperature and the thermal conductivity decreases slightly with temperature. The density is comparable to that of alumina and aluminum nitride. Al-SiC is electrically conductive and may be plated with nickel or aluminum or flame-sprayed with a silver alloy for solderability. Where electrical isolation is required, the aluminum may be anodized. Al-SiC is used primarily as a baseplate for power circuits.

Another MMC with excellent thermal properties is Dymalloy [12], a copper-diamond composite. Dymalloy has 55 percent diamond by volume with a TCE of 5.5 ppm and a thermal conductivity of 420 W/m-K. The TCE of this material is a much better match for silicon or gallium arsenide than Type IIa diamond, with a TCE of 1.7 ppm.

Carbon impregnated with copper, aluminum, and silver [13] has also been developed for thermal applications. The material is anisotropic, with different TCEs in the *x-y* and *z* directions, but the variation is small and can be tailored to closely match silicon or gallium arsenide. The thermal conductivity is also anisotropic and is slightly lower in the *z* direction. The range is from 140 to 183 W/m-K, depending on the concentration.

The use of standard printed circuit board materials, such as FR-4, can be used for low- to moderate-power applications if thermal vias under the die are used to draw away the heat. For improved thermal conductivity, the vias may be selectively plated to increase the surface area, but may not be plated entirely closed due to the risk of trapping plating materials inside. Referring to Fig. 7.15, the thermal resistance of a thermal via is given by

$$R_{\text{TH (via)}} = \frac{t}{9.8\pi(r_2^2 - r_1^2)} \frac{°\text{C}}{\text{W}} \quad (7.25)$$

where t = thickness of board
r_2 = outer radius of via (radius of drilled hole)
r_1 = inner radius of via (hole diameter—plating thickness)

The thermal conductivity of a PC board is highly dependent on the amount of copper area and the thickness. Eq. 3.6 is an empirical formula which may be used for a particular case.

The parameters of the various substrate and baseplate materials are summarized in Table 7.12.

7.5 WIRE BONDING

The electrical connections to power die are made primarily by bonding a wire from the contacts on the die to the substrate or package metallization which, in practice, is achieved in one of two fundamental ways: ultrasonic wedge (stitch) or thermocompression ball and stitch. A third method, thermosonic bonding, is a combination of thermocompression and ultrasonic.

Ultrasonic bonding is accomplished by threading a gold or aluminum wire through a wedge-shaped tip acting as a transducer to couple ultrasonic energy from a

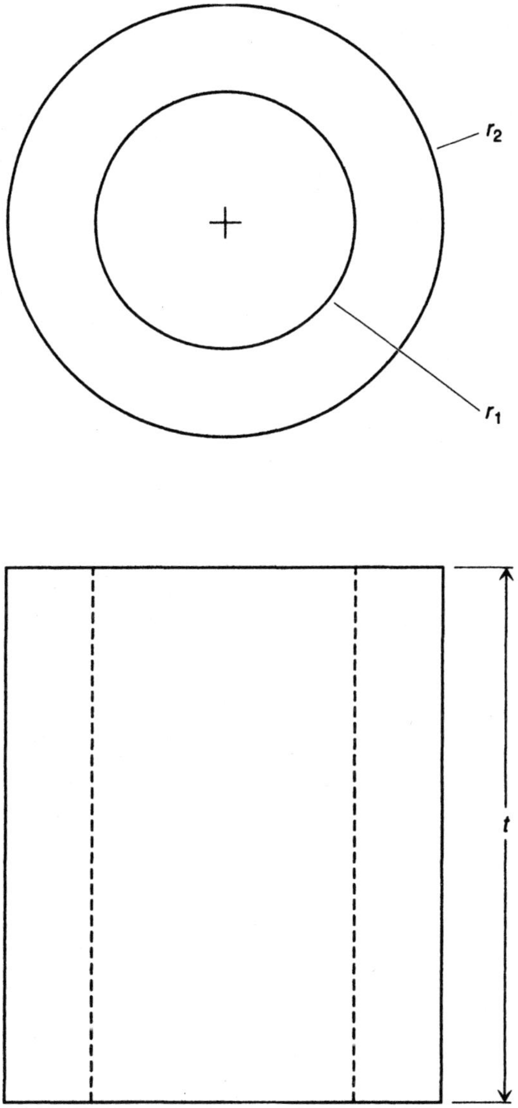

FIGURE 7.15 Dimensions of thermal via.

generator to the bond-surface interface, as shown in Fig. 7.16. The resulting deformation of the two metals produces a shearing action along both surfaces at the interface as depicted in Fig. 7.17, a process not requiring elevated temperature. Ultrasonic bonding primarily uses aluminum wire because the thin oxide coating the wire promotes a cleaning action creating atomically clean surfaces where interatomic forces can bond the two metals. The softer gold wire is not as efficient at cleaning the surfaces unless the bonding surface is also gold.

TABLE 7.12 Thermal Conductivity and TCE of Substrate and Baseplate Materials

Material	Thermal conductivity, W/m-K	Temperature coefficient of expansion, ppm/°C	Modulus of elasticity, GPa
Alumina, 96%	21	6.5	344
Beryllia, 99.5%	250	7.5	345
Aluminum nitride	220	4.4	300
Boron nitride	60	3.8	43
Silicon carbide	270	3.7	407
Thermagon	8*	—†	—†
Aluminum/silicon carbide	270	7	250
Dymalloy	420	5.5	—‡
Carbon-copper	140–183§	5–8§	—‡
FR-4	0.8 (*x,y*) 0.3 (*z*)	16–20 (*x,y*) 50–70 (*z*)	17.4

* Thermal conductivity of the laminate.
† Depends on baseplate material.
‡ Not available.
§ Depends on ratio of carbon to copper.

The bond formed by ultrasonic bonding is called a *wedge* or *stitch* bond. The nature of a wedge bond requires that the second bond is unidirectional. It must continue in the same direction as the first bond to avoid undue stress on the heel of the first bond. This requires that the first and second bond sites be aligned with the direction of the wedge by rotating the substrate mounting platform. After the second bond, the wire is broken by applying a clamp and moving to the next bond site.

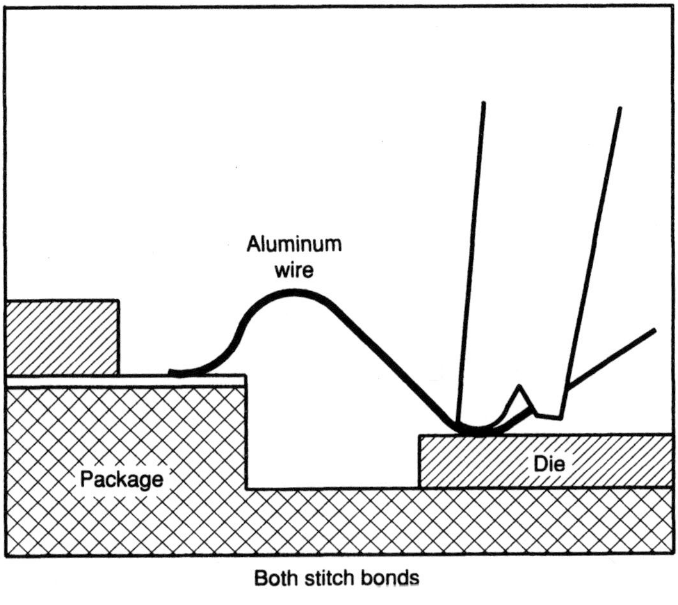

FIGURE 7.16 Ultrasonic aluminum wire bonding.

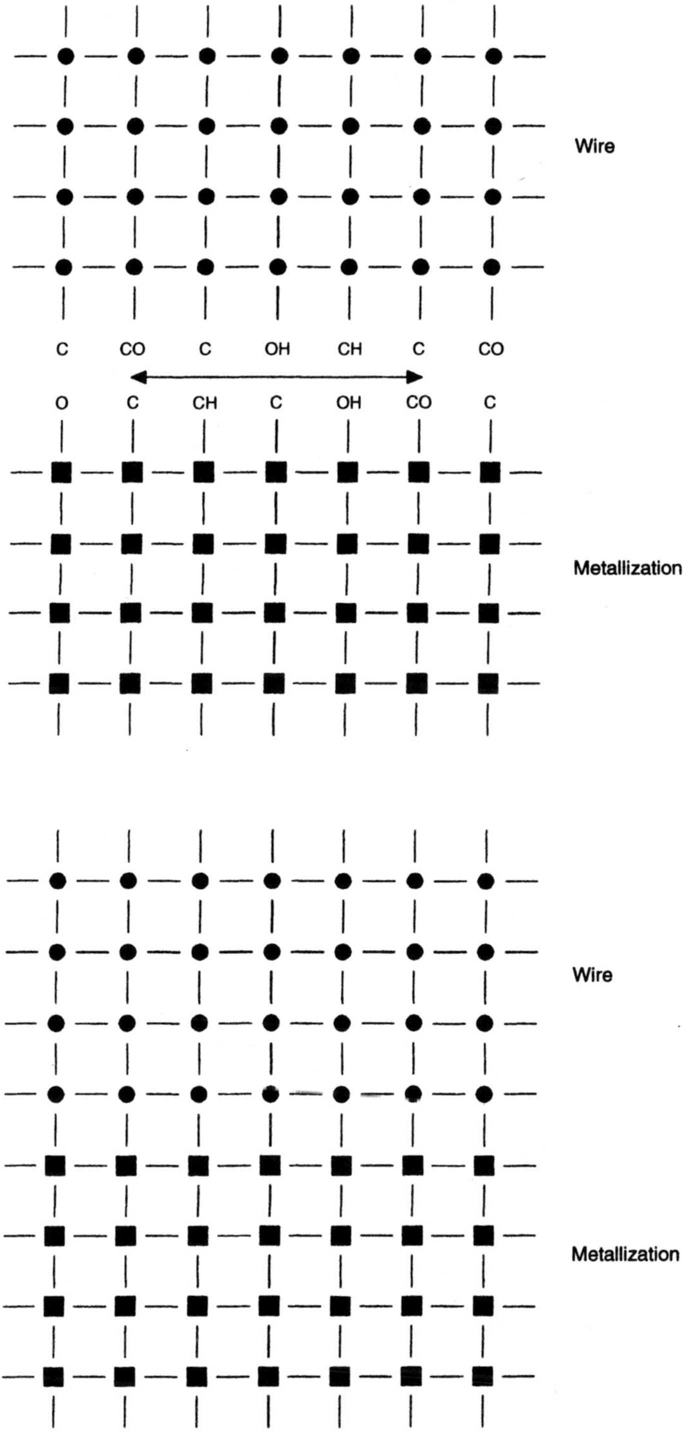

FIGURE 7.17 Ultrasonic bonding mechanism. *(From Johnson, Ref. 14.)*

Ultrasonic bonding is very amenable to large wire sizes, accommodating wires up to 0.022 inches in diameter.

Thermocompression bonding, as the name implies, uses a combination of heat and pressure to form the interatomic bond as shown in Fig. 7.18. This process requires a temperature in excess of 250°C and is used only with gold wire since the oxide coating on aluminum wire prevents the bond from forming. As shown in Fig. 7.18, a wire is fed through a capillary and a ball is formed on the end of the wire by an electric arc. The first bond, usually on the die, is formed by applying heat and pressure for a period of time until the bond is formed. Thermocompression bonding is omnidirectional and the capillary may be moved in any direction, requiring only an *x-y* table for the substrate mounting platform. The second bond is a stitch bond formed by moving the capillary to the next bond site and applying pressure. The wire is clamped and broken at the bond site by moving the capillary.

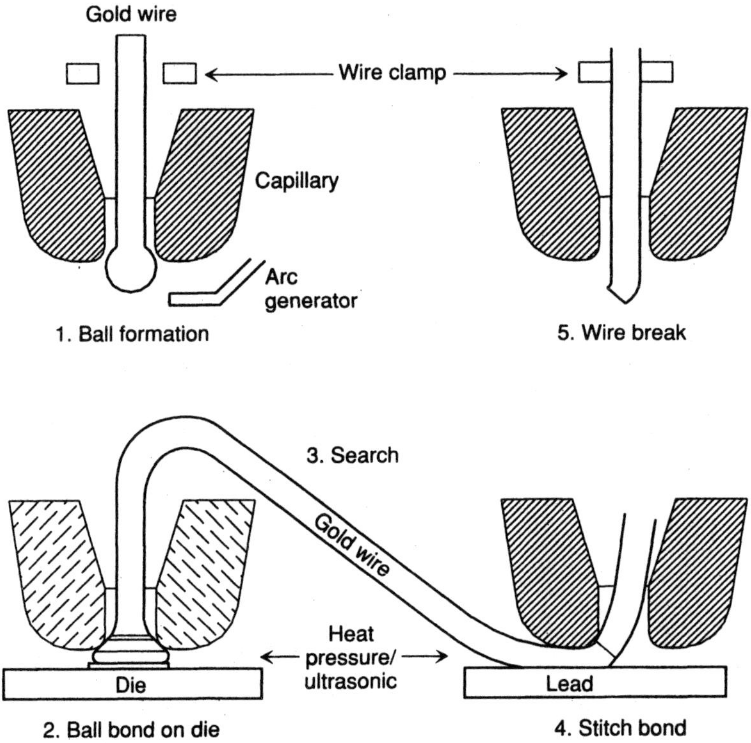

FIGURE 7.18 Thermosonic/thermocompression gold wire bonding.

Thermocompression bonding is rarely used for two reasons:

1. The bonding temperature is above the threshold for intermetallic compound formation between the gold wire and the aluminum bonding pad on the die. The diffusion coefficient of aluminum into gold is substantially higher than that of gold into aluminum. The result is that the aluminum is absorbed into the gold, leaving voids, referred to as *Kirkendall voids,* in the bond site. The presence of Kirkendall voids weakens the bond site and increases the electrical resistance.

2. The lack of scrubbing action does not adequately clean the surfaces, leaving behind contaminants which interfere with the bonding process.

An alternative to thermocompression bonding is thermosonic bonding, also used exclusively with gold wire. In the thermosonic bonding process, the heat is lowered to 150°C and ultrasonic energy is applied. This process compensates for both problems associated with thermocompression bonding and creates a reliable bond without excessive intermetallic compound formation. Thermosonic bonding is limited to gold wires ≤0.003 inches in diameter. Above this size, the weight of gold, coupled with the high ductility, interferes with the wire feeding through the capillary.

Aluminum wire is preferred for power devices for two reasons:

1. Larger wire is required to carry the high currents associated with power devices. The use of gold wire requires multiple bonds to provide the same current-carrying capability.
2. The high junction temperature of power devices accelerates the formation of intermetallic compounds. Aluminum wire provides a monometallic junction at the device level, which eliminates Kirkendall voids. Aluminum bonding to gold traces on the substrate is not as detrimental to reliability. The amount of aluminum constituting the wire is much greater than that which makes up the bonding pad of a die, and voiding is not extreme.

MIL-M-38510 establishes the maximum current density for materials in contact with the substrate as shown in Table 7.13. The current density here may be expressed either as direct current or as RMS AC current. This number does not apply to wire bonds, however, since they are in a free space environment and lose heat only by radiation, convection, or by conduction through plastic encapsulation, which is a relatively poor thermal conductor.

A wire bond must be able to carry the necessary current with an adequate safety factor such that the wire is not stressed during operation. MIL-M-38510 also establishes criteria for the maximum current for different configurations.

$$I_{max} = Kd^{3/2} \tag{7.26}$$

where I_{max} = maximum current in DC amps or RMS AC amps
d = diameter of wire in inches or equivalent diameter for ribbon or other conductor corresponding to the same cross-sectional area
K = applicable constant from Table 7.14

The requirements from Eq. 7.26 are summarized in Table 7.15. Frequently, the maximum current levels calculated from this formula are derated by a factor of 0.5 to provide an additional safety factor.

TABLE 7.13 Maximum Current Density in Wires per MIL-M-38510

Conductor material	Maximum allowable current density A/cm^2	Maximum allowable current density A/in^2
Aluminum (99.99% pure or doped) without glassivation or without glassivation-layer integrity test	2×10^5	1.29×10^6
Aluminum (99.99% pure or doped) glassivated	5×10^5	3.23×10^6
Gold	6×10^5	3.87×10^6
All other (unless otherwise specified)	2×10^5	1.29×10^6

TABLE 7.14 Current Constants for Wire Bonds

	Value of K	
Composition	Length ≤ 0.040 in (0.1 cm)	Length > 0.040 in (0.1 cm)
Aluminum	22,000	15,200
Gold	30,000	20,500
Copper	30,000	20,500
Silver	15,000	10,500
All other	9,000	6,300

The most reliable substrate metallization for aluminum is nickel. [15] The Al/Ni interface is frequently used in automotive applications where ambient temperatures approach 125°C. Aluminum-copper bonds are not considered as reliable at elevated temperatures. Aluminum-gold bonds made with large wire also exhibit higher failure rates. Copper traces, such as those created by the direct bond copper process, are usually nickel-plated to provide a better wire-bonding surface.

7.6 CALCULATION OF CONDUCTOR TRACE WIDTH

The ability of a film conductor to carry large currents is of critical interest to the designer in both the steady state and the transient conditions. In the steady state, the conductor may increase in temperature due to internal resistive heating and contribute significantly to the overall temperature of the circuit, and the voltage drop across the conductor may be detrimental to circuit performance. In the transient case, the conductor may be required to handle currents substantially larger than normal for short intervals of time, and the concern is that the conductor reaches a temperature high enough to evaporate, creating an open circuit. The method described here [16] determines the minimum width based on a predetermined allowable temperature rise in the steady state condition and based on the melting point in the transient case.

TABLE 7.15 Maximum Current for Wire Size

		Maximum current	
Material	Diameter, in	$L < 0.040$ in	$L > 0.040$ in
Gold	0.001	0.949	0.648
	0.002	2.683	1.834
Aluminum	0.001	0.696	0.481
	0.002	1.968	1.360
	0.005	7.778	5.374
	0.008	15.742	10.876
	0.012	28.920	19.981
	0.015	40.417	27.924
	0.022	71.789	49.600

7.6.1 Theoretical Analysis—Steady State

In the steady state condition, the current passing through a conductor will generate heat due to resistive losses in the conductor. The temperature rise of the conductor depends on the thermal path of the system from the conductor through the intervening layers to the heat sink. The term *heat sink* in this context is considered to be the boundary layer, which could be simply air, a liquid-cooled plate, or an actual heat sink. For this case, the substrate will be considered to be mounted on a heat sink of constant temperature and will be the only thermal barrier. This is a common configuration for power hybrids, since it is highly desirable to minimize the thermal path. The principle may be extended to other configurations by simply expanding the thermal analysis to include the entire system.

For this analysis, it is assumed that $K_C >> K_S$, where K_C is the thermal conductivity of the conductor. The implication of this assumption is that the substrate represents the only thermal path.

$$R_{TH} = \frac{T_S}{K_S(W_C)(W_C + 2T_S)} \tag{7.27}$$

where K_S = thermal conductivity of the substrate

The power dissipated in one square of the conductor is given by

$$P = I^2R = I^2\frac{R_B}{T_C} = I^2R_S \tag{7.28}$$

where R_B = bulk resistivity of the conductor
R_S = sheet resistance of the conductor
R = electrical resistance of the conductor
I = steady state DC or RMS current in the conductor

The net temperature rise of the conductor is therefore

$$T_{rise} = PR_{TH} = I^2R_S\frac{T_S}{K_SW_C(W_C + 2T_S)} \tag{7.29}$$

From Eq. 7.29, the maximum current which the conductor can handle for a given width is

$$I_{max} = \sqrt{\frac{T_{rise}K_SW_C(W_C + 2T_S)}{R_ST_S}} \tag{7.30}$$

Solving Eq. 7.30 for the width gives the minimum width for a given current.

$$W_{C_{min}} = \sqrt{\frac{T_SI^2R_S}{K_ST_{rise}} + T_S^2} - T_S \tag{7.31}$$

The results of Eq. 7.31 are depicted in Fig. 7.19.

7.6.2 Theoretical Analysis—Transient

In the transient case, the designer must consider the effect of very high currents for a short period of time. This condition can occur when an output device short circuits,

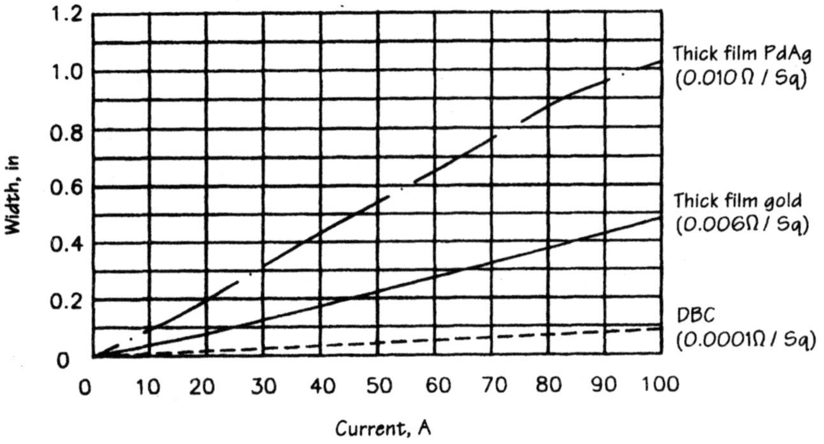

FIGURE 7.19 Minimum width versus current for several conductor materials on a 0.025-in alumina substrate.

for example, or, when the circuit is connected to the AC line, a surge occurs on the line. A common specification for relay and motor control circuits is that they be able to withstand a surge current of up to 20× the rated current for an interval of one cycle, or 0.0167 seconds.

Assume that the conductor is subjected to an impulse of current of magnitude I continuing for the interval $0 \le t \le t_D$, where t_D is the time duration of the pulse. For simplicity, alternating-current signals can be modeled as a pulse with a magnitude of the RMS value of the current. Assume also that the conductor is at a steady state temperature T_0. The power dissipated in the conductor is given by Eq. 7.28.

The thermal analog of Kirchhoff's current law is

$$P = I^2 R_S = P_{R,\mathrm{TH}} + P_{C,\mathrm{TH}} \tag{7.32}$$

which translates to

$$P = \frac{T_{\mathrm{rise}}}{R_{\mathrm{TH}}} + C_{\mathrm{TH}} \frac{d}{dt} T_{\mathrm{rise}}(t) \tag{7.33}$$

Solving Eq. 7.33 for T_{rise} and adding T_0,

$$T(t) = T_{\mathrm{rise}}(t) + T_0 = I^2 R_S R_{\mathrm{TH}}(1 - e^{1/R_{\mathrm{TH}}C_{\mathrm{TH}}}) \qquad (0 \le t \le t_D) \tag{7.34}$$

The thermal time constant is the product of the thermal resistance and the thermal capacitance. It is noteworthy that the only significant dimension is the thickness of the substrate.

$$R_{\mathrm{TH}}C_{\mathrm{TH}} = \frac{T_S}{K_S A_{\mathrm{eff}}} A_{\mathrm{eff}} T_S DS = \frac{T_S^2 DS}{K_S} \tag{7.35}$$

Inserting Eq. 7.34 into Eq. 7.35,

$$T(t) = T_{\mathrm{rise}}(t) + T_0 = I^2 R_S R_{\mathrm{TH}}[1 - e^{-tK_S/T_S^2 DS}] + T_0 \qquad (0 \le t \le t_D) \tag{7.36}$$

At time $t = t_D$, when the power source is removed, the temperature will reach the maximum T_{max} and the temperature will begin to decay toward T_0 at the same rate. The expression for the temperature decay is

$$T(t) = T_{max}e^{(-t + t_D)/R_{TH}C_{TH}} + T_0 \qquad (t > t_D) \tag{7.37}$$

In the transient case, the designer is generally concerned about conductor failure, which may occur where either the adhesion mechanism fails or where the conductor opens as a result of melting or vaporizing. The susceptibility of the adhesion mechanism to heat for a given conductor may vary from lot to lot depending on the quality of the metallizing process, and it is difficult to establish a maximum temperature for this variable. Often, the failure mode is more dramatic: the conductor vaporizes, accompanied by a massive flash of light, and opens, resulting in permanent failure of the circuit.

When a metal is simply melted, requiring somewhat less energy than evaporation, the surface tension of the molten metal will tend to form it into small, isolated spheroids, thereby creating an open circuit and causing a failure. If the melting point of the conductor metal is adopted as the temperature at which failure occurs, Eq. 7.36 may be used to establish criteria for determining the conductor width.

Setting $t = t_D$ and $T(t_D) = T_M$, Eq. 7.36 becomes

$$T_M = I^2R_SR_{TH}[1 - e^{-t_DK_S/T_S^2DS}] + T_0 \qquad (0 \le t \le t_D) \tag{7.38}$$

where T_M = melting point of the conductor

Solving Eq. 7.38 for W_C,

$$W_C = T_S\left\{\left[\sqrt{\frac{I^2R_S}{K_ST_S(T_M - T_0)}(1 - e^{-t_DK_S/T_S^2DS}) + 1}\right] - 1\right\} \tag{7.39}$$

Eq. 7.38 can also be solved for the maximum current for a particular conductor width and time duration.

$$I_{max} = \sqrt{\frac{T_{rise}K_SW_C(W_C + 2T_S)}{R_ST_S}} \tag{7.40}$$

$$t_{D_{max}} = -\frac{T_S^2DS}{K_S}\ln\left[1 - \frac{[(T_S + W_C)/T_S]^2 - 1}{I^2R_S}(K_ST_S)(T_M - T_0)\right] \tag{7.41}$$

The methods described here for determining the minimum conductor width for steady state and transient conditions are based on the approximate methods presented in the section on thermal analysis. They have been verified experimentally and should be applicable in most situations.

By comparison, MIL-M-38510 requires that a conductor dissipate no more than 4 W/in^2. For a gold thick film conductor with a sheet resistance of 0.006 Ω/□ on alumina, this corresponds to a current of 25.82 amperes. Eq. 7.29 predicts a temperature rise of 0.64°C for this structure.

7.7 HEAT EXCHANGE

The process of heat flow progresses as depicted in Fig. 7.20. The heat is generated at the junction of the semiconductor device and flows through the bulk of the silicon,

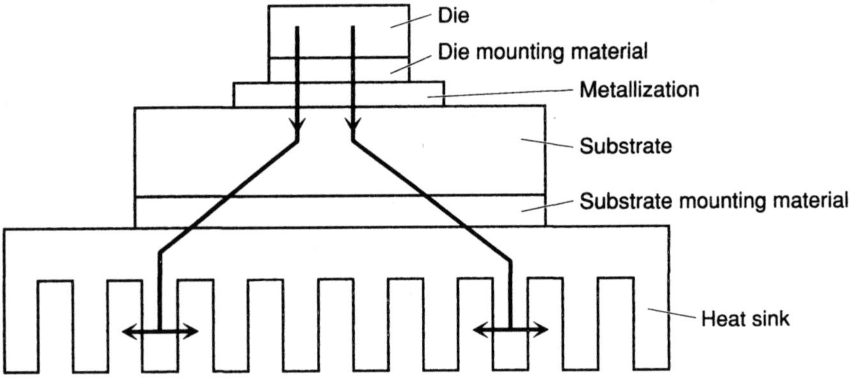

FIGURE 7.20 Heat flow path.

the die bonding material, the substrate metallization, the substrate, the substrate bonding material, and, finally, through the heat sink. The heat sink is the final interface between the heat generated at the junction and the outside world. Heat sinks for the circuit packaging level may be divided into four categories.

1. *Finned heat sinks.* Finned heat sinks consist of a flat plate with a number of fins extending from the surface. The fins may be formed by extrusion, by machining, or by attachment with thermally conductive material. The fins help to dissipate the heat by increasing the surface area.
2. *Cold plates.* Cold plates are fabricated by drilling holes in a block of metal and attaching fittings so that a fluid may be circulated through the block, carrying away the heat to the outside world if the fluid can be discarded, or to a heat exchanger if the system is closed.
3. *Heat pipes.* Heat pipes utilize the heat of vaporization of a liquid to draw the heat away from a source. The heat turns the liquid into a vapor transported by self-convection to a heat exchanger where it is turned back to a liquid. The liquid is transported back to the source by a wick and the cycle is repeated.
4. *Thermoelectric coolers.* Thermoelectric coolers employ the Peltier effect to transport heat. An electric current through a series of p-n junctions creates a temperature differential across the junction, cooling the bottom surface of the substrate.

7.7.1 Finned Heat Sinks

A finned heat sink relies on the principle of convection as defined in Eq. 5.26 to carry heat away from the surface.

$$Q_c = h_c A_s (T_s - T_A) = h_c A_s \Delta T \tag{5.26}$$

where Q_c = heat transferred from a surface to ambient by convection in watts
A_s = surface area in cm^3 or in^3
T_s = surface temperature in °C
T_A = ambient temperature in °C (temperature to which the heat is being transferred)
h_c = convection heat transfer coefficient in watts/cm^3-°C or watts/in^3-°C

The thermal resistance due to convection may be derived from Eq. 5.26.

$$R_{\text{TH(Conv)}} = \frac{\Delta T}{Q_c} = \frac{1}{h_c A_s} \tag{7.42}$$

Convection cooling takes place by transferring the heat from a surface to a fluid, which may be air or a liquid. The heat increases the kinetic energy of the fluid particles, which transport heat away due to the increased motion. There are two basic types of convection: *free convection,* in which the convection is generated by the heat source, and *forced convection,* in which the fluid is forced across the heated surface by an external source. In general, except for a few simple geometries, the solution to Eq. 5.26 is transcendental, no closed form solution can be obtained, and a numerical solution is required. For most cases, the solution requires the determination of the coefficient h_c from which the temperature differential can be calculated. The coefficient is a function of orientation since the heated particles tend to move upward. Referring to Fig. 5.9, the coefficient will be different for a flat plate depending on whether the heated side is facing down or up.

To understand the mechanics of convection, it is necessary to understand the basics of the relationship between fluid flow and heat transport. There are two fundamental types of fluid flow: *laminar flow,* in which flow is due primarily to density differences, and *turbulent flow,* in which the flow is created primarily by external means, such as a pump or fan. The distinction and nature of fluid flow are determined empirically by the values of a series of dimensionless numbers, from which the convection coefficient can be determined. The properties of air and water relating to heat transport are tabulated in Tables 3.9 and 3.10, respectively.

The Reynolds number Re is the ratio of the inertial forces to the viscous forces in a coolant flow and is defined as

$$\text{Re} = \frac{\rho v d}{\mu} \tag{7.43}$$

where ρ = density of the fluid in gm/cm^3
v = velocity of the fluid in cm/sec
d = a critical dimension, such as the edge of the circuit under consideration
μ = viscosity of the fluid in gm/sec-cm

The *Reynolds number* is a function of temperature since both ρ and μ are functions of temperature. For external flow, the value of Re classifies turbulent and laminar flow.

Laminar flow	$\text{Re} < 5 \times 10^5$
Turbulent flow	$\text{Re} > 5 \times 10^5$

For confined flow, as in a tube, the boundary is in the range of 2100 to 8000.

The *Prandtl number* Pr is the ratio of fluid and thermal boundary layer thicknesses and is equal to 0.7 for air over a flat plane. The value of the Prandtl number for air is relatively insensitive to temperature. The Prandtl number for water as a function of temperature is given in Table 3.10.

The *Nusselt number* Nu is defined as

$$\text{Nu} = \frac{hd}{k} \tag{7.44}$$

where h = convection constant in W/m^2-°C
d = critical dimension in m
k = thermal conductivity in W/m-°C

The *Grashof number* Gr is the ratio of buoyancy forces to viscous forces and is defined as

$$\mathrm{Gr} = \frac{\rho^2 g \alpha d^3 (T_S - T_F)}{\mu^2} \tag{7.45}$$

where g = acceleration of gravity = 9.78 m^2/sec
α = coefficient of expansion of the fluid
T_S = temperature of surface
T_F = temperature of fluid

The *Rayleigh number* Ra is the product of the Grashof and Prandtl numbers. The dimensionless numbers defined here are related by a series of empirical equations. For free convection,

$$\mathrm{Nu} = c(\mathrm{Gr\ Pr})^n = c(\mathrm{Ra})^n \tag{7.46}$$

where c and n are constants depending on the value of the Rayleigh number as shown in Table 3.11.

For forced convection,

$$\mathrm{Nu} = c\ \mathrm{Re}^n\ \mathrm{Pr}^m \tag{7.47}$$

where c, n, and m are given in Table 3.11.

For zero to moderate external airflow, heat is transported away from the source by a combination of free and forced convection. The contribution due to free convection is by far the most difficult to calculate analytically, since a trial-and-error solution is required. The Grashof number is a function of the difference between the fluid temperature and the plate temperature. In most cases, the fluid, or ambient temperature, is known and the plate temperature is unknown. To obtain a solution for equations of this type, a value must be estimated for the plate temperature and the system of equations solved for the actual plate temperature based on the estimate. If the estimate is different from the solved value, another estimate is made and the process is repeated until the two converge. The complexity of this process requires a computer routine to obtain an acceptable degree of accuracy. By contrast, the effect of forced convection can be calculated from Eq. 7.47 by determining whether or not the flow is laminar or turbulent from the Reynolds number and using the appropriate form from Table 3.11.

A fin may be considered a flat plate with heat flowing in one end and being dissipated into the surrounding fluid as it travels toward the other. As heat flows down the fin, the combination of the thermal resistance of the heat sink impeding the flow and the heat lost due to convection, the temperature of the fin and, therefore, the heat transfer to the fluid, will vary from one end to the other. This factor may be simplified by defining a term called the *fin efficiency* η [17].

$$\eta = \frac{\text{Actual heat transfer}}{\text{Heat transfer with fin at uniform temperature}} \tag{7.48}$$

The thermal resistance of the fin in terms of the efficiency may be derived from Eq. 7.38.

$$R_{\text{fin}} = \frac{1}{2\eta h A} \tag{7.49}$$

where A = area of the fin

The factor 2 is due to the fact that there are two sides of the fin exposed to the fluid stream.

The efficiency of a rectangular fin is given by

$$\eta = \frac{\tanh ml}{ml} \tag{7.50}$$

$$m = \sqrt{\frac{2h}{k\delta}} \tag{7.51}$$

where l = length of fin
δ = thickness of fin
k = thermal conductivity

The value of h depends on whether the fin is in a state of free or forced convection. Eq. 7.50 is plotted in Fig. 7.21.

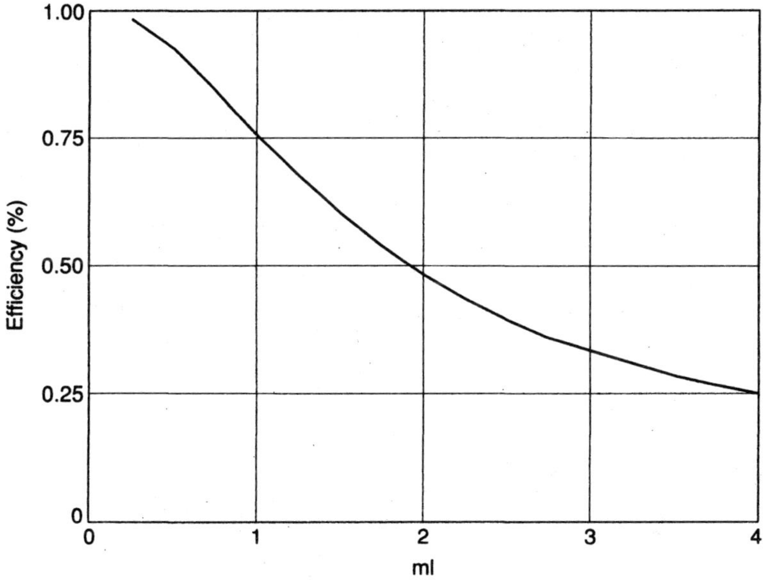

FIGURE 7.21 Efficiency of rectangular heat sink.

The efficiency of a triangular heat sink is given by

$$\eta = \frac{2}{ml}\left[\frac{I_1(ml)}{I_0(ml)}\right] \tag{7.52}$$

where $I_1(ml)$ and $I_0(ml)$ are modified Bessel functions of the first kind.

Eq. 7.52 is plotted in Fig. 7.22.

Example 7.4 Calculate the thermal resistance due to convection of an aluminum fin 0.5 × 0.2 × 0.05 inches in a flow of air at a velocity of 1000 ft/min (200 in/sec) at a temperature of 27°C. The direction of flow is along the 0.5-in dimension as shown in Fig. 7.23.

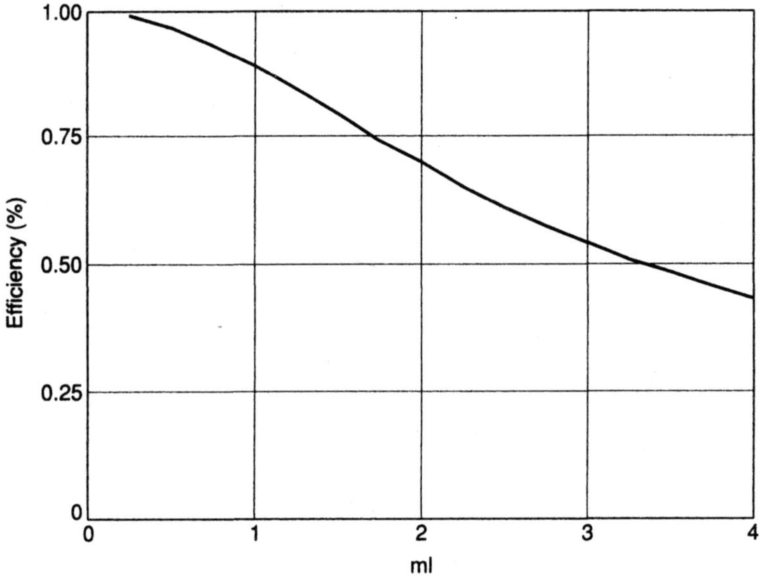

FIGURE 7.22 Efficiency of triangular heat sink.

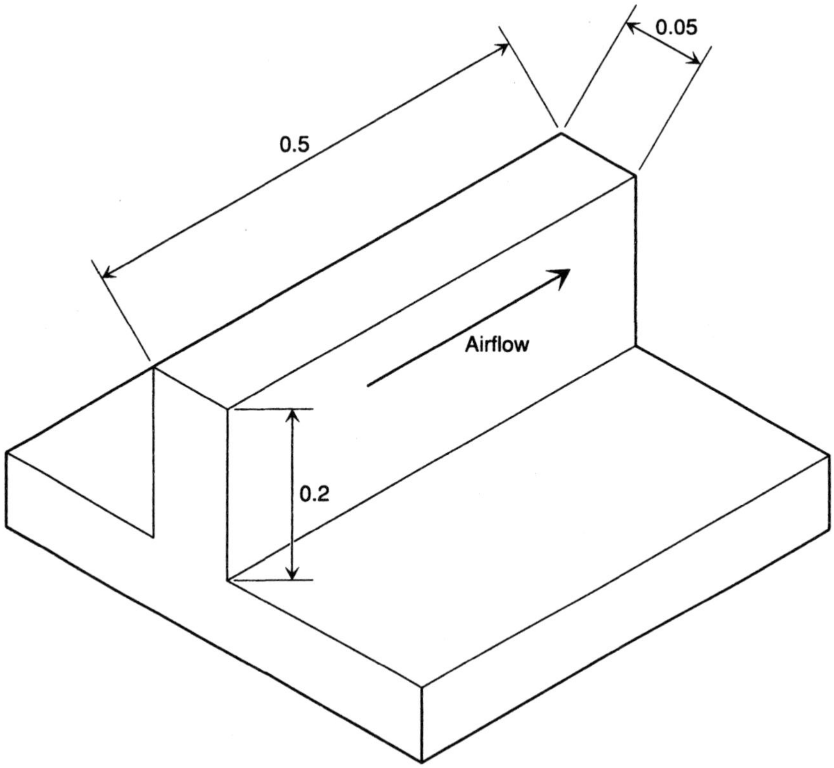

FIGURE 7.23 Fin dimensions for Example 7.4.

The Reynolds number is

$$\mathrm{Re} = \frac{0.0193 \times 200 \times 0.5}{0.000469} = 4.113 \times 10^3$$

which indicates laminar flow.

For laminar flow, the Nusselt number is given by

$$\mathrm{Nu} = (4.113 \times 10^3)^{0.5} \times (0.7)^{.33} = 57.011$$

The convection coefficient is

$$h_c = \frac{57.011 \times 0.000103}{0.5} = 0.012$$

The factor m is calculated from Eq. 7.51

$$m{:} = \sqrt{\frac{2 \times 0.012}{6.02 \times 0.05}} = 0.279$$

from which the parameter ml is

$$ml = 0.279 \times 0.5 = 0.14$$

The efficiency is

$$\eta = \frac{\tanh 0.14}{0.14} = 0.994$$

and the thermal resistance of the fin from the surface to the air is

$$R_{\mathrm{TH}} = \frac{1}{2 \times 0.994 \times 0.012 \times 0.2 \times 0.5} = 428.5 \, \frac{°\mathrm{C}}{\mathrm{W}}$$

The fin described in Example 7.4 is quite small for practical use. Most fins are much larger than this with a correspondingly lower thermal resistance. The thermal resistance can also be lowered by adding multiple fins in parallel.

The distance between the fins is also important. If they are too close together, the airflow is lessened and the thermal resistance is increased. Bar-Cohen and Rohsenow [18] describe a method for calculating the minimum distance between fins when the fin temperatures are known.

Figure 7.24 depicts the thermal resistance of an aluminum fin 0.5 in thick and 2.0 in high as a function of air velocity and length. Note that the thermal resistance drops rapidly as the airflow increases until the velocity reaches about 200 in/min. The knee of the curves marks the point at which the flow changes from laminar to turbulent.

It is very important to understand that it is the mass flow of air and not simply the volume which governs the degree of cooling. At high altitudes, this factor must be taken into account by noting the pressure and density difference of the air. Figure 7.25 is a graph of the density of air up to 20 km (12 mi).

Airflow is developed by a fan or a blower. Fundamentally, a fan moves air in a direction parallel to the fan blade axis and is designed to produce a high flow rate in a low-pressure environment, while a blower delivers air in a direction normal to the blower axis and delivers a low flow rate of air in a higher-pressure environment. The

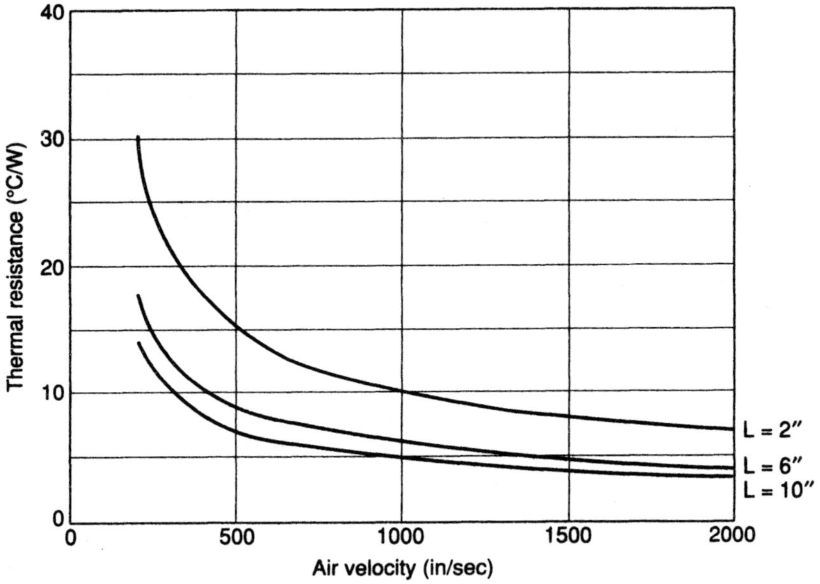

FIGURE 7.24 Thermal resistance versus air velocity for aluminum fin $0.5 \times 2 \times L$.

performance of a fan can be described in a fan curve as shown in Fig. 7.26. The operating point of the fan is determined by the characteristics of a specific system pressure curve. It is preferable to operate toward the high-flow, low-pressure side of the curve for maximum efficiency.

Turner [19] characterizes a fan by four equations (the *fan laws*).

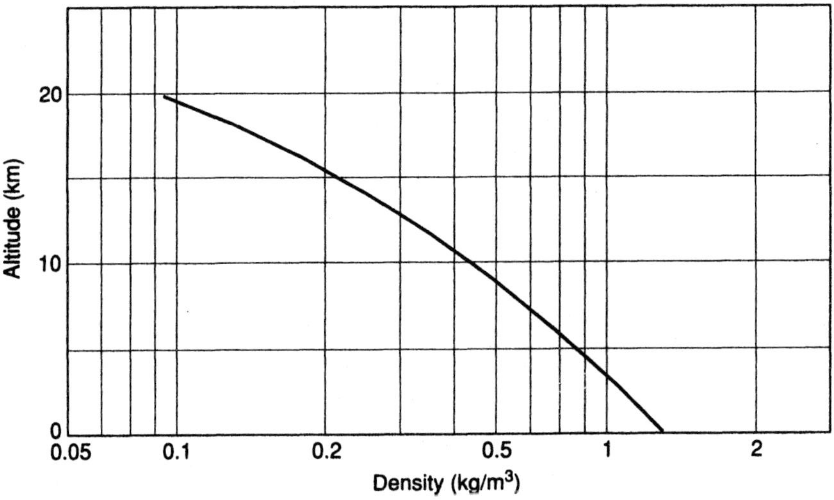

FIGURE 7.25 Density of air versus altitude in km.

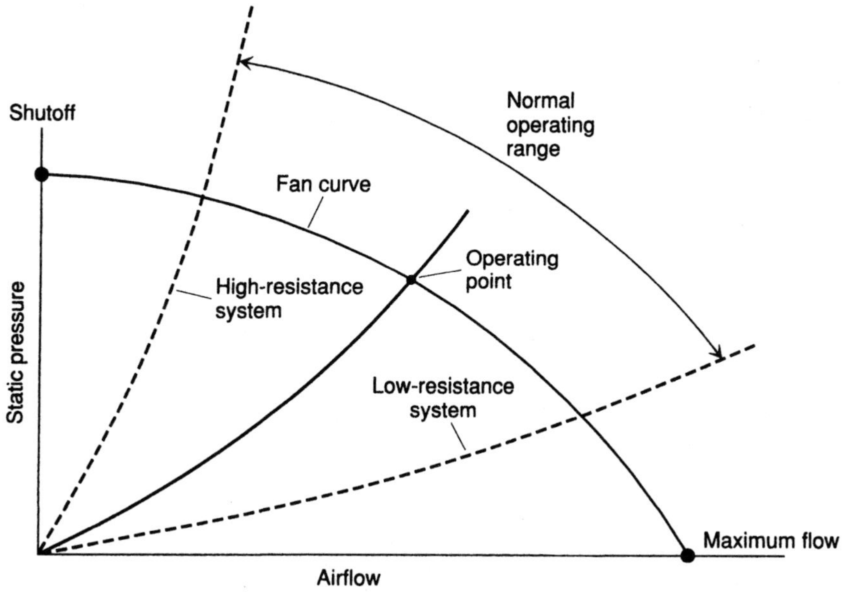

FIGURE 7.26 Typical fan-system curve. *(Source: Ref. 19.)*

$$G = K_q N D^3 \quad \text{Volumetric flow rate} \tag{7.53}$$

$$\dot{m} = K_m \rho N D^3 \quad \text{Mass flow rate} \tag{7.54}$$

$$P = K_p \rho N^2 D^2 \quad \text{Pressure} \tag{7.55}$$

$$HP = K_{HP} \rho N^3 D^5 \quad \text{Power} \tag{7.56}$$

where K = constant for similar operation
G = volumetric flow rate
N = fan speed in RPM
HP = power output
D = fan diameter
ρ = air density
$\dot{m}$ = mass flow rate

These relationships may be used to determine the performance of a fan under other conditions as summarized in Table 7.16.

Heat is transferred to a fluid in proportion to the mass flow rate, the specific heat, and the temperature differential, as shown in Eq. 7.57:

$$\dot{Q} = \dot{m} C_p \Delta T \tag{7.57}$$

where $\dot{Q}$ = heat transferred to system in watts
C_p = specific heat of air in J/kg-K
$\dot{m}$ = mass flow rate of air in kg/s
ΔT = temperature differential, fin to air in K

The mass flow rate is given by

$$\dot{m} = \rho G \tag{7.58}$$

TABLE 7.16 Summary of Fan Laws*

Constants	Variable	Fan laws
		$G_2 = G_1\left(\frac{N_2}{N_1}\right)$
Diameter (D)	Speed (N)	$P_2 = P_1\left(\frac{N_2}{N_1}\right)^2$
Density (ρ)		$HP_2 = HP_1\left(\frac{N_2}{N_1}\right)^3$
		$G_2 = G_1\left(\frac{D_2}{D_1}\right)^3$
Speed (N)	Diameter (D)	$P_2 = P_1\left(\frac{D_2}{D_1}\right)^2$
Density (ρ)		$HP_2 = HP_1\left(\frac{D_2}{D_1}\right)^5$
Diameter (D)		$P_2 = P_1\left(\frac{\rho_2}{\rho_1}\right)^2$
Speed (N)	Density (ρ)	$HP_2 = HP_1\left(\frac{\rho_2}{\rho_1}\right)^5$
Volumetric flow rate (G)		$G_2 = G_1\left(\frac{\rho_2}{\rho_1}\right)$
Diameter (D)	Density (ρ)	$P_2 = P_1\left(\frac{\rho_2}{\rho_1}\right)$
Mass flow rate ($\dot{m}$)		$N_2 = N_1\left(\frac{\rho_2}{\rho_1}\right)$
		$HP_2 = HP_1\left(\frac{D_2}{D_1}\right)^5$

* From Turner, Ref. 19.

where G = volumetric flow rate in m^3/s
ρ = density in kg/m^3

The required airflow at sea level may be calculated by combining Eqs. 7.53 and 7.54.

$$G = \frac{\dot{Q}}{\rho C_p \Delta T} \tag{7.59}$$

The net airflow will, of course, be dependent on the operating point of the fan and the system as illustrated in Fig. 7.26. The pressure drop in a system is

$$\Delta P = K\rho G^N \tag{7.60}$$

where ΔP = system pressure drop
K = system constant
N = constant = 1 for completely laminar flow = 2 for completely turbulent flow

It is important to note that a fan can only deliver one flow at one pressure in a given system.

Turner [19] further offers several suggestions on selecting and locating a fan.

1. Locate components with highest heat dissipation near the enclosure air exits.
2. Size the enclosure air inlet and exit vents at least as large as the venturi opening of the fan used.
3. Allow enough free area for air to pass with a velocity of less than 7 m/s.
4. Avoid hot spots by cooling with a small fan.
5. Locate components with the most critical temperature sensitivity nearest to the air inlet to provide the coolest airflow.
6. Pressurize the cabinet to keep air out.
7. Use the largest filter possible to increase the dust capacity and to reduce the pressure drop.

Air-cooled fins are one of the most commonly used heat sinks and are frequently used in combination with other methods. For example, the heat exchangers required for both the heat pipe and the thermoelectric cooler may be implemented in this manner.

Example 7.5 A total of 25 watts is dissipated into a heat sink. If air at 25°C is to be used to cool the heat sink, what volume of air at sea level is required to keep the fin temperature at 75°C?

$$G = \frac{25}{1177 \times 1.006 \times (75 - 50)} = 4.223 \times 10^{-4} \frac{\text{m}^3}{\text{s}}$$

7.7.2 Cold Plates

Cold plates are essentially blocks of metal with holes or fins at specified locations to enable a liquid to circulate through the block and carry away the heat by convection. The openings may be spread throughout the block such that the block is essentially at the same temperature or may be localized to cool specific areas. The primary coolant is water, although other fluids may be used as well.

The basic principles for cold plate operation in a closed environment are the same as for an open environment. The criteria for laminar and turbulent flow are not as well-defined as for an open environment.

$\text{Re} \leq 2100$	Fully developed laminar flow
$2100 \leq \text{Re} \leq 8000$	Transitional flow
$\text{Re} \geq 8000$	Turbulent flow

The Nusselt number for fully developed laminar flow is a constant as summarized in Table 7.17. There are two conditions of interest: constant heat flux, where the

TABLE 7.17 Nusselt Numbers for Fully Developed Laminar Flow*

Section	Aspect ratio	Constant heat flux	Isothermal walls
Circular	N/A	4.36	3.66
Rectangular	1	3.61	2.98
		4.12	3.39
		5.33	4.44
	∞	8.23	7.54
Triangular	N/A	3.00	2.35

* From Tummala and Rymaszewski, Ref. 17.

power flow rate is a constant, and isothermal walls, where the wall temperature is considered to be a constant.

For fully developed turbulent flow [20],

$$\text{Nu} = 0.023\text{Re}^{0.8}\,\text{Pr}^{0.4}\left(\frac{\mu}{\mu_w}\right)^{0.014} \tag{7.61}$$

where μ = viscosity at the bulk fluid temperature
μ_w = viscosity at the wall temperature

For most cases, the wall and bulk temperatures are virtually the same and the ratio μ/μ_w may be approximated as 1.

For the transitional region [21],

$$\text{Nu} = 0.023\text{Re}^{0.8}\,\text{Pr}^{0.4}\left(\frac{d_h}{l}\right)^{0.014} \qquad 0 < \frac{l}{d_h} < 400 \tag{7.62}$$

where d_h = equivalent hydraulic diameter (Eq. 7.63)
l = length of path

$$d_h = 4\frac{A_c}{P} \tag{7.63}$$

where A_c = area of the opening
P = perimeter of the opening

Example 7.6 Water at 37.8°C is flowing into a 4- × 4-inch cold plate through four holes 0.5 inches (0.01270 m) in diameter oriented as shown in Fig. 7.27 at a rate of 20 gal/hour. What is the maximum power which may be dissipated if the cold plate temperature is not to exceed 50°C? It may be assumed that the cold plate temperature is constant.

The flow rate of 20 gal/hr translates to 2.1×10^{-5} m³/sec. The velocity is

$$v = \frac{G}{A} = \frac{2.1 \times 10^{-5}}{\pi 0.00635^2} = 0.166\,\frac{\text{m}}{\text{s}}$$

The Reynolds number is

$$\text{Re} = \frac{996.4 \times 0.166 \times 0.0127}{0.00325} = 646.4$$

which indicates laminar flow. From Table 7.17, the Nusselt number is 4.36, the condition for constant heat flux.

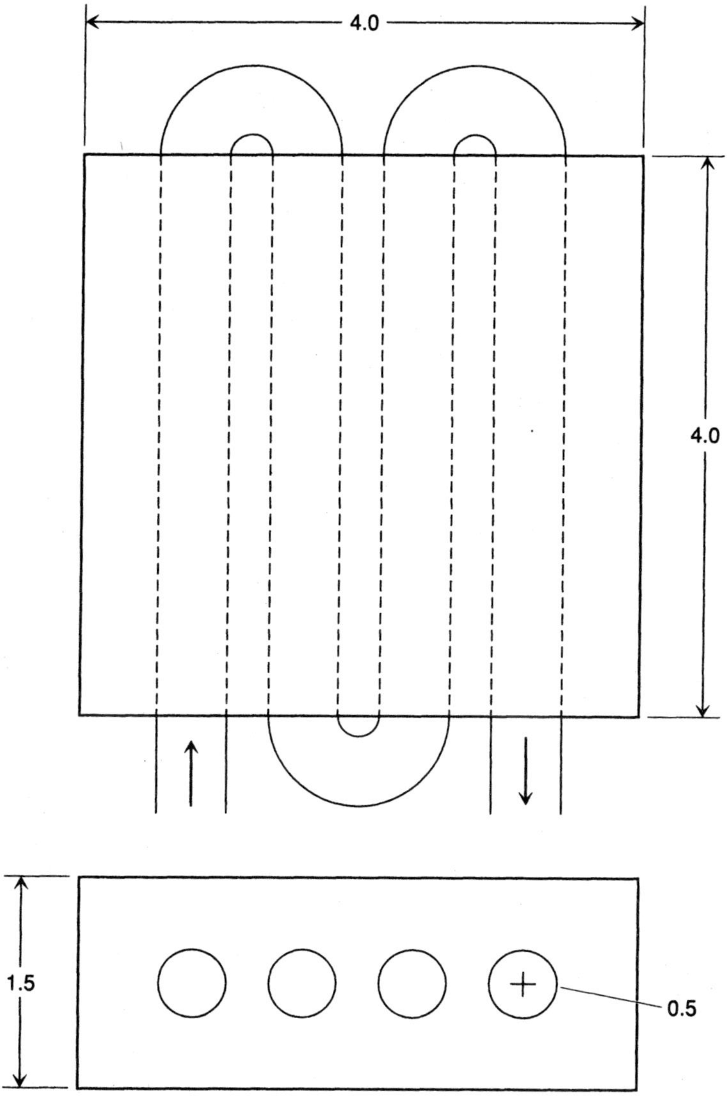

FIGURE 7.27 Cold plate for Example 7.6.

The convection coefficient is

$$h = \frac{4.36 \times 0.631}{0.01270} = 216.7$$

The allowable power dissipated is

$$P = h_c A \Delta T = 216.7 \times 2\pi 0.00635^2 \times 16 \times (50 - 37.8) = 42.85 \text{ W}$$

This analysis does not take into account the increase in water temperature or the drop in pressure as a function of length. This requires a more sophisticated numerical analysis as discussed in Chap. 6. The assumption of uniform cold plate temperature does not consider the temperature gradients due to thermal conductivity differences. This factor makes a copper cold plate, for example, more efficient than one made from aluminum. A more comprehensive analysis is found in Ref. 22.

7.7.3 Heat Pipes

The heat of vaporization of a fluid is defined as the amount of heat required to change a fluid from the liquid state to the vapor or gaseous state. This property can be utilized to efficiently transport the heat from an electronic circuit to a region where it can be safely dissipated.

The principle of operation of heat pipes is to place the heat-generating elements in proximity to a fluid inside a closed system. The heat transforms the fluid into a vapor transported to a heat exchanger by the internal pressure which builds up. The vapor returns to the liquid form and is transported back to the heat source by the capillarity action of a wicking system. A diagram of a heat pipe is shown in Fig. 7.28. With proper selection of materials, heat pipes may be used from cryogenic temperatures out to several hundred °C.

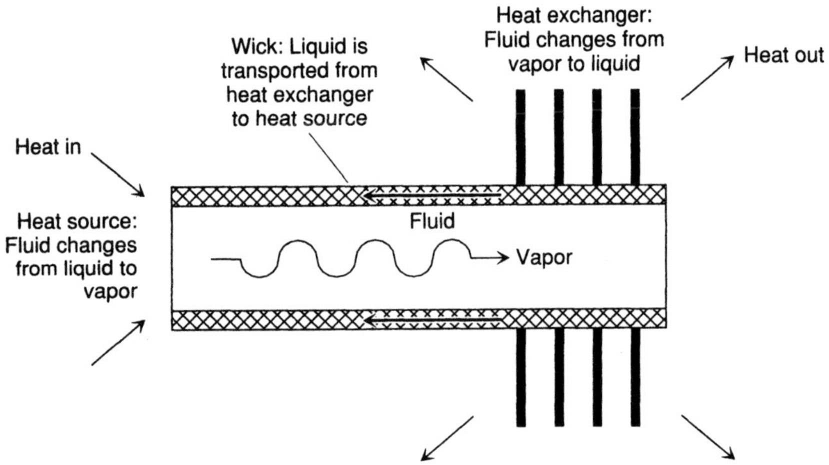

FIGURE 7.28 Basic heat pipe construction.

Typical materials used in commercially available heat pipes include water and certain of the fluorocarbons. Properties of representative materials are given in Table 7.18. Water is one of the more common materials used in electronics for two reasons:

1. The boiling point is in the range of maximum allowable junction temperatures.
2. The high heat of vaporization allows a large amount of heat to be conveyed for a relatively small amount of material.

There are three basic types of wicks used in heat pipes: grooves, screens, and powder metal. A coarse wick, such as a groove, allows a high volume of liquid to be moved, but produces a lower amount of lift and is very sensitive to orientation with

TABLE 7.18 Properties of Selected Materials Used in Heat Pipes

Property	Material		
	Water	FC-72	FC-77
Boiling point (°C)	100	56	97
Vapor pressure (Torr)	760	232	31
Heat of vaporization (kJ/kg)	2,260	94.4	89.9
Density at BP	963	1,680	1,780

respect to gravity. By contrast, a fine wick, such as powder metal, produces a higher amount of lift but lower transport volume. The limitations of the wick make the orientation of heat pipes with respect to gravity critical. Those manufactured from powder metal offer the highest range.

Heat pipes are usually classified by the power rating and the thermal resistance. For most applications, the wick is the limiting factor in the amount of power which can be dissipated, as it will govern the cycle time of the fluid. The thermal resistance has three components: the thermal resistance of the heat source–fluid contact, the thermal resistance of the heat exchanger–air, and the axial thermal resistance of the heat pipe fluid. Typical numbers for a copper-water heat pipe with a powder metal wick [23] are 0.2°C/W-cm^2 for the end contacts and 0.02°C/W-cm^2 for the axial resistance. The overall thermal resistance is extremely low compared even to pure metals.

Heat pipes are commercially available and are widely used to control the temperature of the microprocessors in notebook computers. Typical parameters of the IPC-3 copper-FC72 heat pipe manufactured by Thermacore are shown in Table 7.19.

7.7.4 Thermoelectric Coolers

Thermoelectricity is the science relating heat and electricity and includes three principles, the Seebeck effect, the Peltier effect, and the Thomson effect, as the primary basis. The *Thomson effect* has elements of both the Seebeck and the Peltier effects, but has not been commercialized and will not be discussed in detail.

The *Seebeck effect* states that, when heat is applied to a junction of dissimilar materials, a potential is developed across the junction proportional to the temperature. This principle is stated mathematically in Eq. 7.60 and is depicted schematically in Fig. 7.29.

TABLE 7.19 Parameters of IPC-3 Heat Pipe Manufactured by Thermacore

Parameter	Rating
Maximum power	700 watts
ΔT from ambient at max power	45°C
Thermal resistance	0.06°C/W
Operating temperature	$-25°C \leq T \leq 100°C$
Voltage standoff	9,000 V
Heat input area	65 mm dia. (nominal)
Projection of heat pipe beyond fins	1.0 in
Storage temperature	$-65°C \leq T \leq 120°C$
Operating position	+5° to +90°

Source: Thermacore, Incorporated.

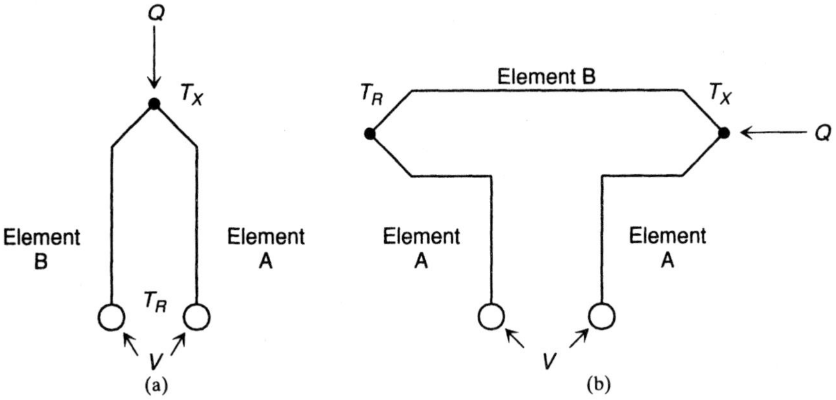

FIGURE 7.29 Seebeck effect: (*a*) single junction; (*b*) dual junction.

$$V = \alpha_S(T_X - T_R) \tag{7.64}$$

where V = Seebeck voltage
α_S = Seebeck coefficient
T_X = temperature at the heated (or cooled) junction
T_R = temperature at the reference junction

There may be second- and higher-order effects as well, and the potential will take the form

$$V = \alpha_1 T + \alpha_2 T^2 + \alpha_3 T^3 \alpha_4 T^4 + \alpha_5 T^5 + \alpha_6 \tag{7.65}$$

but these are generally negligible, except α_6, for certain materials.

The Seebeck effect is the basis for the thermocouple, which is widely used to measure temperature in a variety of applications. The Seebeck coefficient may be positive or negative, depending on the choice of materials. First-order Seebeck coefficients for a variety of materials are presented in Table 7.20.

Thermocouples may be fabricated from one junction or two as shown in Fig. 7.29*a* and *b*, respectively. When only one junction is used, the reference temperature is the ambient temperature, and the apparent temperature at the junction tracks the ambient, up or down. The ambient temperature must be monitored and considered in the calculation. The dual-junction thermocouple, where one junction is maintained at a constant, known temperature, is more accurate. The reference junction

TABLE 7.20 Seebeck Coefficients of Selected Materials

Materials	Type	Seebeck coefficient, V/°C × 10^{-6}
Iron/copper-nickel	J	52.68
Nichrome/nickel-aluminum	K	40.95
Nichrome/copper-nickel	E	63.17
Platinum-10% rhodium/platinum	S	6.45
Platinum-13% rhodium/platinum	R	6.47

may be placed in liquid nitrogen, boiling water, or other materials to obtain a known reference. It is also important to point out the significance of the contact materials where the potential is measured. A voltage is also developed at this junction, and the contact materials for the two leads must be the same to prevent erroneous readings.

The *Peltier effect* is the analog of the Seebeck effect. Referring to Fig. 7.30, if a potential is placed across two junctions, heat will be absorbed into one junction and expelled from the other in proportion to the current. As the carriers go from a lower to a higher energy state across the junction, they absorb heat from the environment. Conversely, as they go from a higher to a lower state, they release heat at the junction. The amount of heat transported is dependent on the number of carriers, which, in turn, make up the electrical current.

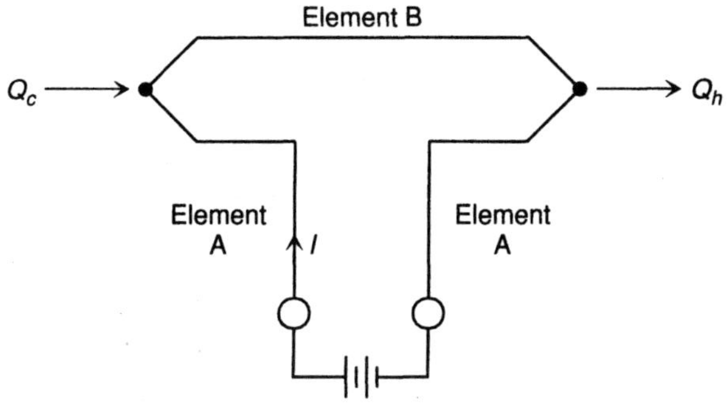

FIGURE 7.30 Peltier effect.

$$Q = \pi_P I \tag{7.66}$$

where Q = heat in or out in watts
π_P = Peltier coefficient in watts/ampere
I = current in amperes

A thermoelectric cooler may also be characterized by the Seebeck coefficient.

The Peltier effect is the basis for the thermoelectric cooler (TEC). While most material combinations will exhibit the Peltier effect to a degree, it is most obvious across a p-n junction, as depicted in Fig. 7.31. The increase and decrease in energy as a carrier transverses the junction is much greater than that of a metal, where the energy states are closer together. For example, as electrons are transported from the p side of a junction to the n side, they are elevated to a higher energy state and absorb heat, cooling the surrounding area, and, when they are transported from the p side to the n side, they release heat. The Peltier effect is reciprocal: if the current is reversed, so is the heat flow.

Several materials have been used to make thermoelectric coolers, including bismuth telluride (Bi_2Te_3), lead telluride (PbTe), and silicon-germanium (SiGe). These semiconductors are doped during fabrication for optimum parameters. Of these materials, bismuth telluride has the best performance at temperatures of interest for electronic components and is the most commonly used. A p-n combination of this type is commonly referred to as a *couple*. Material parameters for bismuth telluride are found in Table 7.21.

A thermoelectric cooler (TEC) is constructed by placing a number of couples, from one to several hundred, electrically in series and thermally in parallel between

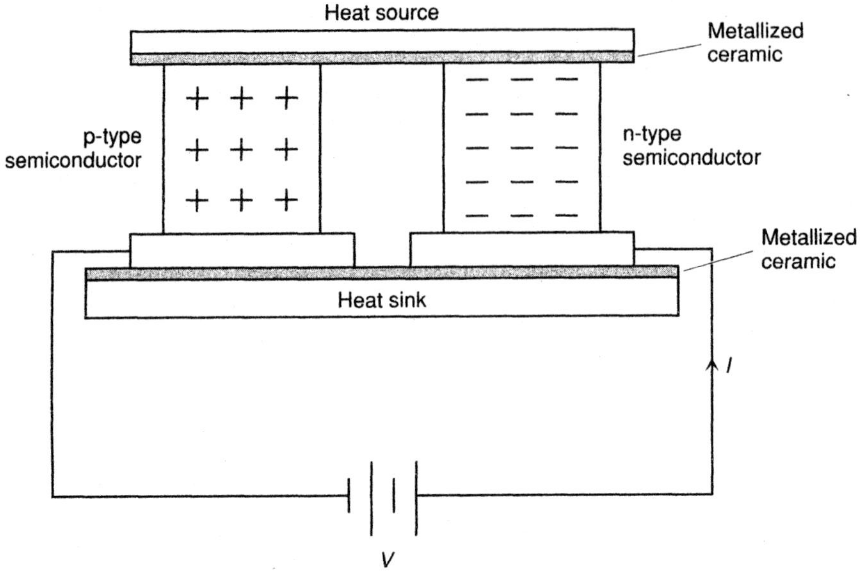

FIGURE 7.31 Diagram of thermoelectric cooler.

two pieces of metallized thermally conductive ceramic acting as an electrical insulator. A TEC should be considered a heat pump as opposed to a heat sink. If the heat is not removed from the hot side by a heat sink, the temperature differential will lessen and the TEC will not be as efficient.

The heat flow through a TEC may be calculated by [25]

$$Q = 2k_{TE}N\Delta TG \tag{7.67}$$

where Q = power in watts
k_{TE} = thermal conductivity of TEC element in W/°C-m
N = number of couples
ΔT = temperature differential in °C
G = area/length ratio of TEC element in m

TABLE 7.21 Material Parameters of Bismuth Telluride as a Function of Temperature

T_{AV}, K	α_s, $\times 10^{-4}$ V/K	ρ_{TE}, $\times 10^{-3}$ Ω-cm	k_{TE}, $\times 10^{-2}$ W/cm-K
273	1.94	0.92	1.61
300	2.02	1.01	1.51
325	2.07	1.16	1.53
350	2.10	1.28	1.55
375	2.00	1.37	1.58
400	1.96	1.48	1.63
425	1.90	1.58	1.73
450	1.86	1.68	1.88
475	1.79	1.76	2.09

A TEC may be rated by a figure of merit as given by

$$\text{FOM} = \frac{\alpha_s^2}{\rho_{\text{TE}} k_{\text{TE}}} \tag{7.68}$$

where FOM = figure of merit in 1/°C
α_s = Seebeck coefficient
ρ_{TE} = resistivity of TEC element
k_{TE} = thermal conductivity of TEC element

The voltage across the TEC is

$$= 2N\left[\left(\frac{I\rho_{\text{TE}}}{G}\right) + (\alpha \Delta T)\right] \tag{7.69}$$

The maximum current is given by

$$I_{\max} = \frac{k_{\text{TE}} G}{\alpha_s} [(1 + (2\text{FOM}T_H))^2 - 1] \tag{7.70}$$

where T_H = hot side temperature in K

and the optimum current is

$$I_{\text{OPT}} = \frac{k_{\text{TE}} \Delta T G[1 + (1 + \text{FOM} \times T_{\text{AV}})^{0.5}]}{\alpha_s T_{\text{AV}}} \tag{7.71}$$

where $T_{\text{AV}} = \dfrac{T_H + T_C}{2}$
T_C = cold side temperature in K

The figure of merit as a function of temperature is given in Table 7.22.

When plotted with current as a parameter, the ΔT versus Q curve for a TEC is linear as shown in Fig. 7.32. The right side of Fig. 7.32 may be considered as the heat capacity of the TEC, the point where it has absorbed sufficient heat such that the temperature differential is zero. However, TECs are rarely used near the capacity because the internal joule heating of the current begins to add significantly to the heat which must be dissipated. Godfrey [26] suggests a limit of 80 percent of maximum current for most efficient operation.

TABLE 7.22 Figure of Merit for Bismuth Telluride versus Temperature

T_{AV}, K	Figure of merit, $\times 10^{-3}$/K
273	2.54
300	2.68
325	2.44
350	2.22
375	1.88
400	1.59
425	1.32
450	1.06
475	0.87

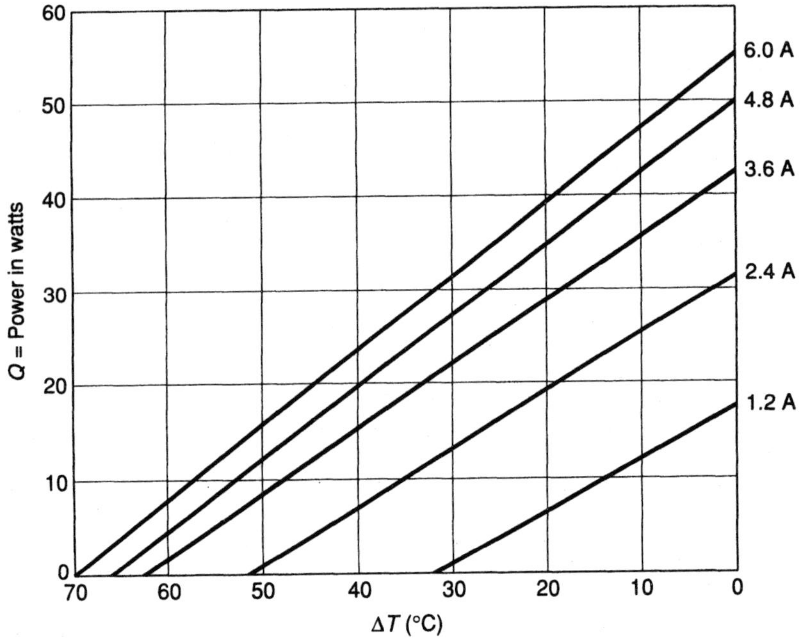

FIGURE 7.32 ΔT versus Q for Melcor CP1.4-127-06L TEC. *(Courtesy of S. Godfrey, Ref. 26.)*

The performance of a TEC is the ratio of the heat pumped to the input power.

$$\text{COP} = \frac{\text{Heat pumped}}{\text{Input power}} = \frac{Q_c}{P_{\text{in}}} \tag{7.72}$$

where COP = coefficient of performance

The COP is a function of many variables, including the temperature differential and the input power, and must be considered accordingly. The optimum COP is given by

$$\text{COP}_{\text{OPT}} = \frac{T_{\text{AV}}}{\Delta T}\,\frac{((1+\text{FOM}\times T_{\text{AV}})^{0.5}-1)}{((1+\text{FOM}\times T_{\text{AV}})^{0.5}+1)} - 0.5 \tag{7.73}$$

Given a set of design criteria, a TEC may be selected from the performance and COP curves. It must be emphasized that the TEC and the heat sink must be selected as an entity. For added heat capacity, TECs may be operated in parallel and, for increased ΔT, in series.

TECs can also be used as heaters. One of their more common applications is in temperature controls. By attaching a thermocouple, the TEC can be used with a closed loop control system to keep a component or circuit at a specific temperature. In this application, it is often necessary to know the time required to bring an object to a specific temperature. Eq. 7.74 may be used as an estimate.

$$t_{\text{eq}} = \frac{mC_p\Delta T}{Q} \tag{7.74}$$

The general specifications of the iTi FerroTec 6301/127/040 are summarized in Table 7.21.

7.8 THERMAL INTERFACE RESISTANCE

The effect of the interfaces between surfaces, such as a substrate and a heat sink, can often be more pronounced than that of the elements themselves. Referring to Fig. 7.33, at the microscopic level, the interface consists of point-to-point contacts surrounding air pockets. Air is a very poor thermal conductor, and some means of eliminating the pockets must be implemented to improve the thermal resistance. One possibility is to apply mechanical pressure to the surfaces to smooth out the interface. This improves the situation to a point when soft metals are used, but the pressure required to offer an acceptable solution may be much greater than the materials can withstand [27], which is approximately 50 psi for most practical cases.

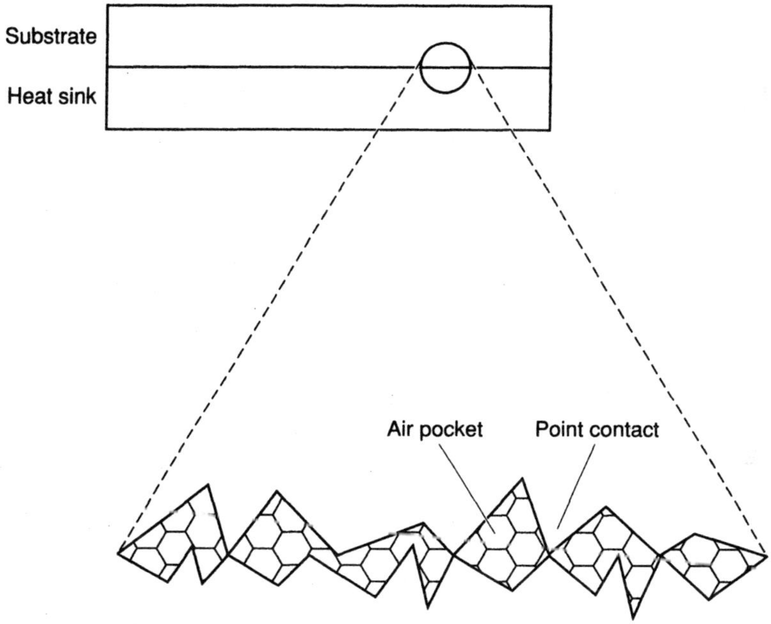

FIGURE 7.33 Interface contacts between surfaces.

Clearly, eliminating the gaps with mechanical pressure is not a complete solution, and a means of filling them with a material with a high thermal conductivity is needed. Solder offers the best thermal contact, but creates other problems.

- The solder temperature may be substantially higher than the materials can withstand. The maximum processing temperature recommended for thermoelectric coolers, for example, is 150°C, well below that of most solders.

- Trapped flux may create large voids in the solder which contribute to a high thermal resistance.
- TCE differences may result in bowing or fracture.

There are a wide variety of commercially available organic materials used as fillers. These generally fall into one of three categories.

1. *Thermal grease.* A *thermal grease* uses silicone or one of the hydrocarbon oils as a base filled with a thermally conductive material, such as aluminum oxide or zinc oxide powder. The thermal resistance is a strong function of the particle size, since larger particles may result in a larger bond thickness. Thermal greases are volatile to a certain degree and may partially evaporate with time. They do not provide adhesion, and some degree of mechanical contact, such as screws, must be used to establish sufficient pressure to minimize the bond thickness and to maintain interface integrity. Thermal grease may contaminate contact or solder areas.
2. *Elastomers.* These are usually in the form of silicone pads filled with thermally conductive material. They are much easier to handle than thermal grease, but require a higher mechanical pressure to completely fill the voids. As the amount of material is fixed, the bond thickness cannot be reduced beyond a certain limit, and the thermal resistance is higher.
3. *Conductive adhesives.* Conductive adhesives may be filled with silver, boron nitride, or ceramic and provide adhesion between the two surfaces after curing. They are furnished as one of two types: liquid and tape. The liquid types may be heavily loaded with material and are capable of achieving a thin bond line. They provide the lowest overall thermal resistance of the available materials. The adhesive tapes are basically B-staged epoxies which may be cut to size for convenience. They are very easy to handle, but, like the elastomers, the bond line thickness is fixed and the thermal resistance may be higher as a result.

It is important to emphasize that mechanical pressure is required for optimum performance of all organic materials to insure that the voids between the two surfaces are completely filled.

7.9 THERMAL MEASUREMENT

The measurement of the temperature at a given point on an electronics circuit is not an exact science. There are four fundamental methods by which this may be accomplished.

1. *Noncontact.* It is known that objects emit infrared (IR) radiation when heated. The amount of heat is proportional to a number of variables, including the temperature and the emissivity. The emissivity is a function of the surface characteristics; smooth, dark surfaces are more efficient radiators than rough, light surfaces. Infrared temperature measurement systems which scan the surface must be calibrated for each material in the circuit to give an accurate reading, which can be a formidable task. A further complication is that the surface may change during the reading as a result of surface oxidation or other reaction. The two major advantages are that the temperature of the entire surface may be scanned in a short time, and there is no direct contact with the surface to disturb the reading.

2. *Contact.* Contact methods use a thermocouple or thermistor probe to measure the temperature at a given point on the hybrid circuit. While there are no surface irregularities to be concerned about, the probe can act as a heat sink to draw heat away from the point in question and affect the reading.

3. *Junction voltage.* The current in a perfect forward-biased diode at low current is given by

$$I = Ae^{qV/kT} \tag{7.75}$$

where I = current in amperes
A = constant
q = charge on one electron
V = voltage across the diode
k = Boltzmann's constant
T = temperature in K

Taking the logarithm of both sides and solving for T,

$$T = \frac{qV}{k \ln\left(\frac{I}{A}\right)} \tag{7.76}$$

Note that T is directly proportional to V. This characteristic can be used to obtain accurate temperature measurements by passing a constant current through the diode and measuring the voltage. Bipolar devices can be measured directly, and FETs can be measured by using the built-in diode. Diodes can also be used as temperature-monitoring devices at various points in a circuit, serving as protection mechanisms for higher-powered devices. Each diode must be calibrated individually when used for this application, as the constant A may vary from device to device.

4. *Thermistors.* Thermistors are basically resistors having a high TCR and are sensitive to temperature changes. A highly doped semiconductor is frequently used for this application. Thermistors may have a positive or negative TCR, depending on the material. The semiconductor types are invariably negative. Thermistors based on resistive materials will usually have a resistance temperature characteristic as expressed in Eq. 7.77

$$R(T) = R_0 \exp\left(\frac{\beta}{T}\right) \tag{7.77}$$

where T = temperature in K
R_0 = base resistance at 25°C
β = thermistor constant in K

The thermistor constant may be positive or negative.

REFERENCES

1. J. E. Sergent and C. A. Harper, eds., *Handbook of Hybrid Microelectronics, 2d ed.,* McGraw-Hill, New York, 1995.
2. J. C. Bolger, "Polyimide Adhesives to Reduce Thermal Stresses in LSI Ceramic Packages," *14th National SAMPE Technical Conference,* Atlanta, Georgia, October 1982.
3. J. Sergent, *Internal Report,* TCA, Inc., Williamsburg, Kentucky.

4. J. C. Bolger and C. Mooney, "Die Attach in Hi-Rel P-Dips: Polyimides or Low-Chlorine Epoxies?" *Proc. IEEE Components Conf.,* 1984, Washington, DC.
5. C. G. M. Van Kessel, S. A. Gee, and J. J. Murphy, "The Quality of Die Attachment and Its Relationship to Stresses and Vertical Die-cracking," *Proc. IEEE Components Conf.,* 1983, Washington, DC.
6. W. D. Callister Jr., *Materials Science and Engineering—An Introduction,* John Wiley & Sons, New York, 1991.
7. J. E. Sergent and G. Marks, "The Electroformed Stencil: A Solution for Fine Pitch Solder Printing," *Proceedings ISHM Symposium,* 1992, Reston, Virginia.
8. S. Fuchs and P. Barnwell, "A Review of Substrate Materials for Power Hybrid Circuits," *International Journal of Microcircuits and Electronic Packaging,* **20**(1), Qtr. 1, 1997.
9. C. A. Harper, *Electronic Packaging and Interconnection Handbook,* McGraw-Hill, New York, 1997.
10. C. D. Iacovangelo and E. C. Jerabek, "Metallizing CVD Diamond for Electronic Applications," *Proceedings International Society for Hybrid Microelectronics Symposium,* 1993, Reston, Virginia.
11. M. K. Premkumar and R. R. Sawtell, "Al-SiC Technology for Reliable and Thermally Demanding Microelectronic Packages—Manufacturing, Applications, and Finishing," *Proceedings International Society for Hybrid Microelectronics Symposium,* 1995, Reston, Virginia.
12. J. A. Kerns, N. J. Collela, D. Makowiecki, and H. Davidson, "Dymalloy: A Composite Substrate for High Power Density Electronic Components," *Proceedings International Society for Hybrid Microelectronics Symposium,* 1995, Reston, Virginia.
13. W. Shih, L. Bentsen, and S. DeVincent, "Metal Impregnated Carbon Composites for Thermal Management Applications," *Proceedings International Society for Hybrid Microelectronics Symposium,* 1995.
14. W. Johnson, "Fundamentals of Multichip Modules," *ISHM monograph,* 1991, Reston, Virginia.
15. G. Harman, "Wire Bonding for Microelectronics," *ISHM monograph,* 1990, Reston, Virginia.
16. J. E. Sergent and G. Shawan, "Determination of the Current-Carrying Capacity of Film Conductors," *Journal of Hybrid Microelectronics, ISHM,* September 1991.
17. R. Tummala and E. J. Rymaszewski, *Microelectronics Packaging Handbook,* Van Nostrand Reinhold, New York, 1989.
18. A. Bar-Cohen, and W. Rohsenow, "Thermally Optimum Spacing of Vertical, Natural Convection Cooled, Parallel Plates," *Journal of Heat Transfer,* February 1984.
19. M. Turner, "All You Need to Know About Fans," *Electronics Cooling,* **2**(2), May 1996.
20. F. W. Dittus and L. M. K. Boelter, Univ. of California, Berkeley, *Pub. Eng.,* **2**, p. 443, 1930.
21. J. P. Holman, *Heat Transfer,* 3d edition, McGraw-Hill, New York, 1972.
22. S. Song, K. P. Moran, D. P. Rearick, and S. Lee, "Thermal Performance Modeling and Measurements of Localized Water Cooled Cold Plate," *Proceedings IEPS Symposium,* 1993, Dallas, Texas.
23. S. D. Garner, "Heat Pipes for Electronics Cooling Applications," *Electronics Cooling,* **2**(3), September 1996.
24. ITI Ferro Tec, Technical Reference Manual.
25. Melcor, Technical Reference Manual.
26. S. Godfrey, "An Introduction to Thermoelectric Coolers," *Electronics Cooling,* **2**(3), September 1996.
27. M. Early, S. Lee, and M. Pellilo, "Thermal Performance of Interface Material in Microelectronics Packaging Applications," *Proceedings IEPS Conference,* 1995, Austin, Texas.

CHAPTER 8
ELECTRONIC DEVICE COOLING

Richard F. Porter
Ohbin Kwon*

8.1 INTRODUCTION

Electronics of all types are becoming more a part of everyday life for everyone. Household electronics of some type represent a large portion of family disposable income and compose approximately a third of the cost of most aircraft. The buyers and users need and expect long-lasting quality in their products. Temperature control is one of the major contributors to the reliability we all seek.

This chapter presents tools, techniques, and some devices that can be used to define, control, and reduce temperatures in commercial and military electronics. The difference between the two is not addressed directly because the material can be applied to either. The difference is in the environments in which the electronics are expected to operate that define the boundary conditions for the thermal engineer.

8.2 HEAT TRANSFER THEORY

The basic theory of heat transfer is found in many references; however, to make this chapter a complete reference, pertinent theory and information is presented in a practical way to ensure the volume can be used without or at least with minimal referral to external sources.

All forms of heat transfer are used in the packaging of electronics and a basic understanding of them is necessary to produce the best cooling system for a given application. To this end, conduction, convection, radiation, and ebullient heat transfer theory is presented in the following sections.

8.2.1 Conduction

Conductive heat transfer in its most general form is defined by the energy balance.

$$k_x\frac{\partial^2 T}{\partial x^2} + k_y\frac{\partial^2 T}{\partial y^2} + k_z\frac{\partial^2 T}{\partial z^2} + q = \rho c_p\frac{\partial T}{\partial t} \tag{8.1}$$

If the material conductivity k is uniform in all directions, this equation reduces to

$$\frac{\partial^2 T}{\partial x^2} + \frac{\partial^2 T}{\partial y^2} + \frac{\partial^2 T}{\partial z^2} + q/k = \frac{\rho C_P}{k}\frac{\partial T}{\partial t} \tag{8.2}$$

and if the system is time invariate (steady state) it becomes

* Northrop Grumman Corporation, Baltimore, Maryland.

$$\frac{\partial^2 T}{\partial x^2} + \frac{\partial^2 T}{\partial y^2} + \frac{\partial^2 T}{\partial z^2} + q/k = 0 \tag{8.3}$$

However, this equation is generally reduced to the Fourier equations, which express the heat transfer in each direction independently. Thus, a set of equations, essentially one-dimensional in nature, can be used to represent a three-dimensional system using a network of conductances between points. This is often called the *lumped parameter method* in which each node of the network is approximated as a single temperature; heat is transferred between nodes at different temperatures, just as current is transferred between nodes of different voltages in an electrical circuit and is, in practice, used in a finite difference format:

$$q_z = -kA \frac{\partial T}{\partial z} \tag{8.4}$$

$$q_y = -kA \frac{\partial T}{\partial y} \tag{8.5}$$

$$q_x = -kA \frac{\Delta T}{\Delta x} \tag{8.6}$$

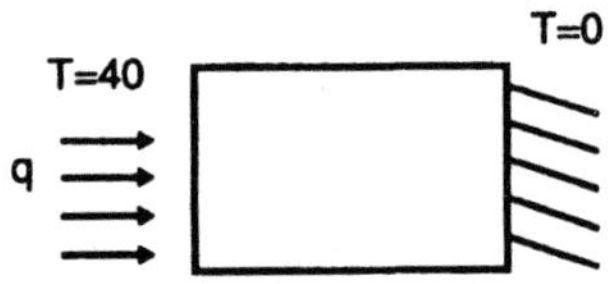

FIGURE 8.1 Plate conduction problem.

The use of the Fourier equation can be demonstrated by a simple plate problem as shown in Fig. 8.1. In this illustration, one side of the plate is at 0°C and the opposite surface is at 40°C.

To calculate the amount of heat transferred through the plate to maintain this temperature at steady state, the following calculation is performed.

$$q = \frac{kA}{x}(40 - 0) \tag{8.7}$$

If the area of the plate, the conductivity of the plate, and the thickness of the plate x are substituted into the equation, the heat flow (in watts) required to maintain this temperature difference can be computed.

The conduction equations presented so far have been in cartesian coordinates. It is often necessary to analyze parts using a cylindrical coordinate system, such as represented in Fig. 8.2. Here, the Fourier equation becomes

$$q = -kA \frac{dT}{dr} \tag{8.8}$$

but now $A = 2\Pi rt$, where t is the thickness of the material. Substituting, the equation becomes

$$q = -2\Pi ktr \frac{dT}{dr} \tag{8.9}$$

and integrating from r_1 to r_2 yields:

$$q = \frac{2\Pi kt(T_1 - T_2)}{\ln (r_2/r_1)} \tag{8.10}$$

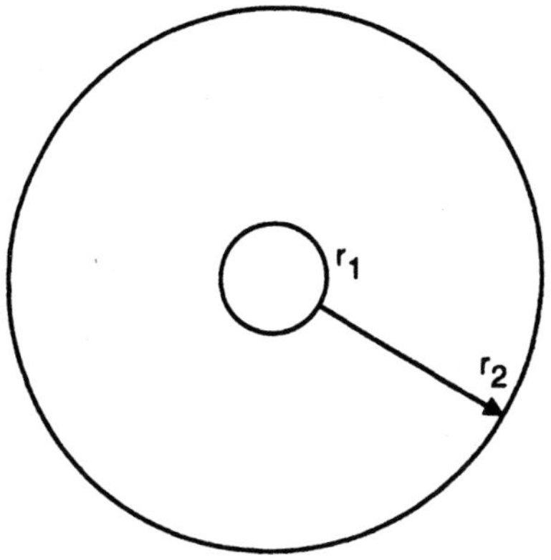

FIGURE 8.2 Radial conduction through a cylinder or annulus.

On some occasions, conduction occurs in only a portion of a cylinder and can be tangential as well as radial. The radial conductance is easily found by integrating from θ_1 to θ_2, instead of over 2Π, as in Eq. 8.10. The tangential conductance is found by defining the area $A = tdr$ and defining x as $rd\theta$. The resulting integration yields

$$q = kt \frac{\ln (r_2/r_1)}{\theta_2 - \theta_1} (T_1 - T_2) \qquad (8.11)$$

One element of conduction that is treated very differently is contact conductance. *Contact conductance* is defined as the conductance across an interface of two surfaces in contact. The value of this conductance is a function of the pressure applied, flatness, surface finish (smoothness), hardness, and malleability of the two surfaces. Although contact conductance can be predicted, so much information is needed about the surfaces in contact that it is not practical to use. As a result, contact conductance is determined empirically. Because many practical equations used in heat transfer are based on empirical data, this isn't a detriment. The contact conductance G is expressed in units of watts/°C-m^2 and the temperature drop across the interface is then:

$$\Delta T = \frac{Q}{GA} \qquad (8.12)$$

Data exists for many combinations of materials and is usually expressed in terms of conductance versus pressure as in Fig. 8.3.

The actual pressure over the area in contact is often difficult to determine. The pressure applied can be determined by computing the tensile force on the bolt or screw:

$$P = \frac{T}{KD} \qquad (8.13)$$

Where P = tensile force or tension (lb$_f$)
T = torque applied to the bolt (ft-lb$_f$)
D = diameter of the bolt (ft)
K = friction coefficient = 0.15

However, the pressure reduces with distance from the bolt or screw. A good general rule is that the pressure can be considered to apply over an area defined by a two bolt-head-diameter radius.

Many materials have been developed to enhance and worsen the contact conductance. Materials that improve contact conductance are composed of different types of soft materials that conform to the irregularities of the surface, which is never perfectly smooth and to fill gaps caused by imperfect flatness. Of course, the smoother and flatter the two surfaces are, the better the conductance will be. Some interstitial materials can be used to reduce the conductance in cases where increased resistance is desired; usually, a hard porous material, such as steel screens, is used.

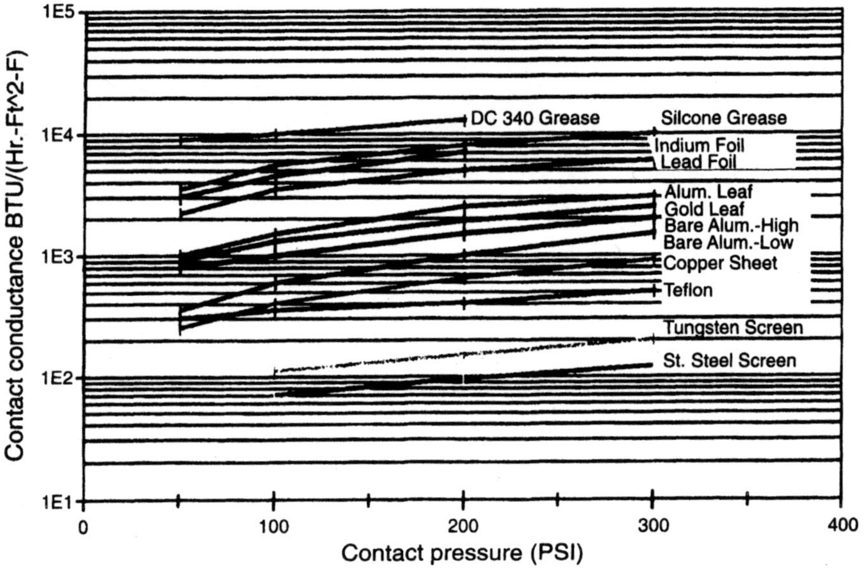

FIGURE 8.3 Contact conductance versus pressure for several interstitial materials.

Some new materials change phase (i.e., melt and are trapped in the interface, which is reportedly as good or better than thermal grease, which has been the best when applied correctly). These phase-change fillers wet the surfaces of the interface materials to provide high-transfer areas over very small distances. They can also have (although not yet proven) the advantage of being less sensitive to pressure, which eliminates the need to control this parameter in thermal designs.

8.2.2 Convection

Convection is extremely important in cooling electronics because it represents a very efficient method of removing heat that can be easily designed into the system. The challenge in convection heat transfer analysis is the determination of the convective heat transfer coefficient h. Convection cooling in electronics can occur through either natural convection or forced convection. Natural convection involves the movement of a fluid (usually air) over a warm body or surface induced by the buoyancy effect of the air being heated. It is, therefore, free, whereas forced convection requires the inclusion in the design of a forcing mechanism, such as a fan or pump.

Forced Convection. Forced convection can occur externally (e.g., over the outside of an electronic box) or internally (e.g., flow through a tube). The flow can be either laminar or turbulent. This is important because of the difference in the boundary layer (where the heat transport from the surface actually takes place and changes the characteristics of heat transfer) which affects the coefficient h which expresses this mechanism.

Several dimensionless parameters are important in convection:

Reynolds number, $\mathrm{Re} = \rho V d/\mu$

Prandtl number, $\mathrm{Pr} = C_p\mu/k$

Nusselt number, $\mathrm{Nu} = hd/k$

Stanton number, $\mathrm{St} = \mathrm{Nu}/(\mathrm{RePr})$

The transition from laminar to turbulent flow takes place at Re values of 5×10^5 for external flow and 2100 for internal flow. The velocity profile of the fluid in the boundary layer is different for laminar and turbulent flows. Figure 8.4 illustrates the laminar boundary layer velocity profile and Fig. 8.5 shows the turbulent profile. Although the turbulent boundary layer has a laminar sublayer, it is much thinner than pure laminar boundary layers; therefore, heat transfer is improved. The transition from laminar to turbulent is shown in Fig. 8.6.

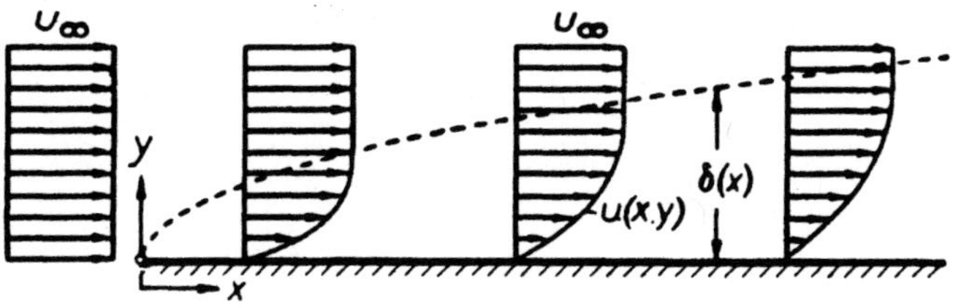

FIGURE 8.4 Fluid velocity profile in the laminar boundary layer.

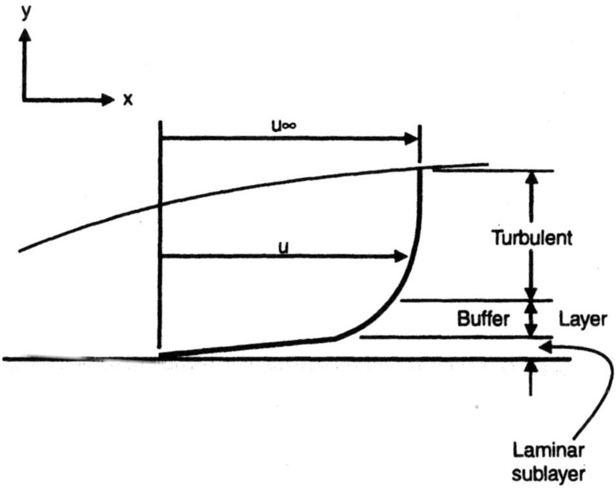

FIGURE 8.5 Fluid velocity profile in the turbulent boundary layer.

Although the turbulent boundary layer is thicker than the laminar boundary layer, the laminar sublayer is thinner, which improves the transport of heat into the fluid; heat transfer coefficients are thus higher.

The laminar heat transfer coefficient for external flow over a flat plate is the most often used and useful expression for computing external flow heat transfer. For a plate heated over its entire surface, the Nusselt number is

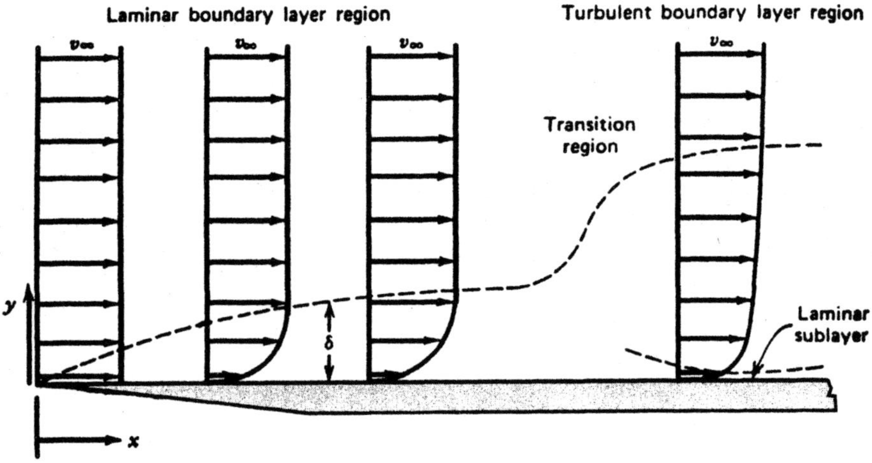

FIGURE 8.6 Fluid flow transition from laminar to turbulent.

$$Nu_x = 0.332 Pr^{1/3} Re_x^{1/2} \tag{8.14}$$

where x is measured along the flow direction on the plate. The average heat transfer coefficient for the plate is obtained by integrating over the length of the plate to obtain

$$Nu = \frac{hL}{k} = 0.664 Pr^{1/3} Re_L^{1/2} \tag{8.15}$$

For turbulent flow, the local Nusselt number is

$$Nu_x = 0.0288 Re_x^{4/5} Pr^{1/3} \tag{8.16}$$

and the average is

$$Nu = 0.036 Re_L^{4/5} Pr^{1/3} \tag{8.17}$$

Heat transfer coefficients and how to compute them are detailed in *Engineering Heat Transfer* by James Welty [1] and *Heat Transfer* by J. P. Holman [2]. These include cylinders and spheres, as well as banks of tubes, which can be useful in the design of heat exchangers.

Internal heat transfer is most often defined by flow through a tube. Seider and Tate [3] have correlated experimental data to the expression for laminar flow:

$$Nu = 1.86 \left(RePr \frac{D}{L} \right)^{1/3} \left(\frac{\mu_m}{\mu_0} \right)^{0.14} \tag{8.18}$$

Where μ_m is evaluated at the mean fluid temperature and μ_0 is evaluated at the wall temperature. Then for turbulent flow in a tube, the most general expression is given by McAdams [4] as

$$St = 0.023 Re^{-0.2} Pr^{-2/3} \left(\frac{\mu_m}{\mu_0} \right)^{0.14} \tag{8.19}$$

where:

1. All fluid properties are evaluated at the average fluid temperature, except μ_0, which is evaluated at T_0, the wall temperature.
2. $\mathrm{Re} > 10^4$
3. $0.7 < \mathrm{Pr} < 17{,}000$
4. $L/D > 60$

These tube expressions are accurate for circular tubes (Fig. 8.7), but can be used for noncircular tubes with reasonable accuracy. Information on corrections for noncircular tubes can be found in *The Handbook of Heat Transfer* by Rohsenow and Hartnett [3].

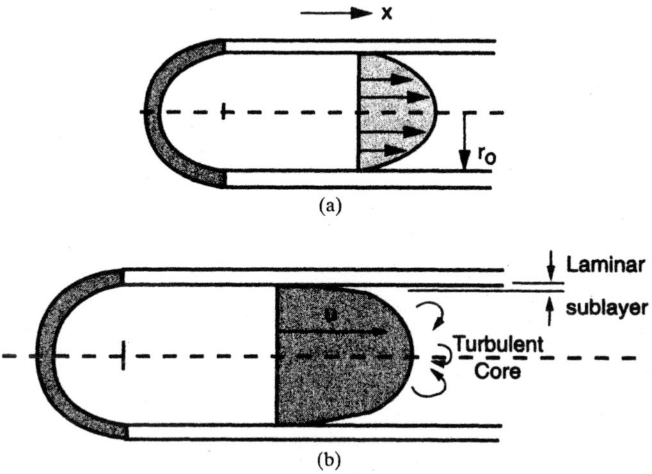

FIGURE 8.7 A comparison of laminar-flow and turbulent-flow velocity profiles in a tube.

Many abbreviated expressions have been defined for convection heat transfer coefficients when the cooling fluid is air at or near sea level conditions. Some of these are presented here. All are in units of BTU/hr-°F-ft². In these equations, the following units must be used:

ρ: Density in lb/ft³
V: Gas velocity in ft/hr
L or D: Length or diameter in ft
C_p: Specific heat in BTU/lb-°F

1. Laminar flow over a flat plate with velocities less than 15 ft/s:

$$h = 0.8 + \frac{V}{322} \tag{8.20}$$

2. Turbulent flow over a flat plate with velocities greater than 15 ft/s:

$$h = \frac{V^{0.75}}{830\,L^{0.25}} \tag{8.21}$$

3. Turbulent flow in a tube:

$$h = 0.0144 \frac{C_P(V\rho)^{0.8}}{D^{0.2}} \tag{8.22}$$

Natural Convection. Natural convection occurs because the movement of fluid is caused by buoyancy effects as a result of the heating of the fluid. Obviously, buoyancy depends on the external action of body forces, such as gravity or some sort of acceleration, so that no natural convection can occur in the microgravity environment of space or in orbiting spacecraft. Whereas forced convection is characterized by the Reynolds number, in natural convection, the Grashof number is the defining parameter, which is involved in defining the laminar versus turbulent boundary flow. The critical Grashof number for a vertical flat plate is approximately 4×10^8 and the Grashof number is defined as

$$\mathrm{Gr}_x = \frac{g\beta(T_w - T_\infty)x^3\rho^2}{\mu^2} \tag{8.23}$$

where g = acceleration of gravity
β = coefficient of volumetric expansion, which is $1/T$ for a perfect gas
T = absolute temperature of the gas
T_w = wall temperature
T_∞ = fluid bulk temperature
x = distance from the start of flow or heating
ρ = fluid density
μ = fluid viscosity

The average free convection heat transfer coefficient can be expressed as

$$\overline{\mathrm{Nu}} = C(\mathrm{GrPr})^m \tag{8.24}$$

where all the fluid properties are evaluated at the film temperature, $T_f = (T_\infty + T_w)/2$, T_∞ is the free stream or ambient fluid temperature, and T_w is the wall temperature. Values for C and m are given in Table 8.1 for isothermal surfaces. For vertical plates, the characteristic dimension for calculating the Grashof number is the height or length of the plate; for horizontal cylinders, the diameter should be used (see Figs.

TABLE 8.1 Coefficient for Eq. 8.24 for Isothermal Surfaces*

Geometry	GrPr	C	m
Vertical planes and cylinders	$10^{-1} - 10^{4}$	Use Fig. 8.8	Use Fig. 8.8
	$10^4 - 10^9$	0.59	¼
	$10^9 - 10^{13}$	0.021	⅖
	$10^9 - 10^{13}$	0.10	⅓
Horizontal cylinders	$0 - 10^{-5}$	0.4	0
	$10^{-5} - 10^4$	Use Fig. 8.9	Use Fig. 8.9
	$10^4 - 10^9$	0.53	¼
	$10^9 - 10^{12}$	0.13	⅓
Upper surface of heated plates or lower surface of cooled plates	$2 \times 10^4 - 8 \times 10^6$	0.54	¼
	$8 \times 10^6 - 10^{11}$	0.15	⅓
Lower surface of heated plates or upper surface of cooled plates	$10^5 - 10^{11}$	0.58	⅕

* From Holman, Ref. 2.

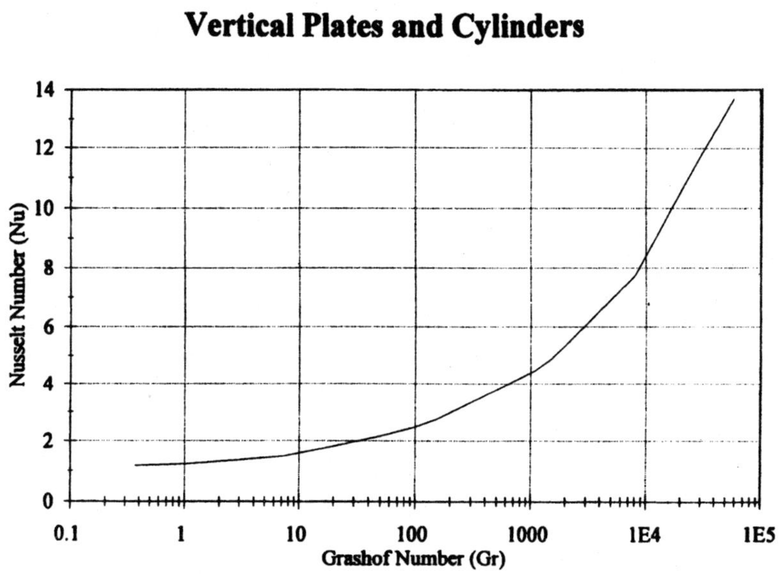

FIGURE 8.8 Grashof and Nusselt numbers for vertical plates and cylinders. *(From Ref. 2.)*

8.8 and 8.9). Often, the vertical plate coefficient is used for vertical cylinders if the boundary-layer thickness is small compared to the cylinder diameter. The criterion for determining whether the plate equation can be used is

$$\frac{D}{L} \geq \frac{35}{\mathrm{Gr}_L^{1/4}} \tag{8.25}$$

For vertical surfaces with constant heat flux instead of isothermal surfaces, the Nusselt number can be expressed in terms of a modified Grashof number:

$$\mathrm{Gr}^{*x} = \mathrm{Gr}_x N_x = \frac{g\beta\rho^2 q_w x^4}{k\mu^2} \tag{8.26}$$

where q_w is the wall heat flux in energy per unit area.

For laminar flow, the Nusselt number expressed in terms of the modified Grashof number is:

$$N_{xf} = \frac{hx}{k_f} = 0.60(\mathrm{Gr}_x^*\mathrm{Pr}_f)^{1/5} \tag{8.27}$$

$$10^5 < \mathrm{Gr}_x^* < 10^{11}$$

$$q_w = \text{constant}$$

For turbulent flow, the Nusselt number is

$$\mathrm{Nu}_x = 0.17(\mathrm{Gr}_x^*\mathrm{Pr})^{1/4} \tag{8.28}$$

$$2 \times 10^{13} < \mathrm{Gr}_x^* < 10^{16}$$

$$q_w = \text{constant}$$

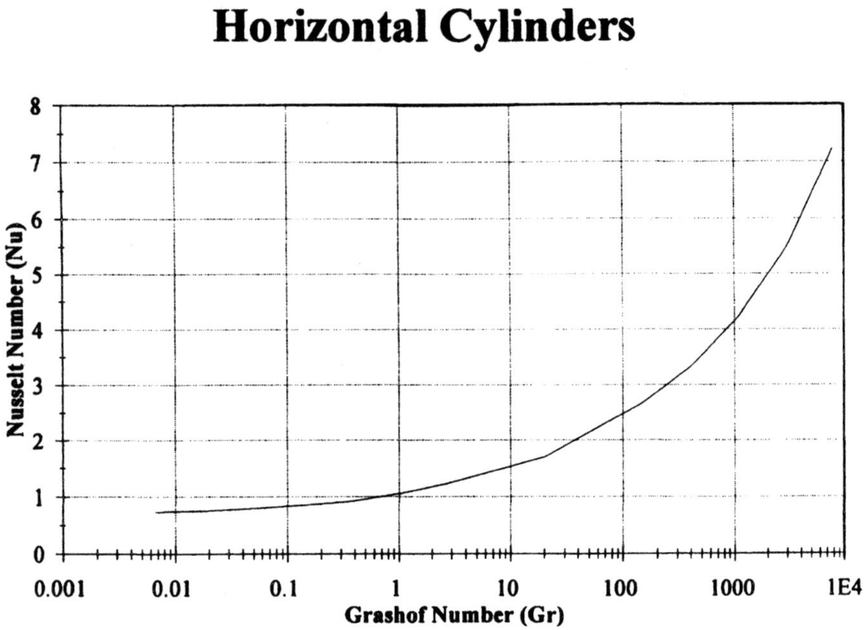

FIGURE 8.9 Grashof and Nusselt numbers for horizontal cylinders. *(From Ref. 2.)*

Again, in most cases the designer is interested in the average heat transfer coefficient of the surface and this can be obtained for laminar flow by integrating Eq. 8.28 to get:

$$\bar{h} = \frac{5}{4} h_{x=L} \tag{8.29}$$

and for turbulent flow, Holman [2] shows that the coefficient is constant with x.

For use with air at one atmosphere, some simplified expressions have been developed and are shown in Table 8.2.

TABLE 8.2 Simplified Equations for Natural Convection in Air at One Atmosphere*

Surface	Laminar $10^4 < Gr_f Pr_f < 10^9$	Turbulent $Gr_f Pr_f > 10^9$
Vertical plate or cylinder	$h = 1.42(\Delta T/L)^{1/4}$	$h = 0.95(\Delta T)^{1/3}$
Horizontal cylinder	$h = 1.32(\Delta T/d)^{1/4}$	$h = 1.24(\Delta T)^{1/3}$
Horizontal plate:		
Heated plate facing upward or cooled plate facing downward	$h = 1.32(\Delta T/L)^{1/4}$	$h = 1.43(\Delta T)^{1/3}$
Heated plate facing downward or cooled plate facing upward	$h = 0.61(\Delta T/L^2)^{1/5}$	$h = 0.61(\Delta T/L^2)^{1/5}$

* From Holman, Ref. 2.
h = heat transfer coefficient, W/m²-°C
$\Delta T = T_w - T_\infty$, °C
L = vertical or horizontal dimension, m
d = diameter, m

Time Constant. The time constant of a system is defined as the length of time required for the system to reach 63.2 percent of its final, steady state temperature from some initial condition. In very complex systems, many time constants are involved and different parts of the system are reaching steady state at different times. However, it is very often useful to treat a system as a single degree of freedom lumped mass. The time constant of a single degree of freedom system is defined as

$$\tau = WC_pR \tag{8.30}$$

where W = weight
C_p = weight specific heat
R = thermal resistance of the lumped mass to the sink

The time constant equation for a system at an initial temperature exposed to a sudden sink temperature change is

$$\frac{T - T_i}{T_s - T_i} = (1 - e^{-t/\tau}) \tag{8.31}$$

where T_i = initial temperature
T_s = sink temperature
t = time

and if $t = \tau$, then:

$$\frac{\Delta T}{\Delta T_{\text{final}}} = (1 - e^{-1}) = 0.63212 \tag{8.32}$$

However, in the design of electronics cooling systems, the question is most often related to how fast the equipment will heat up as a result of its own internal power dissipation. In this situation, the following equation is more helpful.

$$WC_p\frac{dT}{dt} + G(T - T_s) = Q \tag{8.33}$$

This can be rewritten as

$$\frac{dT}{dt} + \frac{(T - T_s)}{WC_pR} = \frac{Q}{WC_pR} \tag{8.34}$$

which can be solved by separation of variables to yield

$$T = (T_i - T_s - RQ)e^{-t/\tau} + T_s + RQ \tag{8.35}$$

Radiation. Radiation heat transfer predicts energy exchange between two surfaces. The exchange is determined by the surface properties of the two surfaces, the temperatures of the surfaces, and how well each can see each other, called the *view factor.* Theoretical approaches usually present black body theory, followed by gray body theory, and some information about real surfaces. Radiation heat transfer, in practice, uses gray body theory and only in extreme cases makes correction for real surfaces. In fact, most surfaces approximate gray bodies reasonably well. The ability to radiate energy from a surface is determined by its emissivity ε which is the ratio that the surface can radiate compared to a theoretical black body which is 1. Thus, all real emissivities are less than 1. Absorption of energy is defined by α, the absorptivity. The fraction of energy reflected from the surface is defined by ρ, the reflectivity.

If the surfaces in question are transparent, energy can also be transmitted through the surface. In most cases of electronic cooling situations, transmittance is not a consideration and is not addressed in any detail here. Detailed examinations can be found in textbooks, such as *Heat Transfer* by J. P. Holman [2] and *Engineering Heat Transfer* by James R. Welty [1].

The total energy that can be radiated by a black body is $E_b = \sigma T^4$, where σ is the Stefan-Boltzmann constant defined as

$$\sigma = 5.669 \times 10^{-8} W/m^2\text{-}K^4 (1.714 \times 10^{-8} BTU/hr°R^4) \tag{8.36}$$

Gray bodies radiate energy per unit area, according to $E = \sigma\varepsilon T^4$ (i.e., ε percent of a black body). A comparison of emissive power E of black, gray, and real surfaces is shown in Fig. 8.10.

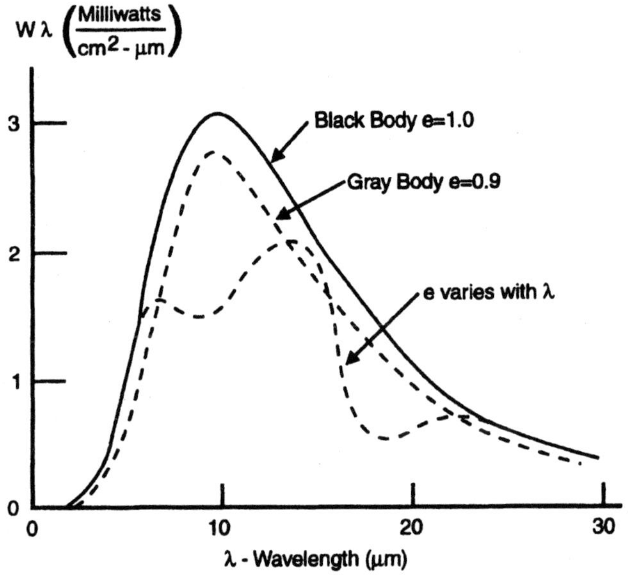

FIGURE 8.10 Comparison of emissive power for black, gray, and real surfaces.

The difference between black body, gray body, and real surfaces has a more fundamental difference than the emissive power. Both black body and gray body theory presume that the surface radiates using Lambert's cosine principle (i.e., the radiative intensity varies with the cosine to the normal to the surface). This gives a smooth hemispherical intensity relationship, which can be contrasted with real surfaces (Fig. 8.11). Notice that organic or nonconductive surfaces reasonably approximate the cosine distribution, except at large θ, whereas metallic surfaces differ greatly.

The spectral range of radiant energy from gray bodies is represented in Fig. 8.12. An object begins to glow red hot at approximately 800°C. Where electronics are involved, radiation occurs below 200°C. To illustrate the radiative properties of real materials, Figs. 8.13 and 8.14 (taken from *The Handbook of Infrared Radiation Measurement* by Barnes Engineering, Ref. 5) show the transmissibilities of several

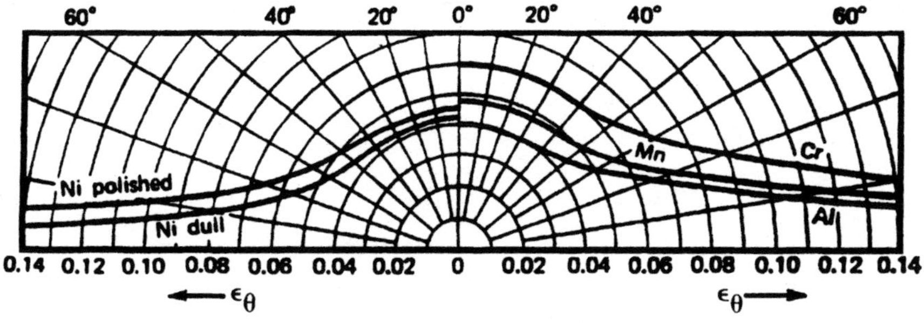

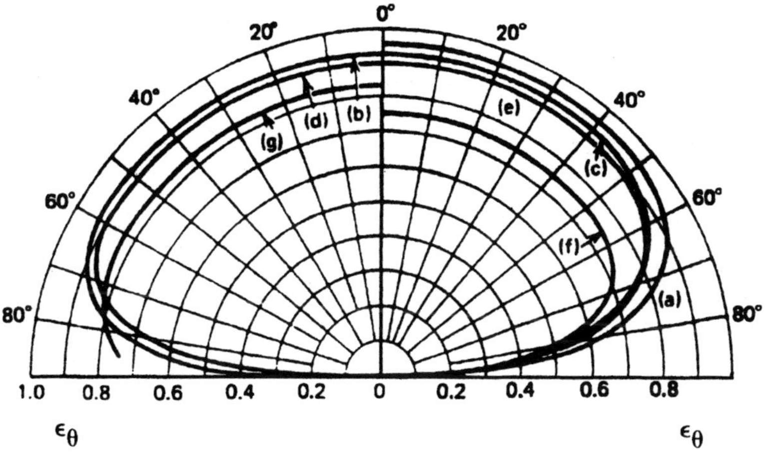

FIGURE 8.11 Radiative intensity distributions for real surfaces (*a*) ice; (*b*) wood; (*c*) glass; (*d*) paper; (*e*) clay; (*f*) copper oxide; (*g*) aluminum oxide.

organic materials as a function of wavelength. These materials have low emissivities where their transmissibility is high and, because their reflectance is very low, have high emissivities ($\alpha = \varepsilon$) in the high-absorption bands. One technique to keep a surface cool in sunlight is to have a low α_s/ε_t, where α_s is the absorptivity to the solar spectrum and ε_t is the emissivity to the thermal spectrum. Using this technique, one of the best surfaces for reflecting solar radiation while having a high emissivity is a second surface mirror, using quartz instead of glass. Figure 8.15 shows the transmissibility of quartz and glass. The transmissibility of quartz, especially in the 2.75- to 5-micron range allows a high percentage of solar radiation to pass through and reflect off the silvered back surface. However, at thermal wavelengths above 5 microns, the quartz has a high emissivity to effectively radiate energy.

Where two surfaces are involved, all the energy leaving one surface might not strike the other (Fig. 8.16). This geometric relationship between the two surfaces is defined by the view factor *F.* The view factor between two surfaces F_{1-2} is defined as the fraction of the total energy leaving surface 1 which strikes surface 2. Mathematically, the view factor for an area dA_1 to a surface dA_2 is presented as

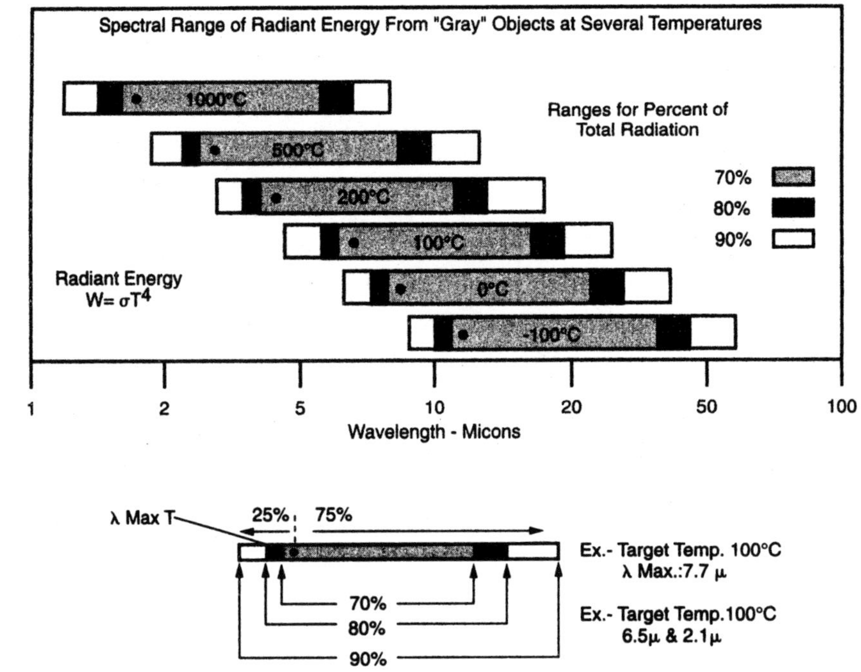

FIGURE 8.12 Gray body radiation bandwidths.

$$dA_1 dF_{dA^1 - dA^2} = \frac{\cos \theta_1 \, dA_1 \cos \theta_2 \, dA_2}{\Pi r^2} \tag{8.37}$$

The view factor from dA_1 to A_2 is thus:

$$dA_1 F_{1-2} = dA_{1\int A2} \frac{\cos \theta_1 \cos \theta_2 \, dA_2}{\Pi r^2} \tag{8.38}$$

Then integrating over A_1 gives the total view factor from surface 1 to surface 2:

$$A_1 F_{1-2} = \int A_1 \int A_2 \frac{\cos \theta_1 \cos \theta_2 \, dA_1 \, dA_2}{\Pi r^2} \tag{8.39}$$

In practice, numerical integration is easily accomplished by replacing the integration in Eq. 8.23 with a summation of finite subdivisions or nodes over surface 2. This is done for each subdivision of surface 1 and summed. With the reciprocity theorem:

$$A_1 F_{1-2} = A_2 F_{2-1} \tag{8.40}$$

the view factor from either surface is known. The reciprocity theorem is a consequence of both surfaces radiating with the same directional intensity distribution (i.e., gray or black body theory). If real surfaces are involved, it is still true if the distribution is the same; however, if one surface is bare metallic and the other painted, the effect of non-Lambertian radiation should be considered, but is beyond the scope of this chapter.

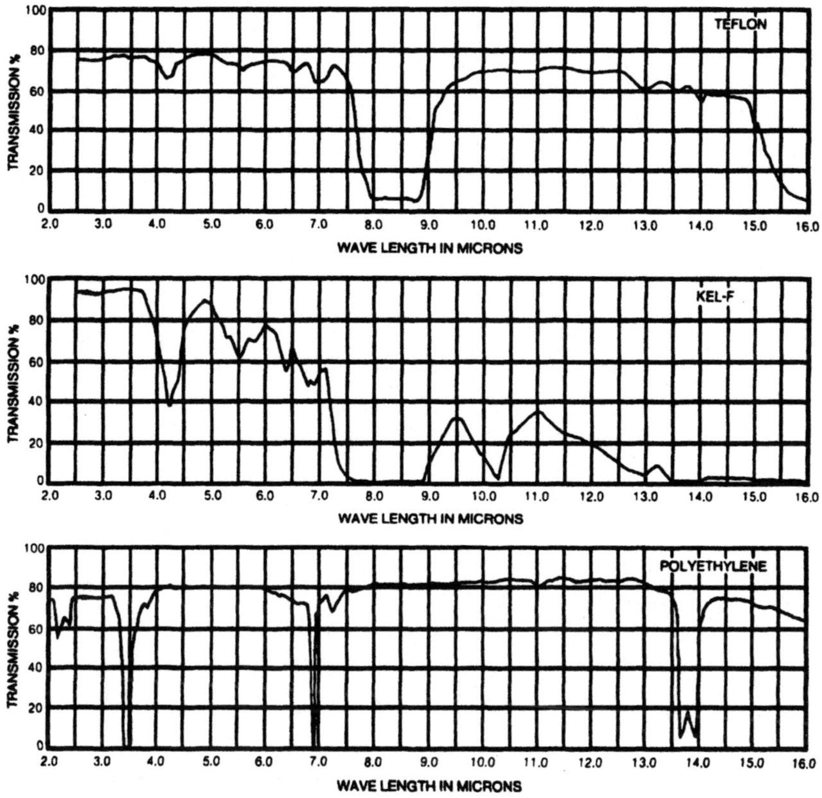

FIGURE 8.13 Transmissibility variation with frequency. *(From Ref. 5.)*

Here, it is convenient to consider the total energy exchange per unit time Q instead of the energy exchange per unit time per unit area E. The computation of the energy exchange

$$Q = \sigma A_1 F_{1-2}(T_1^4 - T_2^4) \tag{8.41}$$

between two surfaces therefore depends on the surface properties ε, α, ρ, τ whose sum must equal one, the temperature of the two surfaces, and the view factor of the two surfaces. If the surfaces are black bodies, this interchange can be expressed as, and F_{1-2} is called, the *black body view factor.*

However, when nonblack surfaces are involved, the emissivity and other properties must be considered and is usually done by altering the black body view factor to a "script F" $\mathcal{F}$. For example, consider two infinite surfaces: the energy leaving surface 1 can strike surface 2, where some is absorbed and some is reflected back to surface 1. Surface 1 then absorbs part of this reflected energy back to surface 2, which reflects it again. This result is an infinite series, which represents an *effective emissivity* between the two surfaces expressed as

$$\mathcal{F}_{1-2} = \frac{1}{(1/\varepsilon_1 + 1/\varepsilon_2) - 1} \tag{8.42}$$

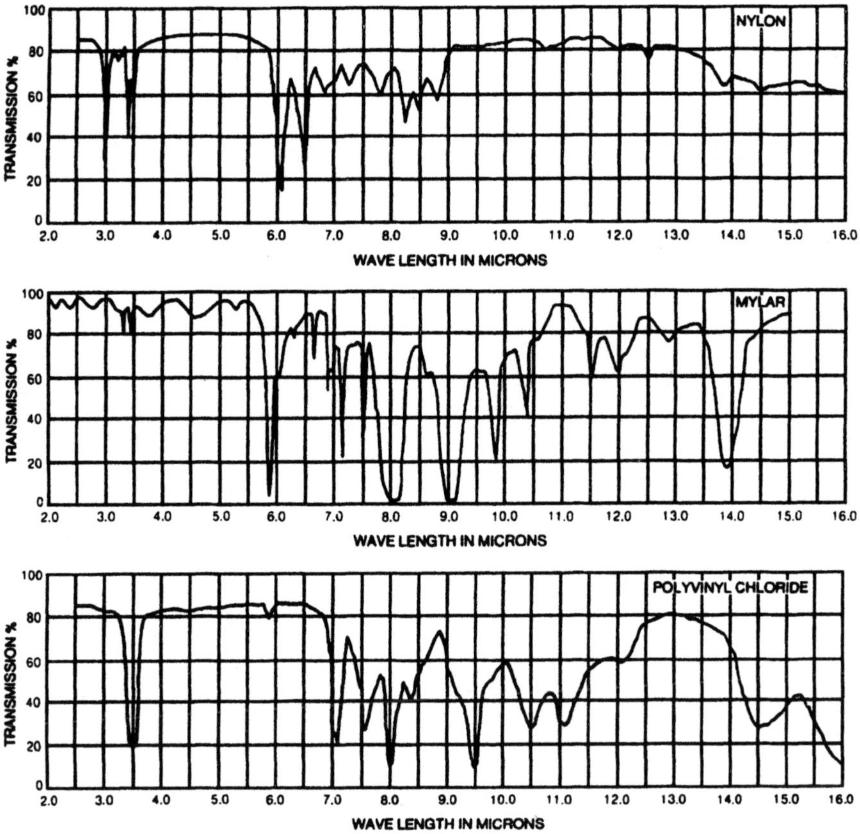

FIGURE 8.14 Transmissibility variation with frequency. *(From Ref. 5.)*

and the energy exchange between the surfaces is

$$Q = \sigma A_1 \mathscr{F}_{1-2}(T_1^4 - T_2^4) \tag{8.43}$$

Table 8.3 lists $\mathscr{F}$ expressions for some commonly encountered surface geometries. Complicated geometries will require actual integration and infinite series evaluation. Computer programs available to do this are TRASYS and NEVADA.

Other geometries have been evaluated in NACA Technical Note 2836 [6]. Holman [2] presents the equation $F_{A1-A2} = D^2/(4R^2 + D^2)$ for a small area A_1 to a larger disk A_2, corresponding to Fig. 8.17.

Figures 8.18, 8.19, and 8.20 from Holman's *Heat Transfer* [2] define black body view factors for parallel rectangles and disks and for perpendicular rectangular surfaces. A particularly useful view factor is given in the NACA Technical Note 2836 [6] (Fig. 8.21) reproduced from that document for a small surface dA_1 relative to any infinite plane A_2 where the plane of A_2 and dA_1 intersect at an angle. For this situation, the view factor is

$$F_{1-2} = \tfrac{1}{2}(1 + \cos\theta)$$

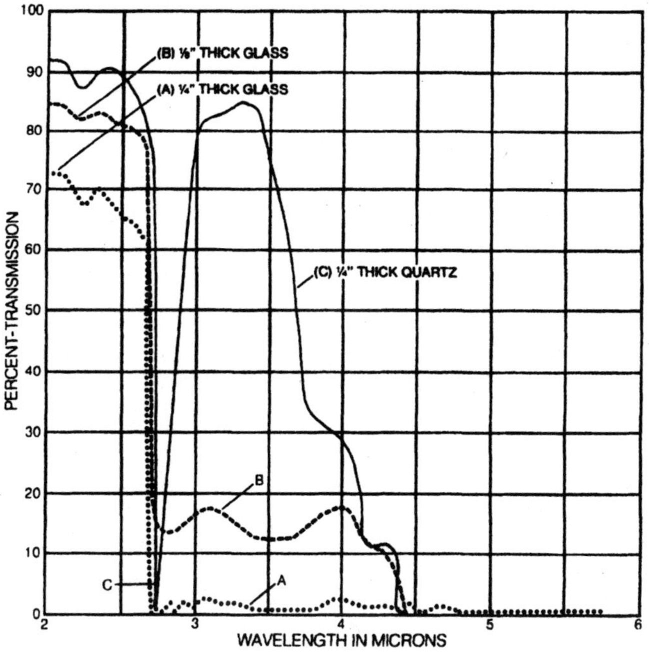

FIGURE 8.15 Transmissibility of glass and quartz. *(From Ref. 5.)*

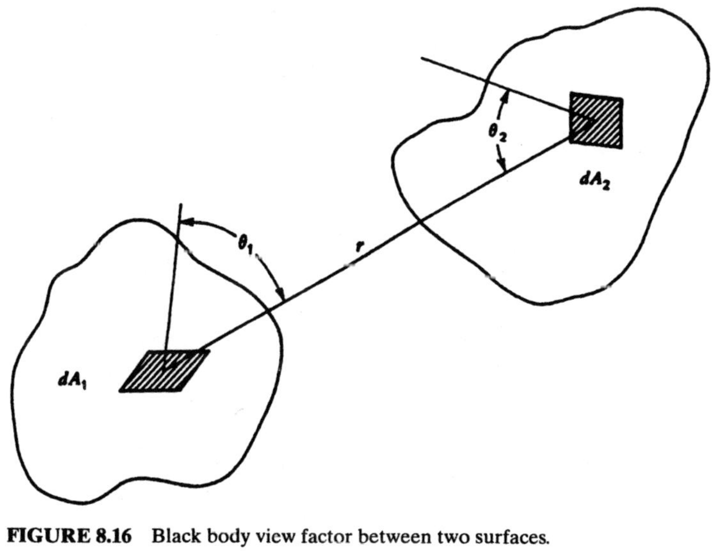

FIGURE 8.16 Black body view factor between two surfaces.

TABLE 8.3 Values for Commonly Encountered Script F Geometries Data*

Surfaces	Area A	$\mathcal{F}$
Infinite parallel plates	Either area	$1/(1/\varepsilon_1 + 1/\varepsilon_2 - 1)$
Small body A_1 enclosed by a large body A_2	A_1	ε_1
Large body A_1 completely enclosed by a larger body A_2	A_1	$1/(1/\varepsilon_1 + \varepsilon_2 - 1)$
Concentric spheres, A_1 inside A_2	A_1	$1/[1/\varepsilon_1 + A_1/A_2(1/\varepsilon_2 - 1)]$
Infinite concentric cylinders, A_1 inside A_2	A_1	$1/[1/\varepsilon_1 + A_1/A_2(1/\varepsilon_2 - 1)]$

* From *Cooling Electronic Equipment,* Ref. 7.

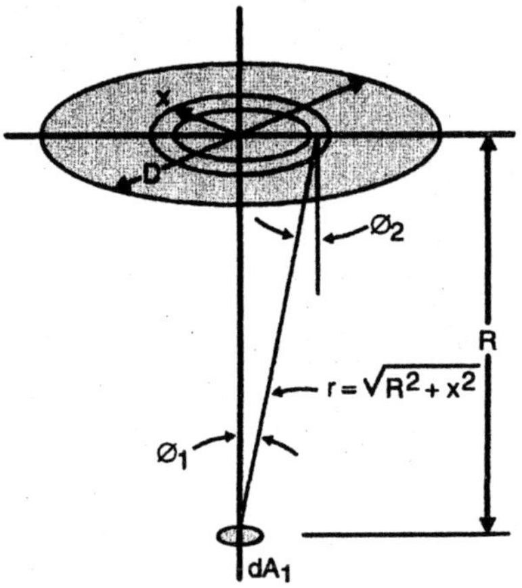

FIGURE 8.17 Radiation from a small area to a disk.

Space-Radiation Considerations. The radiation between surfaces of different temperatures, as presented to this point, presumes that the radiation is of approximately the same wavelength, in which case $\varepsilon = \alpha$. This is usually the case and is always the case inside of electronic enclosures. When considering problems involving sunlight, the effect of the high spectral intensity of visual and ultraviolet radiation must be considered because many materials can have different absorptivities to this spectrum than they do to infrared or thermal radiation (i.e., $\varepsilon \neq \alpha$). In this case, it is convenient to express the interchange of energy, not in terms of $(T_1^4 - T_2^4)$, but as energy radiated, εT_1^4, minus the energy absorbed, $H\alpha \cos\theta$, as in Fig. 8.22.

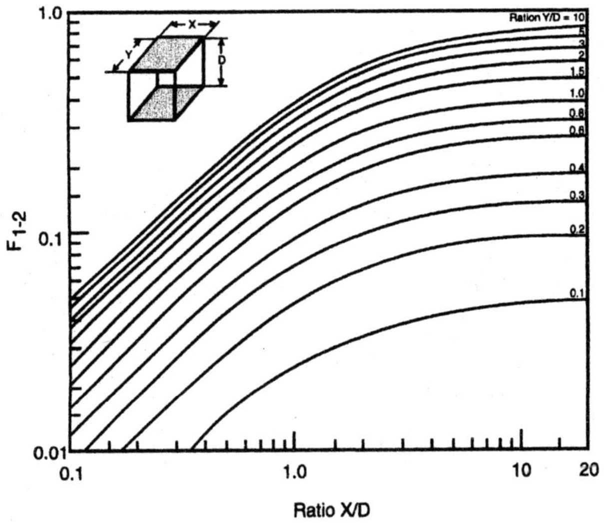

FIGURE 8.18 View factors for parallel rectangles.

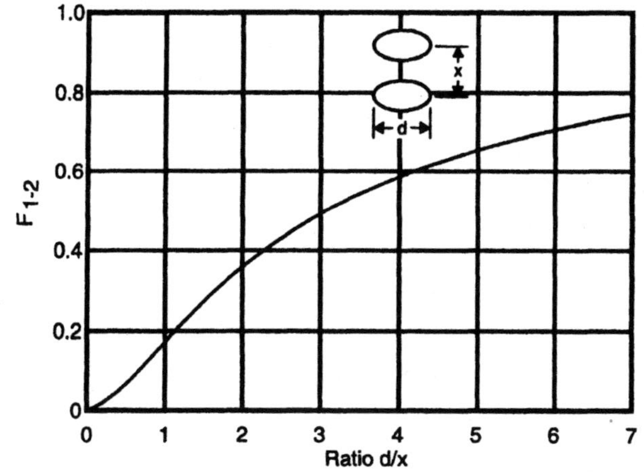

FIGURE 8.19 View factors for parallel disks.

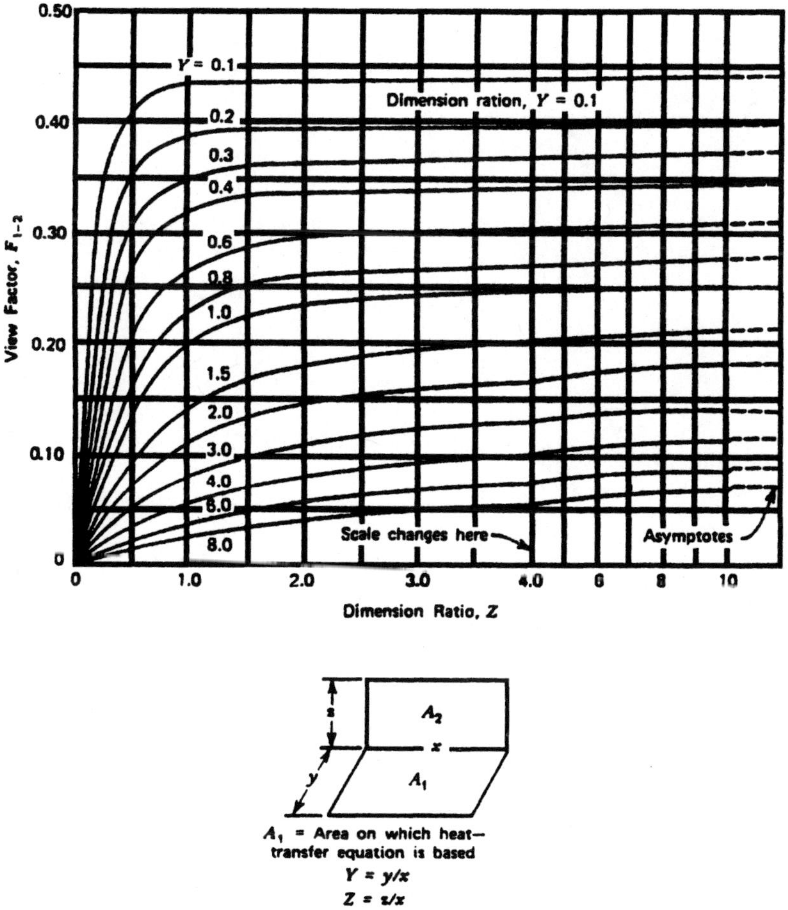

FIGURE 8.20 View factors for perpendicular rectangular surfaces.

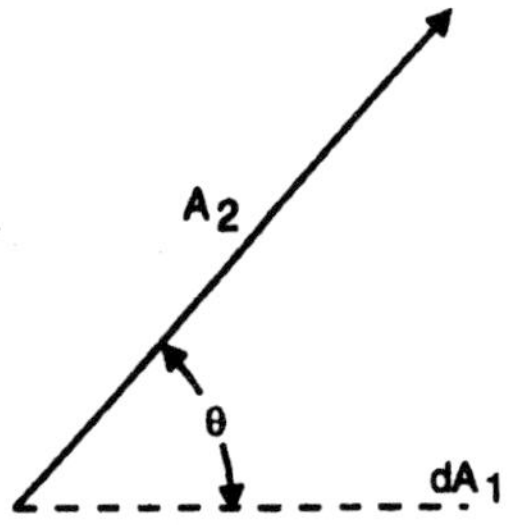

FIGURE 8.21 The view factor from a small plane surface to an infinite plane whose places intersect at an angle θ.

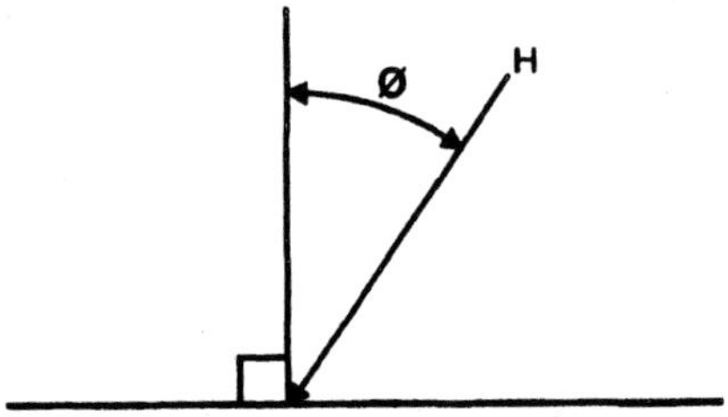

FIGURE 8.22 Solar radiation striking a flat plate.

To illustrate the effect of wavelength on emissivity, values for a few materials are presented in Table 8.4. At any wavelength, $\varepsilon = \alpha$, so the values in this table can be interpreted as the thermal emissivity at the surface temperature and the solar radiation absorptivity at the solar wavelength.

The following computer program computes the view factor for two plane surfaces oriented at any angle to each other. Data input is the center of each subarea of the two surfaces, the areas of each subsurface, and coefficients defining the planes of the surfaces to describe an equation in the form

$$Ax + By + Cz = 0$$

```
50 DIMENSION X1(225),Y1(225),Z1(225),X2(225),Y2(225),Z2(225)
100 DIMENSION AREA1(225),AREA2(225),F(225)
110 DATA TAREA1/0./,TAREA2/0./,AF/0./
120 DATA F/225*0./
150 CALL OPENF(1,"COORD")
C N1 and N2 are the number of sub-areas on each surface
200 READ (1,)N1,N2
250 READ(1,)(X1(I),I=1,N1)
300 READ(1,)(Y1(I),I=1,N1)
350 READ(1,)(Z1(I),I=1,N1)
400 READ(1,)(X2(I),I=1,N1)
450 READ(1,)(Y2(I),I=1,N1)
500 READ(1,)(Z2(I),I=1,N1)
550 READ(1,)(AREA1(I),I=1,N1)
600 READ(1,)(AREA2(I),I=1,N2)
650 READ(1,)A1,B1,C1,A2,B2,C2
700 DO 20 J=1,N1 750 DO 10 I=1,N2
800 R1=X2(I)-X1(J)
850 R2=Y2(I)-Y1(J)
900 R3=Z2(I)-Z1(J)
950 RMAG=SQRT(R1**2+R2**2+R3**2)
1000 COSB1=(R1*A1+R2*B1+R3*C1)/RMAG
1010 R1=-R1 1020 R2=-R2 1030 R3=-R3
1050 COSB2=(R1*A2+R2*B2+R3*C2)/RMAG
1052 IF(COSB1.LT.0.)COSB1=0.
```

TABLE 8.4 Emissivities of Materials at Different Wavelengths

Material	9.3 μ 100°F	5.4 μ 500°F	3.6 μ 1000°F	1.8 μ 2500°F	0.6 μ Solar α_s
Asbestos paper	0.93	0.93			
Red brick	0.93				
White paper	0.95				0.28
Water	0.96				
Carbon, T-carbon	0.82	0.8	0.79		
Polished aluminum	0.04	0.05	0.08	0.19	0.3
Oxidized	0.11	0.12	0.18		
Anodized	0.86	0.42	0.6	0.34	
Polished chromium	0.08	0.17	0.26	0.4	0.49
Polished copper	0.04	0.05	0.18	0.17	
Oxidized	0.87	0.83	0.77		
Polished gold	0.02	0.018	0.035		0.14
Polished iron	0.06	0.08	0.13	0.25	0.45
New galvanized	0.23			0.42	0.66
Dirty galvanized	0.28			0.9	0.89
Oxide	0.96		0.85		0.74
Rusted		0.69			
Lead, gray anodized	0.28				
Magnesium	0.07	0.13	0.18	0.24	0.3
Molybdenum			0.09	0.15	0.2
Polished silver	0.01	0.02	0.03		
Polished stainless steel	0.15	0.18	0.22		
Weathered	0.85	0.85	0.85		
Bright in		0.043			
Paints					
Aluminum lacquer	0.65	0.65			
Cream	0.95	0.88	0.7	0.42	0.35
Flat black (lacquer)	0.96	0.97	0.97	0.97	0.97
Red	0.96				0.74
White (ZnO)	0.95				0.18
White	0.8–0.95				0.19–0.33
White epoxy	0.9–0.95				0.3–0.35
Gray	0.84–0.91				0.5–0.57
Glass	0.9–0.95				

```
1055 IF(COSB2.LT.0.)COSB2=0.
1100 10 F(J)=F(J)+COSB1*COSB2*AREA2(I)/(3.14159*(RMAG**2))
1150 20 CONTINUE 1200 DO 30 M=1,N1
1250 30 AF=AF+F(M)*AREA1(M)
1300 PRINT,"AREA-VIEW FACTOR PRODUCT=",AF
1350 DO 50 K=1,N2
1400 50 TAREA2=TAREA2+AREA2(K)
1450 PRINT,"AREA2=",TAREA2
1500 F21=AF/TAREA2
1600 PRINT,"F(2 TO DA OF 1)=",F21
1700 DO 60 K=1,N1
1750 60 TAREA1=TAREA1+AREA1(K)
1800 PRINT,"AREA1=",TAREA1
```

```
1850 F12=AF/TAREA1
1900 PRINT,"F(1 TO DA OF 2)=",F12
9000 END
```

Another useful computer program is presented to solve the radiative heat transfer equation. This is in FORTRAN and uses the Newton iteration method to solve the equation.

```
C     NEWTON S METHOD ITERATION TO SOLVE NON-LINEAR EQUATIONS
C     Q+H*Area*alpha=sigma*Area*epsilon*T**4+Con*(T-T0)
C     AREA IS IN INCHES, A IS ALPHA, EP IS EPSILON, CON IS CON-
      DUCTANCE
      DATA EP/.03/,A/.1/,AREA1/28.1/
      DATA SUN/430./,T2/733./,POW/0./
      DATA AREA/134.55/,CON/.0105/
      F(E)=1.713E-9*EP*AREA*E**4/144.+CON*(E-T2)-SUN*A*AREA1/
      144.-POW
      FPRIME(E)=4.*1.713E-9*EP*AREA*E**3/144.+CON
      FPRIM2(E)=12.*1.713E-9*EP*AREA*E**2/144.
      WRITE(*,*)'Enter an initial temperature as a starting
      point'
5     READ(*,*)X
      IF(ABS(F(X)*FPRIM2(X))/FPRIME(X)**2-1.)10,15,15
10    WRITE(*,50)
      I=0
      Y=F(X)
      WRITE(*,20)I,X,Y
      DO 11 I=1,30
      X1=X-(Y/FPRIME(X))
      Y=F(X1)
      WRITE(*,20)I,X1,Y
      IF(ABS((X1-X)/X1).LT.1.E-5)GOTO 5
11    X=X1
      WRITE(*,40)
      GO TO 5
15    WRITE(*,30)X
      GOTO 5
20    FORMAT(7X,I2,8X,E14.8,2XE14.8)
30    FORMAT(3H X=,F12.4,27H WILL NOT CAUSE CONVERGENCE)
40    FORMAT(/36H FAILED TO CONVERGE IN 30 ITERATIONS)
50    FORMAT(//41H ITERATION X1 F(X1))
      END
```

8.3 PRESSURE DROP ANALYSIS

The two terms to consider in predicting the pressure drop of a system are the friction and the losses caused by expansions, contractions, and turns. Liquids and gases are handled identically, but friction losses are much higher in liquids and the effect of viscosity is much more important. In either fluid, there is a source of motion (such as a blower or pump) and a source of impedance, which is discussed in this section. In designing a system, these two forces must be matched to obtain the desired flow rates.

The pressure losses covered in this section refer to system energy losses that are not recoverable. The total pressure is composed of static and dynamic pressure and the values of these can change through a system of ducts by converting static pressure to dynamic pressure by contracting the flow and the reverse by expanding the flow. Pressure loss for electronic equipment, even though not specified in some cases, is meant to be loss in total pressure.

Friction Loss. The two commonly used equations for friction loss are identical, except for the definition of the friction factor.

Using the Fanning friction factor, the equation is

$$\Delta P = \frac{fL\rho V^2}{2gd} \tag{8.44}$$

Using the Darcy friction factor, the equation is

$$\Delta P = \frac{4fL\rho V^2}{2gd} \tag{8.45}$$

where, in both equations

L = length in inches
ρ = density of the fluid in lb/in^3
V = velocity of the fluid in in/s
g = acceleration of gravity (384 in/s^2)
d = hydraulic diameter (in)

It is obvious that the Fanning friction factor is four times the Darcy friction factor. This is mentioned as a caution for the designer to use the correct friction factor in the equation. The Fanning friction factor is plotted in Fig. 8.23. It is useful in most cases to define the friction factor for laminar flow (i.e., $R_e < 2100$ as $64/R_e$ for Eq. 9.44 and $16/R_e$ for Eq. 9.45).

Expansions, Contractions, and Turns. When flowing fluid is forced to change direction by turning or, to a lesser extent, by contractions and expansions, additional energy is lost in the form of pressure and is expressed as

$$\Delta P = \frac{K\rho V^2}{2g} \tag{8.46}$$

where K is a loss coefficient, defined empirically and dependent on the severity of the turn. Some useful definitions are given in Figs. 8.24 through 8.26 and Table 8.5. The total pressure loss in a duct is thus defined as

$$\Delta P = \frac{\rho V^2}{2g}\left(\frac{fL}{d} + K\right) \tag{8.47}$$

An examination of the values of K in Figs. 8.24 through 8.26 leads to the conclusion that the turns should be as gradual as possible and changes in area should be gradual. In fact, many tests and references talk about diffusers and recommend that transitions between two duct areas be made at 12° to 15° angles. Although true, there is usually not room for this gradual transition inside of electronic units.

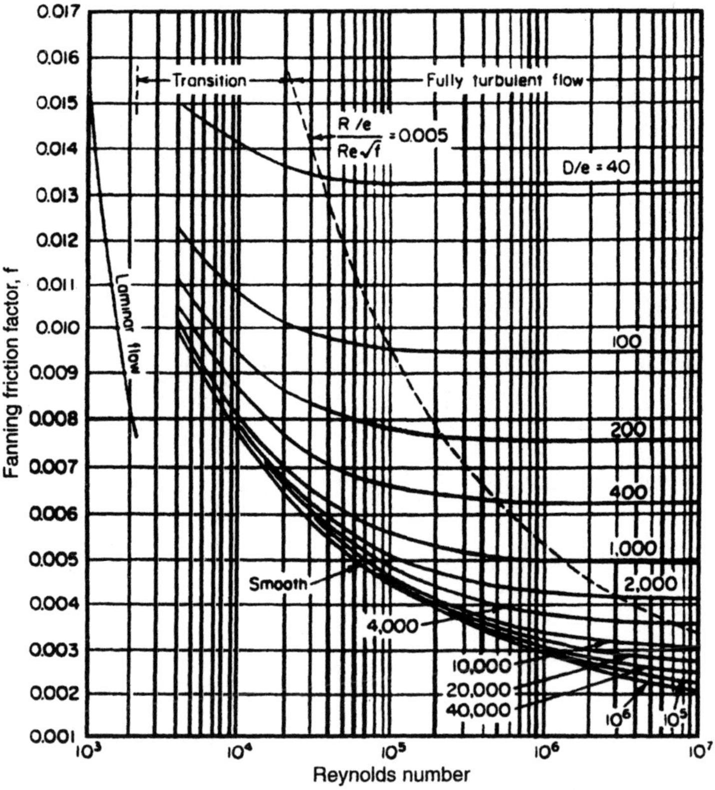

FIGURE 8.23 Fanning friction factor chart (Moody diagram) for smooth and rough tubes.

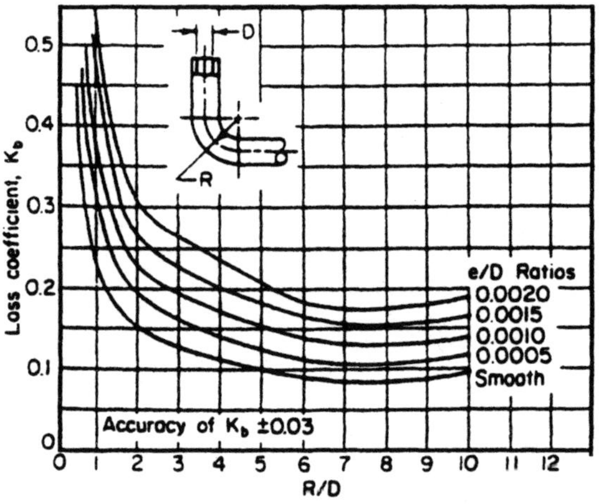

FIGURE 8.24 Loss coefficients for smooth 90° bends.

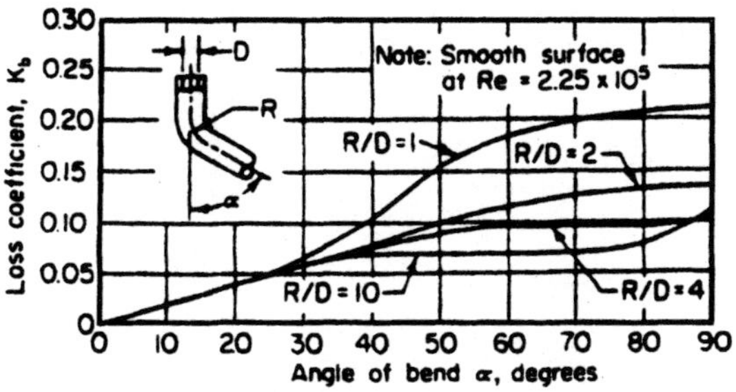

FIGURE 8.25 Loss coefficients for smooth bends in smooth tubes versus bend angle.

8.4 CIRCUIT CARD ASSEMBLY (CCA) COOLING

The CCA is the basic building block of most electronic systems. The exceptions are usually high-power transmitters (built around traveling-wave tubes) or high-voltage power supplies (built around isolated floating decks, where arc lengths and voltage isolation usually require immersion in a dielectric liquid or gas). The thermal designer is therefore most often concerned with removing heat from heat-dissipating components mounted to a CCA. Radiation is of little use and is used only as a safety factor except in some low-power space applications. Heat is removed from the components either by conduction or convection or a combination of the two. Where heat is removed by conduction to the circuit card or a heat sink on the circuit card, methods of removing the heat from the card must be utilized.

Typically in the analysis of CCAs, the components are characterized as a one-dimensional heat source and resistance connected to the card (Fig. 8.27). The resistance is called θ_{jc}, which is a misnomer because θ refers to temperature difference in most heat transfer texts. However, the term originated in the electronics industry and now is tradition.

This one-dimension characterization is usually adequate, but it must be recognized that it is not in all cases and detailed analysis of the chip and its mounting and conduction to the component case are often necessary (covered in Sec. 8.5). However, as system-level thermal designers, there is often little knowledge of how the component is constructed; we have to use either the manufacturer's stated θ_{jc} or a MIL-SPEC value. The MIL-SPEC values usually give conservative (too high) temperature estimates because they are set at a level to cover all manufacturers' values, thus, requiring a design for the worst-case manufacturer. In an increasingly more commercial environment, it is becoming more acceptable to use the manufacturer's values and limit the use of certain parts to acceptable suppliers when necessary.

8.4.1 Conduction-Cooled CCAs

Conduction is used both as the primary cooling method for CCAs and to augment convection cooling. CCAs designed to be cooled by conduction generally are of three different types:

1. Heat sinks under the board.
2. Heat sinks on top of the board.
3. No heat sink: the board is used to conduct heat to a sink.

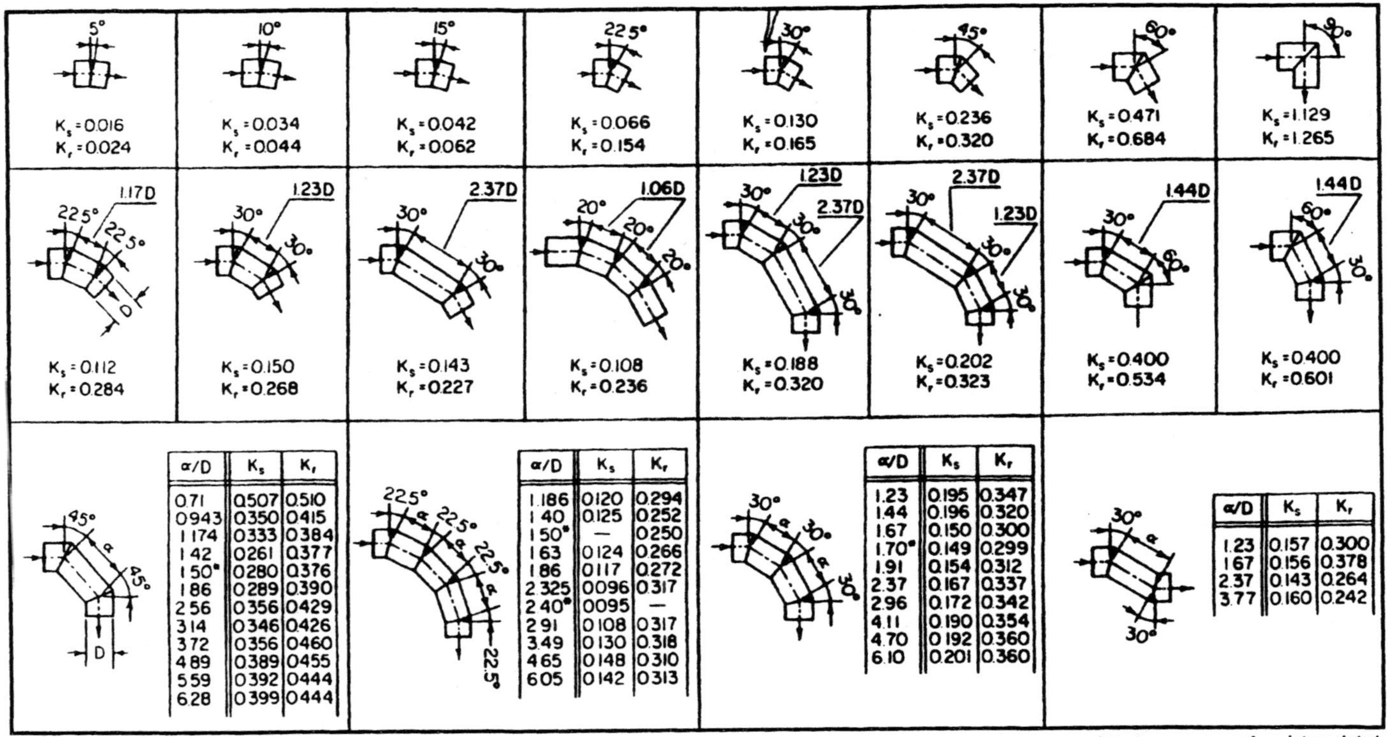

α/D	K_s	K_r
0.71	0507	0510
0943	0350	0415
1174	0333	0384
142	0261	0377
150*	0280	0376
186	0289	0390
256	0356	0429
314	0346	0426
372	0356	0460
489	0389	0455
559	0392	0444
628	0399	0444

α/D	K_s	K_r
1.186	0120	0.294
140	0125	0252
150*	—	0250
163	0124	0266
186	0117	0272
2325	0096	0317
240*	0095	—
291	0108	0317
3.49	0130	0.318
465	0148	0310
605	0142	0313

α/D	K_s	K_r
1.23	0.195	0.347
1.44	0.196	0.320
1.67	0.150	0.300
1.70*	0.149	0.299
1.91	0.154	0.312
2.37	0.167	0.337
2.96	0.172	0.342
4.11	0.190	0.354
4.70	0.192	0.360
6.10	0.201	0.360

α/D	K_s	K_r
1.23	0.157	0.300
167	0.156	0.378
2.37	0.143	0.264
3.77	0.160	0.242

K_s = Resistance coefficient for smooth surface
K_r = Resistance coefficient for rough surface, e/D = 0.0022

*Optimum value of α interpolated

FIGURE 8.26 Loss coefficients for miter bends of various angles and straight sections.

TABLE 8.5 Pressure Drop for Turbulent Flow Through Screwed Valves and Fittings

Type of valve or fitting	Pressure drop*	
	Equivalent L/D	Equivalent velocity head, K_v
Couplings	2	0.04
Unions	2	0.04
Gate valves†		
Open	7	0.1
¾ open	40	0.8
½ open	200	4
¼ open	800	15
Globe valves, bevel seat†		
Open	350	6
½ open	550	10
Composition disk		
Open	330	6
½ open		
Plug disk	500	9
Open	500	9
¾ open	700	13
½ open	2,000	10
¼ open	6,000	110
Angle valves†		
Open	170	3
Y or blow-off valves†		
Open	170	3
Check valves‡		
Swing	110	2
Disk	500	10
Ball	3,500	65
Water meters		
Disk	400	8

* Pressure drop is expressed as a straight pipe of same nominal size as the fitting. Where the pressure drop is expressed as L/D, L is the equivalent length (feet) of straight pipe and D is the inside diameter (feet). Where the pressure drop is expressed as velocity heads, the velocity in the pipe is based on the nominal diameter of the fitting.

† Flow direction through valves has negligible effect on pressure drop.

‡ Values apply only when check valve is fully open, which generally is attained at pipe velocities over 3 ft/s for water.

Source: Reprinted from "Velocity Head Simplified Flow Computation," by C. E. Lapple, by permission from *Chem. Engr.,* 56, May 1949.

Heat sinks under the board are used with surface-mount components, such as flat packs, quad packs, and multichip modules (Fig. 8.28). Either a single CCA can be bonded to the sink or a CCA on each side of the heat sink can be used. This type of electronic module packaging has been defined in specific sizes called *standard electronic modules* (*SEMs*), as illustrated in Fig. 8.29. The heat sink is a solid sheet of material, which extends beyond the edges of the circuit card to be clamped into the chassis to provide a good conductive path to the chassis. A common clamping mechanism made by Wedge-Lock is shown in Fig. 8.30. Resistance through this interface depends

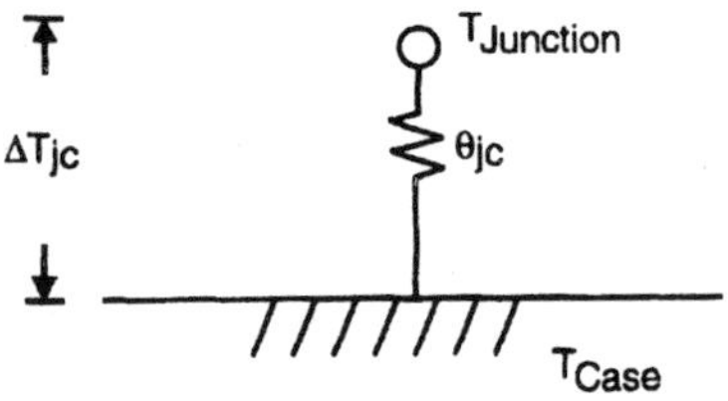

FIGURE 8.27 One-dimensional characterization of an electrical component.

on the number of sections of Wedge-Lock used in the mechanism. Values for 3- and 5-piece mechanisms are defined by Table 8.6.

Providing a uniform heat distribution is assumed, the conduction heat sink provides a parabolic temperature distribution according to Eq. 8.48 and is illustrated in Fig. 8.31:

$$\Delta T = \frac{q'x^2}{2kA} \tag{8.48}$$

where q' is in watts per inch.

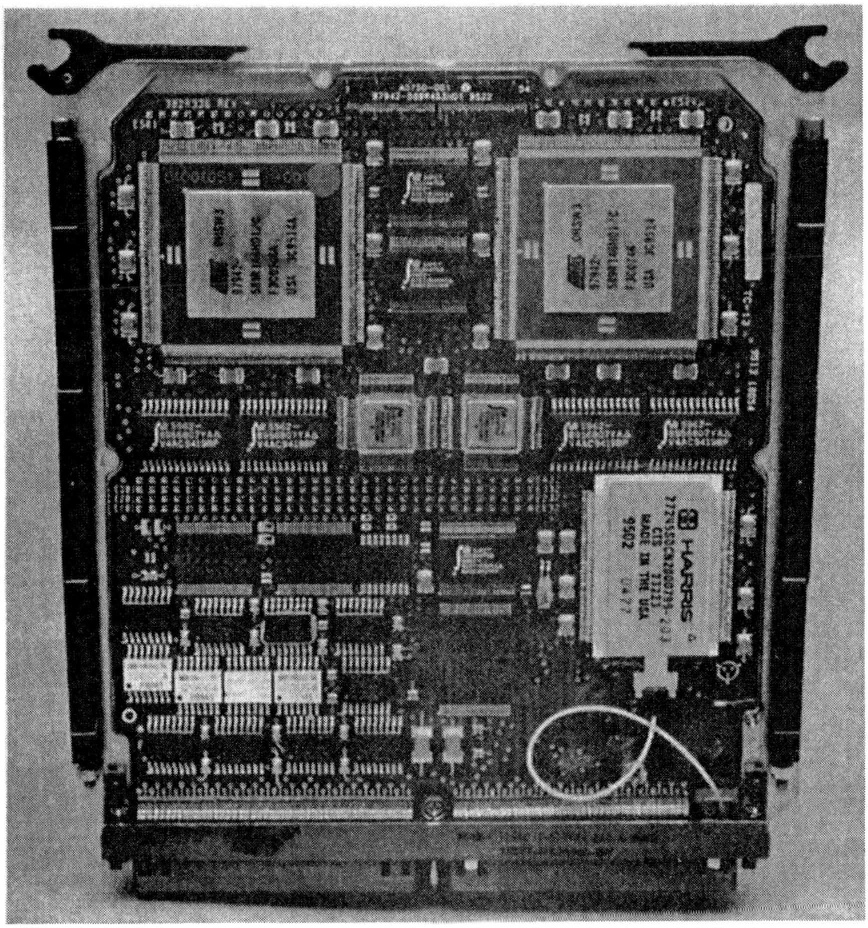

FIGURE 8.28 A conduction-cooled surface-mount-type CCA.

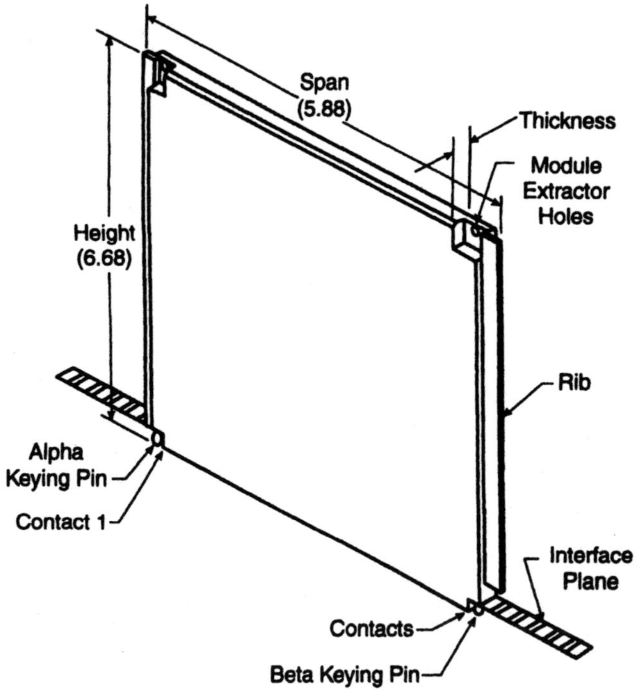

Inches	Metric
5.88	149.35
6.68	169.67

Notes:
1. Dimensions are in inches.
2. Metric equivalents are given for general information only.

FIGURE 8.29 A standard electronic module: E size.

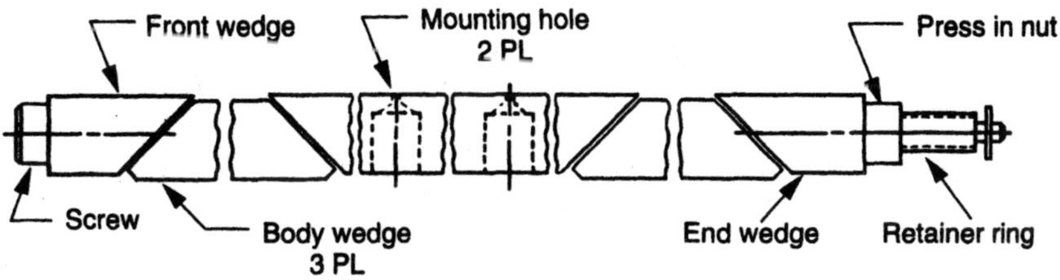

FIGURE 8.30 A CCA clamping device made by Wedge-Lock to provide conduction from the CCA to the electronic chassis.

TABLE 8.6 Contact Resistance Provided by the Wedge-Lock Clamp at the Edge of the Module*

Wedge-Lock type	°C-m²/watt	°C-in²/watt
3 Segment, chemical film finish per MIL-C-5541, Class III	0.0003459	0.5
3 Segment, Type II, Class 2 anodize finish per MIL-A-8625	0.0003079	0.45
5 Segment, chemical film finish per MIL-C-5541, Class III	0.0002864	0.417

* From Ref. 1.

The average temperature is ⅔ of the maximum T and is only ½ of the temperature computed when all the heat is applied at the center. Thus, the maximum temperature can be modeled by putting all of the heat at ¼, the length of the plate from each edge.

Obviously, the temperature of the components on the CCA depends on the conductive path through the circuit card to the heat sink and the conduction in the heat sink to the chassis. These values can be affected favorably by good design or negatively by poor design. The conduction through the board can be enhanced by the addition of vias. In some extreme cases, the board is cut away so that the part can sit directly on the heat sink. The material of the heat sink is also important. Aluminum and copper are commonly used, but new composite materials are starting to be used. Composite materials have the definite advantages of being lighter, having higher conductivities, and having adjustable thermal expansion coefficients, which stress the CCA less. Their disadvantage is cost, but this is coming down. Composite Optics compiled figures to show the cost of building composite heat sinks, compared to the 1993 cost of raw composite (carbon) fiber (Fig. 8.32). Though still more expensive than aluminum or copper, where weight or particularly good thermal performance is needed, the cost is low enough to merit serious consideration (Fig. 8.33). Properties of some materials are shown in Table 8.7.

The effective conductivity of laminated composite heat sinks can be adjusted through fiber orientation. Figure 8.34 shows this effect for P-100/Epoxy made for a SEM-E CCA heat sink. Also note that a similar graph could be plotted for thermal expansion.

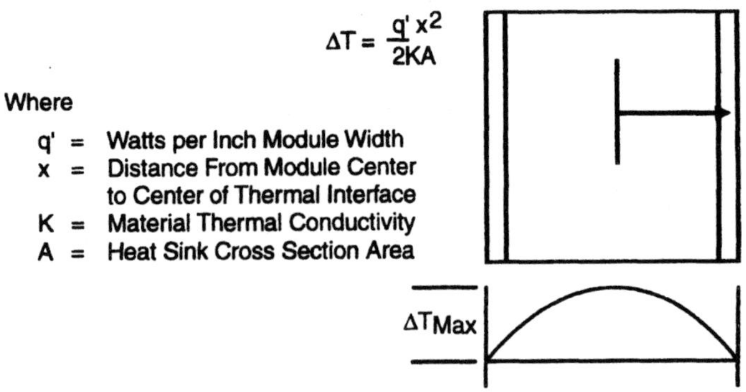

FIGURE 8.31 Conduction-cooled CCAs provide a parabolic temperature distribution.

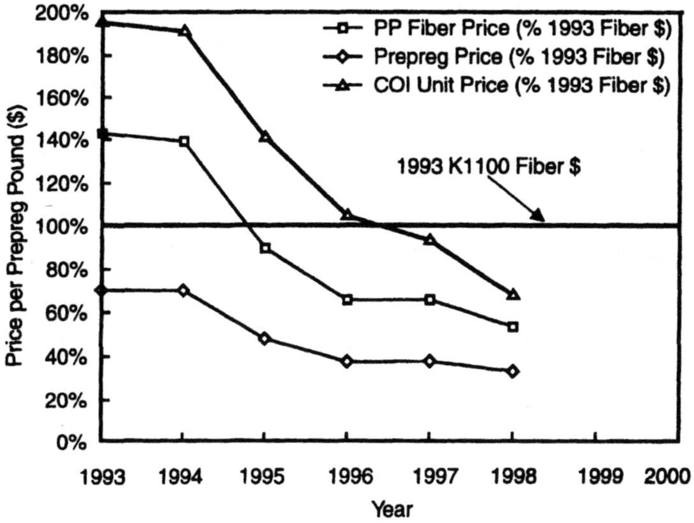

FIGURE 8.32 Reducing cost trends of fabricated advanced thermal composite hardware relative to 1993 cost of raw material. *(Courtesy of Composite Optics Inc.)*

FIGURE 8.33 A composite-material VME-size conduction module core. *(Courtesy of Composite Optics Inc.)*

Heat sinks on top of the board are referred to as *rail* or *ladder cooled* because the configuration of the heat sink has parallel rails or rungs, like a ladder. A CCA of this type is shown in Fig. 8.35. This type of heat sink requires through-hole mounting components, such as dual-inline packages, and is generally used in military equipment, where direct contact of air on the CCAs isn't desirable because of contaminants and moisture in the air or where natural convection isn't adequate and conduction to the electronic chassis is needed.

Like the SEM-type CCAs, the board is actually smaller than the heat sink so that the edges of the sink can be clamped into the chassis to provide a conductive path to the chassis. The temperature profile on each rail can be approximated as a uniformly distributed heat load to again give a parabolic temperature distribution. The SEM conductive plate has the advantage of conducting in two dimensions, but the rail has the advantage of having the component mounted directly onto it so that the problem of conducting through the circuit card is eliminated.

Conduction in the heat sink wall of a chassis can also be improved by using advance thermal composites. Figure 8.36 is a chassis wall developed by Composite Optics, which produced temperatures 20°C cooler than the aluminum equivalent with 400 watts heat sunk to the wall.

TABLE 8.7 Comparison of Advanced Composite Heat Sink Materials to Aluminum

			Thermal conductivity		
Material	Type	Density, g/cm^3	One direction, W/m-K	Two directions, W/m-K	Three directions, W/m-K
Aluminum	2024, isotropic metal	2.7	170	170	170
K13710	CFRP, quasi-isotropic	1.75	40	40	2.1
P120	CFRP, quasi-isotropic	1.75	190	190	2.1
K1100	CFRP, quasi-isotropic	1.75	300	300	2.1
Thermal graph 8000	Unidirectional carbon-carbon	1.76	700	20	20
Thermal graph EWC 600	High k fabric	1.65	160–200	160–200	~1.5
TPG	Pyrolytic graphite, grade T	1.9	1,300	1,300	<20
TC 1050	TPG encapsulated in aluminum	2.5	1,050	1,050	<20
TPG/resin composite	TPG bonded into conductive CFRP	1.85	~1,100	~1,100	~2

8.4.2 Forced Convection–Cooled CCAs

Direct Impingement of Air on CCA Components. Most applications of forced convection cooling to CCAs involves blowing air over the components on the circuit card. Although very effective, accurate prediction of the heat transfer is difficult because of the irregular surface presented to the airflow. Flow can be quite turbulent, even when Reynolds numbers are low. Tall components can shield downstream components. Typically, you can use undeveloped flow over each component and account for temperature rise as the air or cooling fluid passes down the length of the CCA. A correlation by Wills [8] has been shown by experiment to be accurate in the absence of component shielding.

$$h_x = 0.22\text{Re}_x \frac{\text{Pr}^{1/3}k}{x} \tag{8.49}$$

where x = location in the flow direction along the CCA in inches
k = conductivity of air evaluated at the local air temperature in BTU hr-ft-°F

Numerical thermal analysis is today's standard and many finite element and finite difference programs are available. Only through numerical analysis can a high level of accuracy be achieved. A computer program using the method of Direct Numerical Simulation (DNS) was used to analyze a forced convection CCA. The DNS method is explained in detail in Sec. 8.5. An example of the use of DNS using Eq. 8.49 to define the heat transfer coefficient is shown in Figs. 8.37 through 8.41. Figure 8.37 shows the CCA with many components mounted on its surface. The following four figures show the junction temperatures based on the manufacturers θ_{jc}

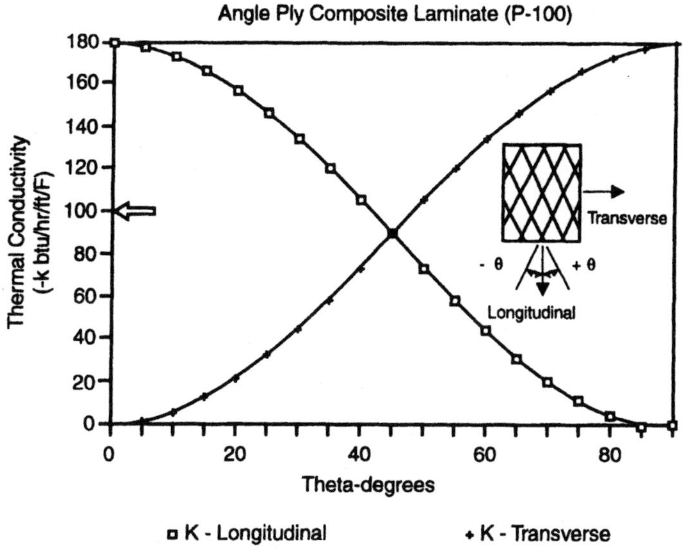

FIGURE 8.34 Effective thermal conductivity of a P-100 epoxy-composite CCA heat sink.

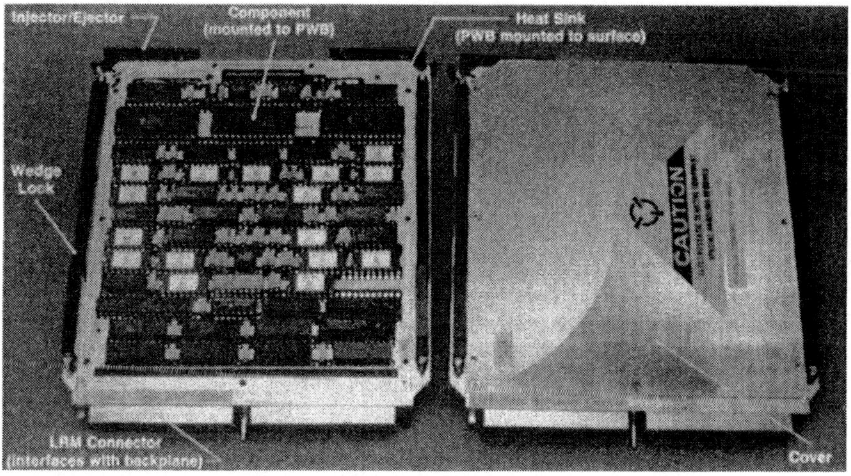

FIGURE 8.35 Conduction-cooled through-hole-type CCA.

values, each component case temperature, the board temperature under the components, and finally the air temperature as it passes across the CCA in the x direction.

Flow-Through-Board Pair Cooling. Forcing coolant to flow between a pair of CCAs is a method used extensively in military electronics. This method is effective for several general reasons: it keeps the coolant from depositing contaminants on

- Design & Fab by COI
- Size: 165mm x 297mm
- Configuration: Single Pass Air Cooled Heat Exchanger, 12 Card Capacity
- Weight: 890 g (Aluminum)
  527 g (COI Composite)
  41% Weight Reduction
- Materials: P75S/954-3 Gr/CE Tape
  K1100/ERL1999 Tape
  6061-T6 Aluminum Rails
- Tested by NAWC: 400 Watt Total Power Input at All 14 Aluminum Rails
- Performance: 20 deg. Lower Operating Temperature than Aluminum

FIGURE 8.36 Advanced composite chassis wall designed to replace aluminum walls *(Courtesy of Composite Optics Inc.).*

the board and components, it gets the coolant very close to the heat source and the electronic part, and liquid or gas can be used as a coolant. In field conditions, air can contain salt, fuel vapors, water vapor, dirt, and other contaminants. Keeping this contamination off the electronic circuitry can add many hours to the reliability of complex equipment. The construction consists of coolant fins sandwiched between two thin faceplates to which two CCAs are bonded so that the components are facing outward. A thin layer of insulation, such as polyimide or Mylar, is used to insulate the CCA circuitry from the metal faceplates. The cooling method and construction is cold plate, which is described in Sec. 8.6. As such, this type of cooling lends itself to more accurate temperature prediction than direct air impingement, which, in turn, allows less-conservative designs, thus saving weight and power.

The DNS method was again used to analyze a board using the equations defined in Sec. 8.7. This analysis is shown in Figs. 8.42 through 8.45. Figure 8.42 shows one CCA on one side of the cold plate. The following three figures show the junction temperatures, the coolant temperature as it passes in the y direction, and a temperature cross-section plot showing temperatures at different levels under the components.

Microchannel and "Macrochannel" Cooling. *Microchannel* cooling defines a technology dealing with very small fins. The absolute dimensions that define microchannel versus simply small fins are not well defined. For general purposes, when the space between fins is 0.005 inches (0.127 mm) or less, the fins can be called *microchannels.* It is always the case that cooler heat source temperatures can be achieved by getting the coolant closer to the heat source. In electronic packaging, the heat source is a chip, resistor, or capacitor, and getting the coolant directly to the part has historically been difficult to impossible. There are several reasons for the difficulty such as contamination of the chips and circuitry; possibility of shorts where liquids are concerned; corrosion, space, and lack of access to modify chip designs by the higher-level packaging engineers. Thus, many chips have to be used as purchased from the foundry. The development of microchannel fin structures [9, 10] has shown the possibility of placing fine-fin structures on the chip or on the substrate on which the chip is mounted.

The large number of fins inherent in microchannel cooling enables a very large heat transfer area at or very near the heat source. It is still not practical to place the fins actually at the heat source because the fins are usually placed on the bottom of the chip while heat is dissipated on or very near, within 2 or 3 microns, of the top of the chip; still, it is a great improvement.

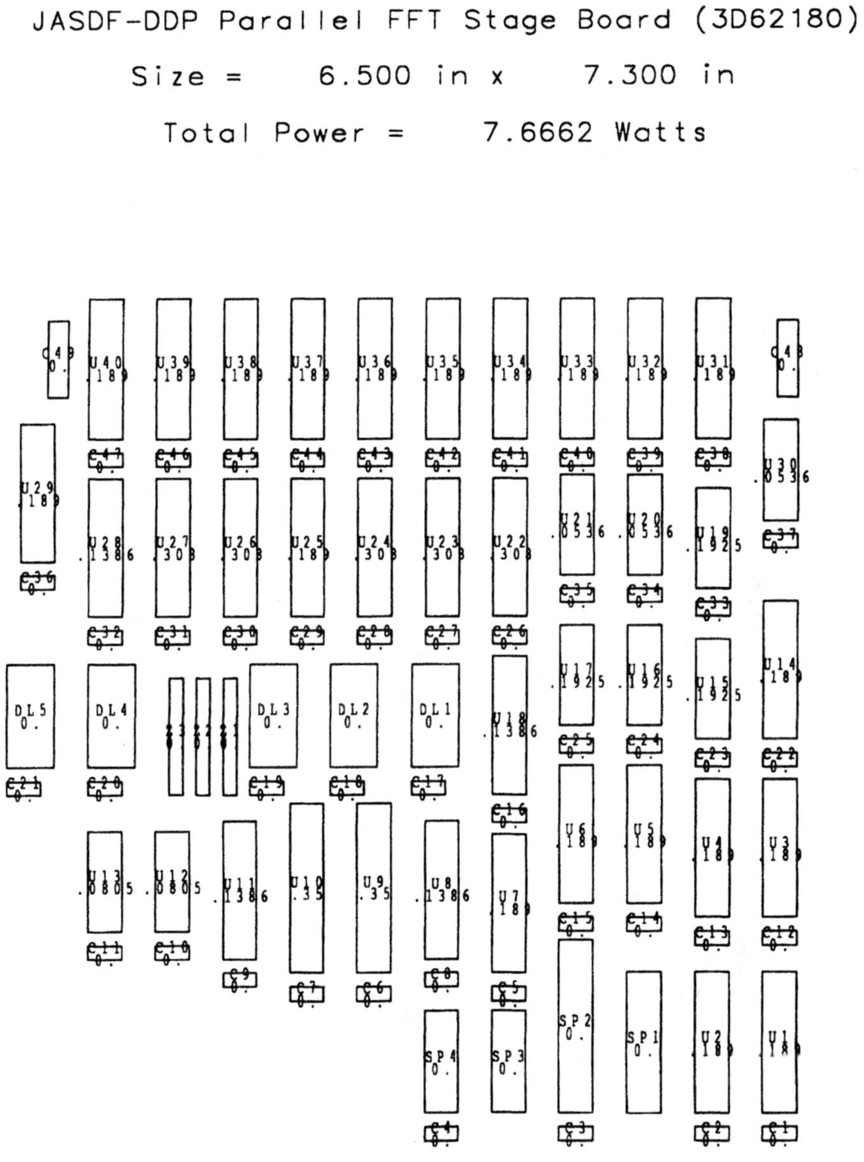

FIGURE 8.37 A CCA used for DNS analysis of direct-impingement cooling of parts on the board.

Design problems must be overcome to successfully use microchannel cooling. The distribution of the liquid to the chip locations in multichip assemblies, such as CCAs and MCPs, can make the assemblies very expensive. The possibility of leaks inside electronic assemblies might not be desirable and can affect yields. Pressure drops can be high. Fluids must be kept very clean to avoid clogging or fouling the small passages and reducing the effectiveness of the fins.

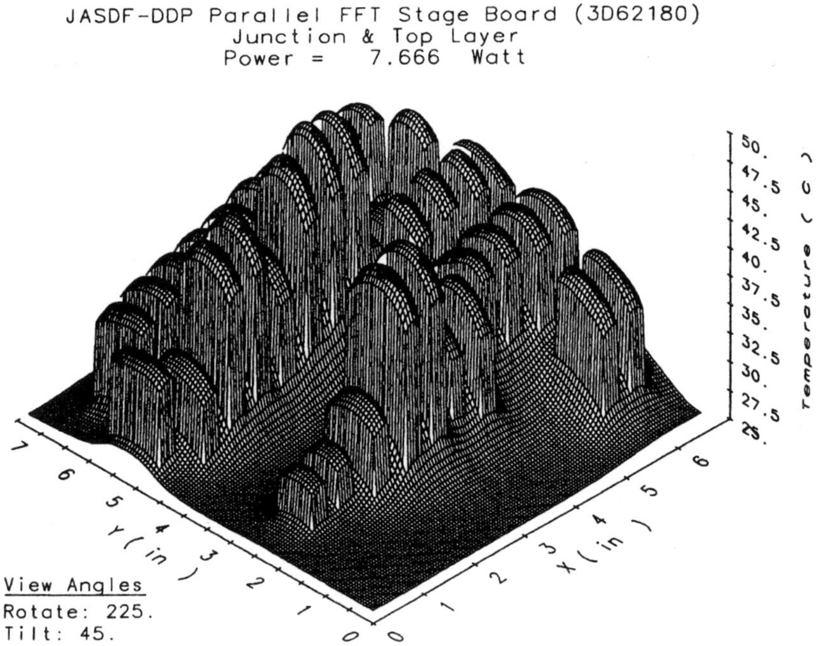

FIGURE 8.38 Computed part junction or hot-spot temperatures.

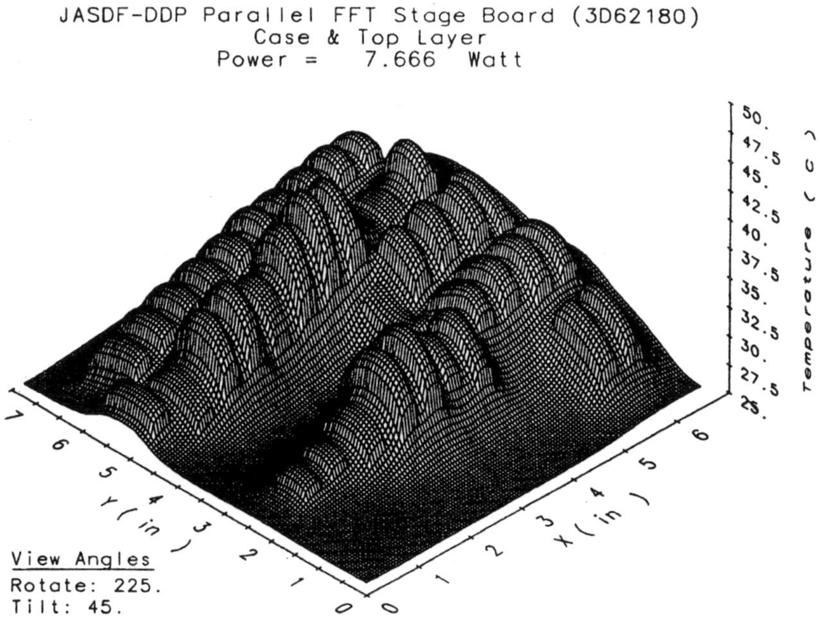

FIGURE 8.39 Temperature of part cases and the CCA board surface.

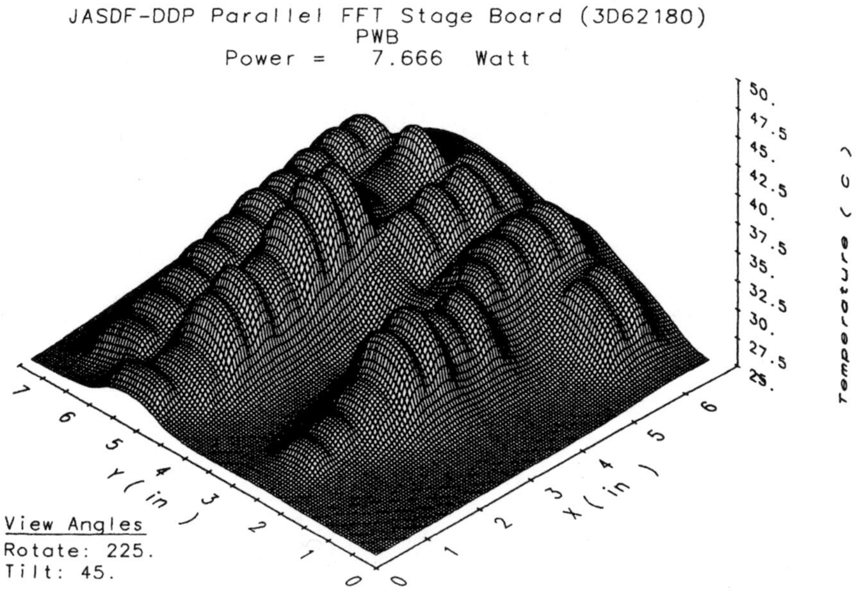

FIGURE 8.40 Temperature of the circuit card surface under the electronic parts.

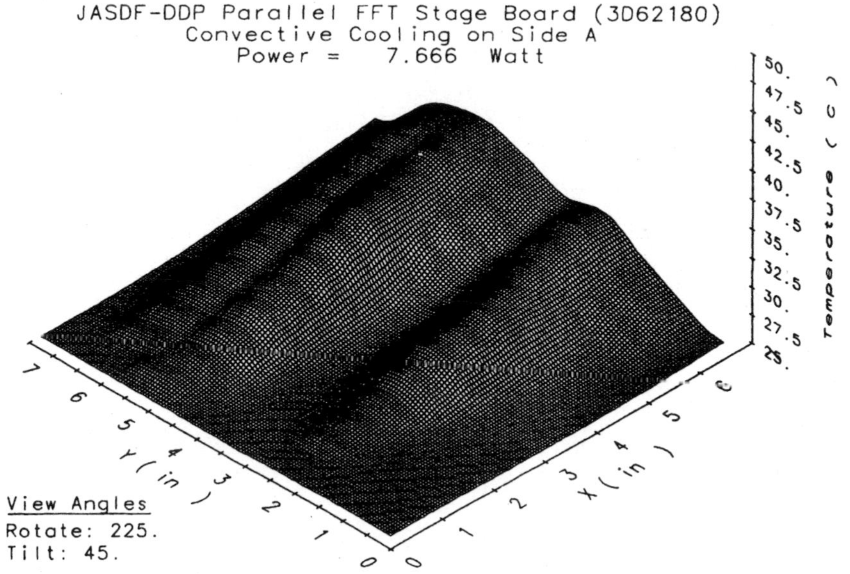

FIGURE 8.41 Air-temperature increase as it cools the electronics passing in the x direction.

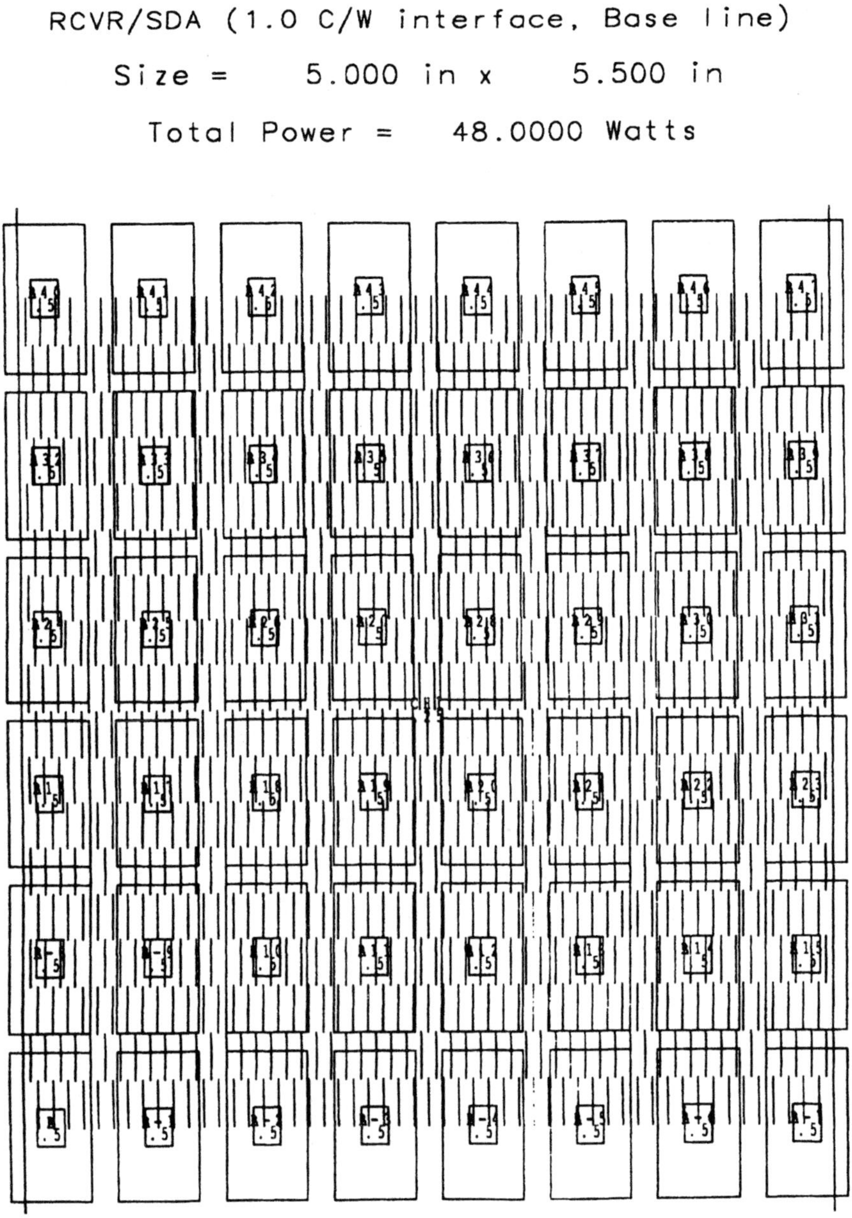

FIGURE 8.42 One side of a CCA pair used to analyze a flow-through-cooled module using DNS.

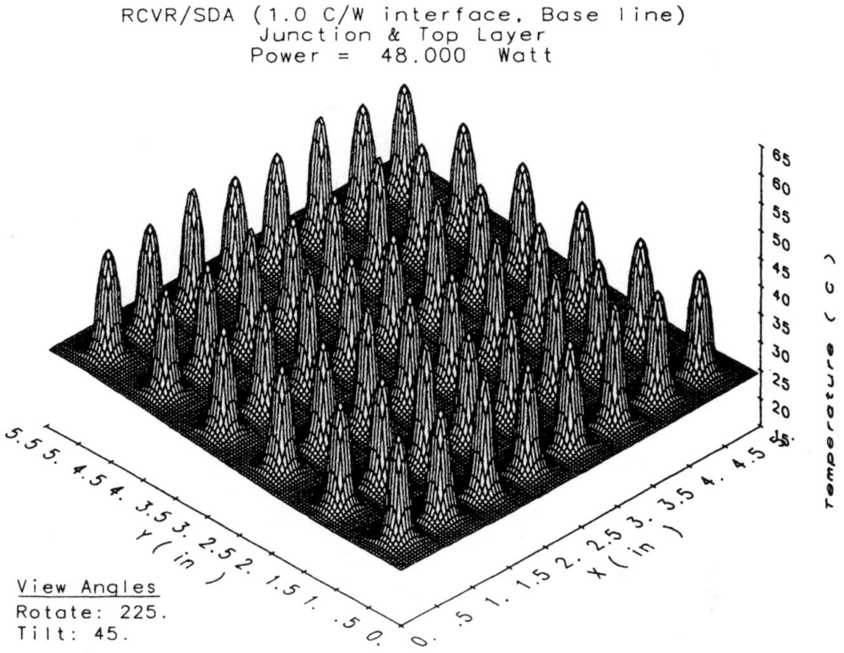

FIGURE 8.43 Junction/hot-spot temperatures of electronic parts on a surface-mount type, flow-through-cooled CCA pair.

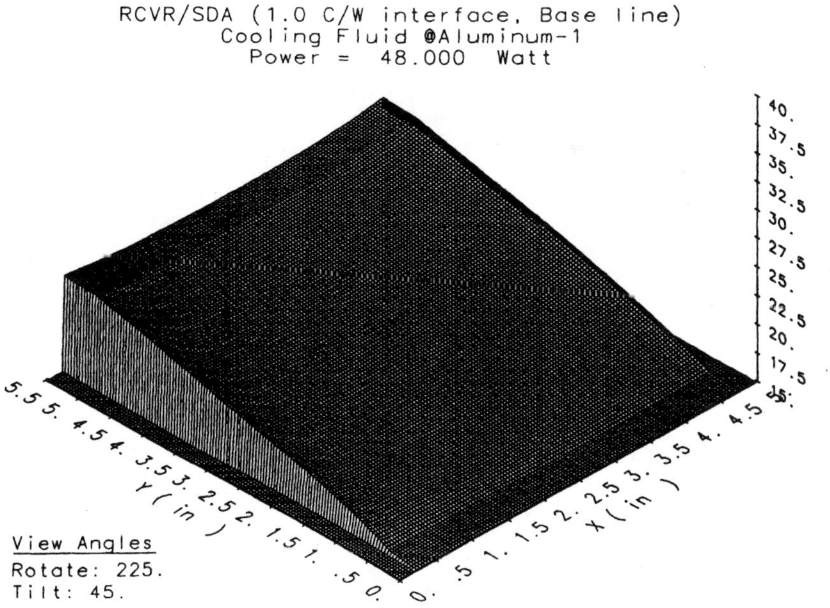

FIGURE 8.44 Temperature increase of air as it passed through the flow-through module.

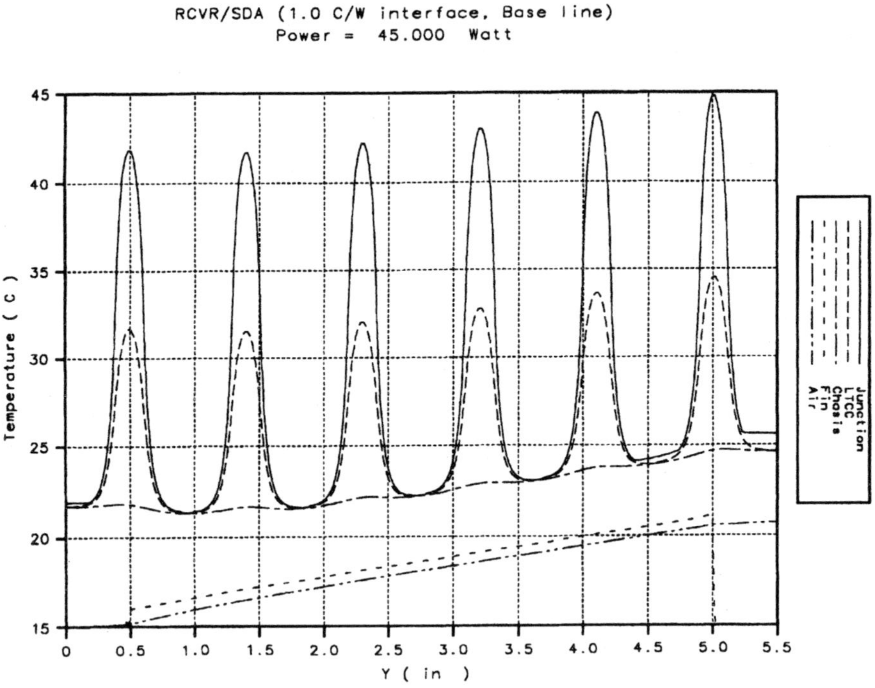

FIGURE 8.45 Temperature profiles at different levels in the CCA.

One class of fins is larger than microchannel (characterized by over 100 fins per inch) and is characterized by more than 25 standard fins per inch. This class is called *macrochannel* for convenience. These fins are too large to be on a chip or an individual device, but are more practical for the package level designer to implement. An example of this type of structure is presented in Figs. 8.46 through 8.49. These figures show the layout of the parts on a power-supply heat sink, which is dissipating 606 watts, and the flow path and fin structure inside the heat sink. Figures 8.48 and 8.49 show the temperatures of the parts and the refrigerated liquid coolant as it passes through the heat sink and absorbs heat. These results were again the result of a DNS analysis.

A promising, but as yet not demonstrated, technology to implement in microchannel cooling is the use of micromachines, such as micropumps (to place the entire coolant loop, including the pump, inside the multichip package or on a circuit card). One concept is shown in Fig. 8.50, where a micropump the size of a dual-inline package provides the work to circulate coolant through a high-power device and the heat exchanger required to cool the liquid after it absorbs the heat from the device.

8.4.3 Natural Convection–Cooled CCAs

Circuit card assemblies are often mounted vertically to induce and enhance natural convection flow of air over the assembly. The natural convection equations presented can be used to define the heat transfer coefficient at any location on the

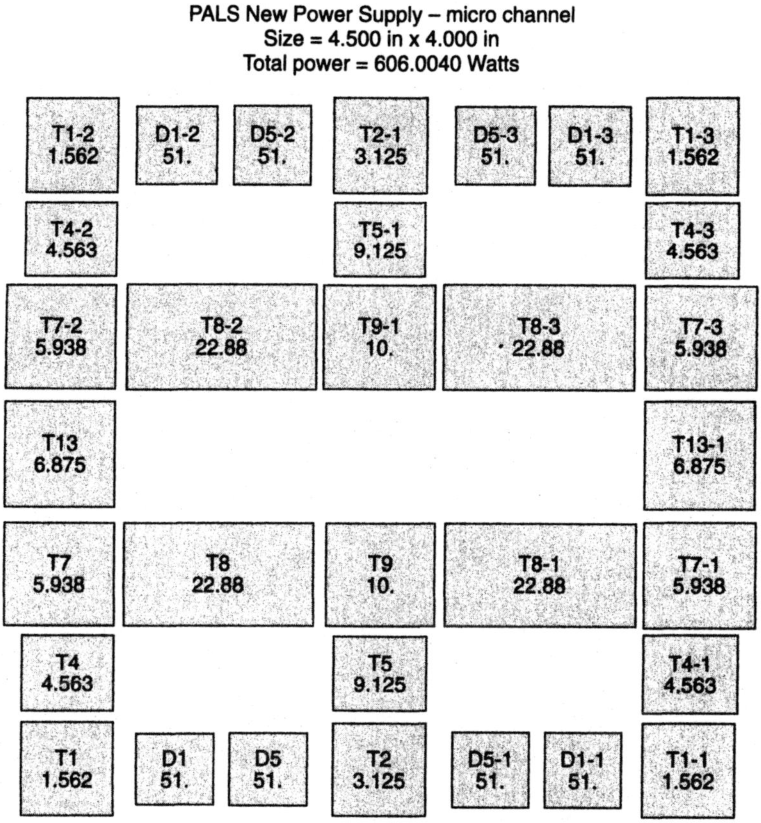

FIGURE 8.46 Parts layout of a microchannel-cooled power supply.

assembly, but presume a knowledge of the air bulk temperature. Most electronic boxes contain multiple circuit cards and air trapped between them increases in temperature as it rises, according to the expression:

$$\Delta T_x = \frac{Q_x}{\dot{m} C_p} \tag{8.50}$$

where Q_x = total heat on the assembly upstream of position x
T_x = air temperature at location x
$\dot{m}$ = mass flow rate of air
C_p = specific heat of air

The temperature rise depends on the mass flow rate, which is driven by the temperature difference between the air within the circuit cards and the outside surrounding air. The air flowing over a single circuit card can be estimated by the expression:

$$\dot{V} = 4.06\mu^{0.5} T_f^{0.5} (T_\infty + 460)^{-0.25} \mathrm{Pr}^{-0.5} (0.952 + \mathrm{Pr})^{-0.25} \Delta T_w^{-0.25} L^{0.75} \tag{8.51}$$

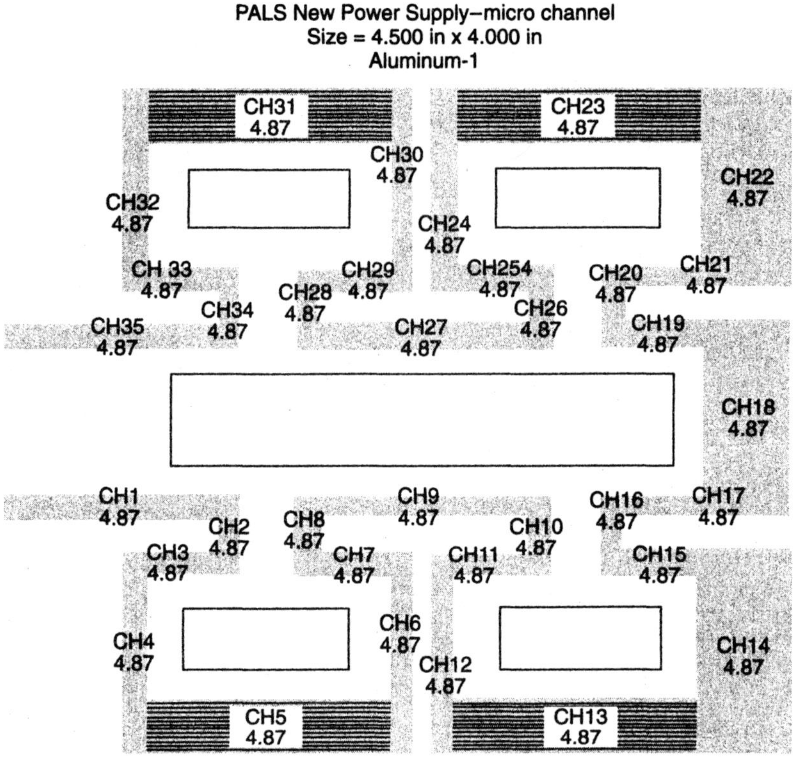

FIGURE 8.47 Fluid channels and microchannel fin structure in the power-supply heat sink.

where $\dot{V}$ = volume flow rate in ft³/min
T_f = average film temperature in °F
T_w = temperature rise of the surface above the ambient
T_∞ = ambient air temperature in °F

However, when air is channelled between boards, the flow rate is determined by a balance between the buoyancy forces and the friction forces. Thus, the actual flow rate is determined by the pressure developed by buoyancy in Eq. 8.3 balanced against the resistance to flow:

$$\Delta P = 2.67\rho L\left[1 - \frac{T_\infty + 460}{\Delta T}\ln\left(1 + \frac{\Delta T}{T\infty + 460}\right)\right] \tag{8.52}$$

where P = pressure difference in inches of water
L = height of the flow channel in inches
T = rise in air temperature through the channel in °F
T_∞ = ambient air outside the channel or at the entrance of the channel in °F

A simple computer program to define the temperature rise between two circuit cards is presented. Note that the expressions presented and used in the computer program are for flat plates, which are smooth, compared to circuit cards, covered

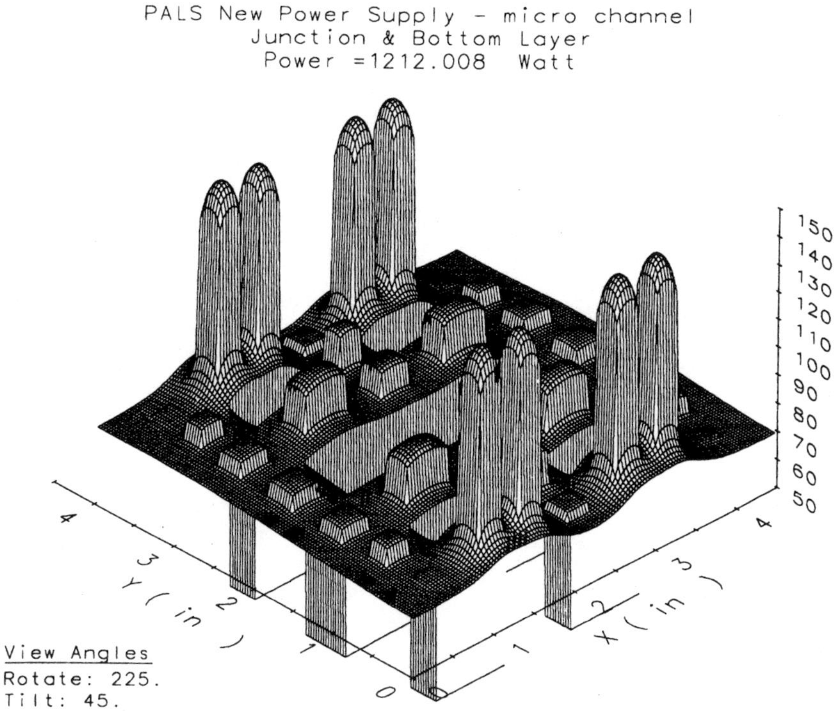

FIGURE 8.48 DNS method–computed part temperatures.

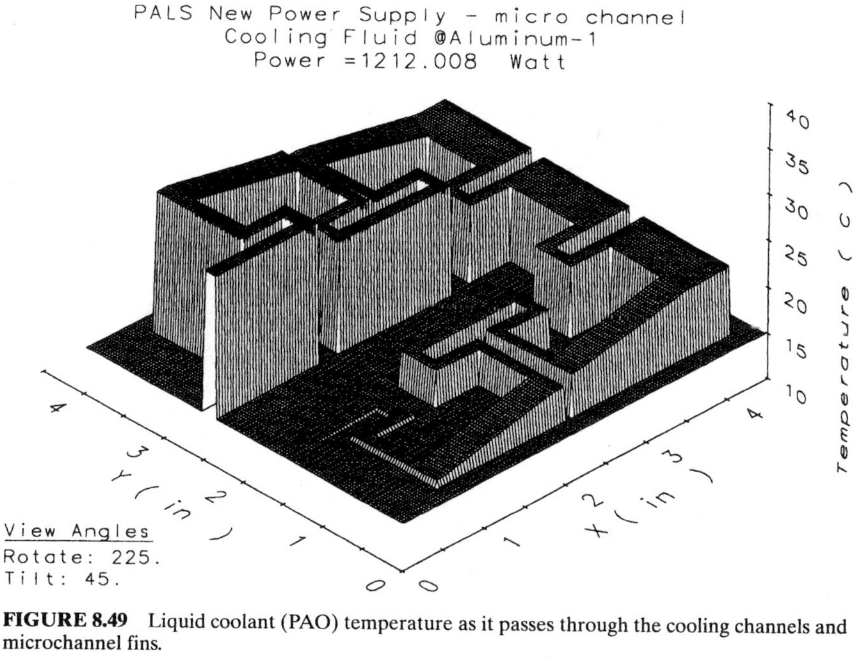

FIGURE 8.49 Liquid coolant (PAO) temperature as it passes through the cooling channels and microchannel fins.

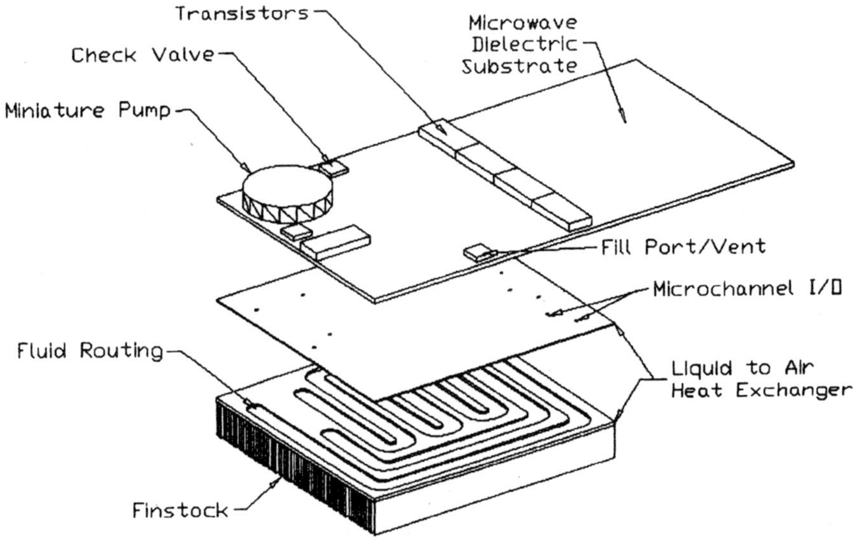

FIGURE 8.50 Micromachine pumping using microchannel cooling.

with components of varying heights and sizes. It is therefore an approximation. Some studies have been done for the specific case of component-covered circuit card assemblies and a perusal of *Advances in Thermal Modeling of Electronic Components and Systems* by Bar-Cohen and Kraus [11] gives additional information. Also be aware that the computer program is not self-limiting: If the circuit cards are moved too far apart, the program will provide flow rates that are too high. The maximum flow rate should be limited to Eq. 8.51.

```
Program is to calculate flow rate and temp. rise of air in natural
'convection between closely space PC boards, which are uniformly
heated.
INPUT "Input ambient temp.(C)"; T
INPUT "Input heat dissipation of board (watt)";
Q Cp = 1 / .1316
Pr = .72
'Estimated film temperature
Tf = T * 1.8 + 32 + 15
INPUT "What is the flow height (inch)"; H
INPUT "What is the air flow space between boards (inch)"; b
INPUT "What is the width of the boards (inch)"; L
A = b * L
Dh = 4 * A / (2 * (L + b))
DT = 10
'k is in watts/inC
1 k = (.0132 + .000024 * Tf) * 1.8 * 12 / 3.413
'Visc is in #/ft.hr.
visc = (3.378E-07 + 6E-10 * Tf - 2.4E-13 * Tf ^ 2) * 115812
'converting visc to #/min.inch
visc = visc / 12 / 60
```

```
'rho is in #/in^3
rho = (.086 * 460 / (Tf + 460)) / 1728!
mdot = Q / Cp / DT
Tf = (T + DT / 2) * 1.8 + 32
Re = mdot * Dh / A / visc
f = 16 / Re
'Buoyancy pressure available
DP1 = 27.67 * rho * H * (1 - (Tf + 460) / (DT * 1.8) * LOG(1 + (DT *
1.8) / (Tf + 460)))
'Friction flow resistance
DP2 = 27.67 * 2 * f * H * mdot ^ 2 / Dh / 386.4 / 3600 / rho / A ^ 2
IF ABS(DP1 - DP2) < .00001 THEN GOSUB pout
IF DP1 > DP2 THEN DT = DT - .05 ELSE DT = DT + .06
GOTO 1
END
pout:
PRINT "Pressure drops =", DP1, DP2
form1$ = " ####.## ###.## #.### ###.##"
PRINT " Height Width Spacing Tambient"
PRINT USING form1$; H; L; b; T
form2$ = " ##.### ##.####### ####.## "
PRINT " mdot DT; Tf"
PRINT USING form2$; mdot; DT; Tf
STOP
END
SUB pout
END SUB
```

Enhancing Natural Convection. There are cases where natural convection is not adequate to cool the circuit card assembly and a cooling enhancement is required without resorting to the installation of a fan or completely changing the design to liquid. One method has been devised and marketed (AAVID) to provide localized enhancement by sandwiching a liquid-filled bag (Fig. 8.51) between the circuit card and the chassis wall or cover or an adjacent circuit card, which has a means of dissipating the heat to a sink. Because natural convection is buoyancy driven, the higher density of liquid provides improved circulation. In addition, the thermal conductivity can be much higher in liquids, thus providing a higher heat transfer coefficient. These bags are very flexible, so good contact is made with each component on the board. The convective heat transfer in the enclosed space of these liquid-filled bags can be computed, but depends on whether they are oriented vertically or horizontally. Also, heat is being transferred to the liquid and then to the other, cooler surface, so the *effective* heat transfer coefficient is different from the coefficient from a single surface to a fluid only. Also, if the bag is oriented horizontally and the upper surface is warmer, no convection currents can be developed and energy transfer must be through conduction only.

For the case of the vertically oriented liquid-filled bags, the Grashof number is calculated as

$$Gr_\delta = \frac{g\beta(T_1 - T_2)\delta^3\rho^2}{\mu^2} \tag{8.53}$$

where δ is the internal space between the two surfaces of the bag (assuming a uniform distance).

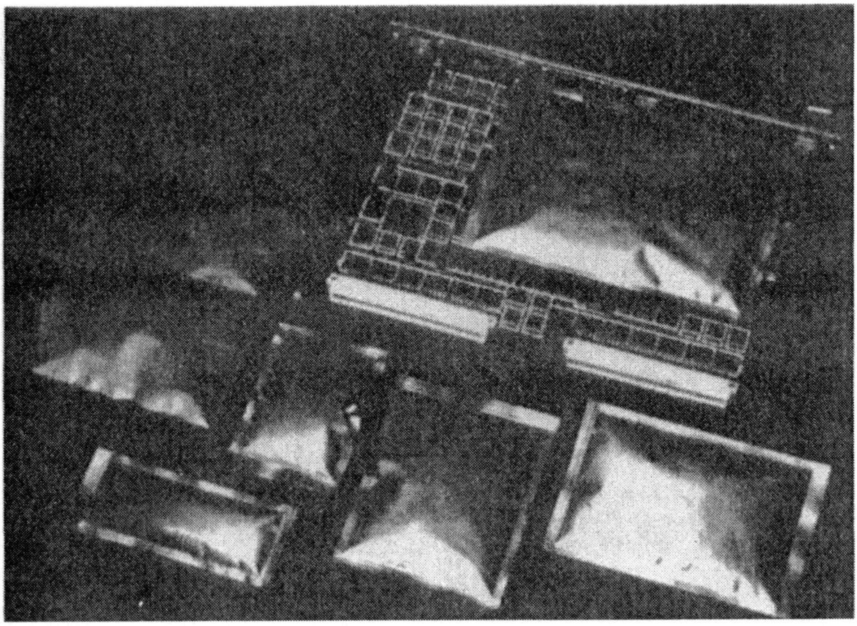

FIGURE 8.51 Liquid-filled heat sink bag used between a circuit card assembly and a chassis wall to provide heat transfer when natural convection is not adequate. *(Courtesy of AAVID, Inc.)*

At low Grashof numbers, the circulation is weak and the temperature difference between the two surfaces is nearly linear through the fluid and is largely a conduction phenomenon. At any Grashof number between 3×10^4 and 1×10^6, a high-velocity circulation is established. For a given geometry, the Grashof number is a function of ΔT; for a desired temperature difference, the Grashof number can be increased by increasing δ or ρ or decreasing μ. The Nusselt number is defined from empirical correlations [12] as

$$N_\delta = 0.42(\mathrm{Gr}_\delta \mathrm{Pr})^{1/4} \mathrm{Pr}^{0.012} (L/\delta)^{-0.03} \tag{8.54}$$

for constant heat-flux conditions and where $10^4 < \mathrm{Gr}_\delta \mathrm{Pr} < 10^7$

$$1 < \mathrm{Pr} < 20{,}000$$

$$1 < L/\delta < 40$$

$$N_\delta = 0.046(\mathrm{Gr}_\delta \mathrm{Pr})^{1/3} \tag{8.55}$$

where

$$10^6 < \mathrm{Gr}_\delta \mathrm{Pr} < 10^9$$

$$1 < \mathrm{Pr} < 20$$

$$1 < L/\delta < 40$$

Then, the heat transfer between the two walls is calculated as

$$Q = A\mathrm{Nu}_\delta \frac{k}{\delta} (T_1 - T_2) \tag{8.56}$$

where A is the area of the heat-dissipating surface.

When the orientation of the two surfaces is horizontal, the following equation can be applied.

$$N_\delta = 0.012(\mathrm{Gr}_\delta \mathrm{Pr})^{0.6} \tag{8.57}$$

When $1700 < \mathrm{Gr}_\delta \mathrm{Pr} < 6000$ and $1 < \mathrm{Pr} < 5000$

$$N_\delta = 0.375(\mathrm{Gr}_\delta \mathrm{Pr})^{0.2} \tag{8.58}$$

When $6000 < \mathrm{Gr}_\delta \mathrm{Pr} < 37{,}000$ and $1 < \mathrm{Pr} < 5000$

$$N_\delta = 0.057(\mathrm{Gr}_\delta \mathrm{Pr})^{0.333} \tag{8.59}$$

When $37{,}000 < \mathrm{Gr}_\delta \mathrm{Pr} < 10^8$ and $1 < \mathrm{Pr} < 20$

$$N_\delta = 0.13(\mathrm{Gr}_\delta \mathrm{Pr})^{0.3} \tag{8.60}$$

When $\mathrm{Gr}_\delta \mathrm{Pr} > 10^8$ and $1 < \mathrm{Pr} < 20$.

The liquid-filled bags can also be used in a two-phased cooling mode (i.e., boiling and condensing usually referred to as *ebullient cooling*). This is covered in the section on ebullient cooling.

Another product designed to aid the transfer of heat to a cover or other structural plate is a sheet of flexible foam-type material (Fig. 8.52). The standard material made by the Bergquist Company of Minneapolis comes in standard widths of 0.02 to 0.125 inches in thickness and is designed to be sandwiched between a CCA and chassis structure that can reject heat to the outside ambient. It has a thermal conductiv-

FIGURE 8.52 The Bergquist Gap Pad material designed to conduct heat from CCA components to a cover or structure. *(Courtesy of The Bergquist Company.)*

ity of 1.75 W/m-°C (at 10 psi) and conforms to the component shape to provide intimate contact to conduct heat to the cover.

8.4.4 Ebullient Cooling of Electronic Devices

Although less often used, ebullient or boiling of a coolant is used in high-power situations to provide very efficient heat transfer. This type of heat transfer is very appealing because of the high heat transfer coefficients it provides, but its implementation requires some special considerations. The simplest method is to totally immerse the electronics in the liquid. This requires that the liquid be a dielectric; in fact, high-voltage applications often use this method because the immersion in a dielectric for arc suppression is already a requirement. The correct liquid must be one with the right dielectric and heat transfer properties. Usually, the heat transfer via boiling is so efficient, compared to other methods, that the liquid is not critical; however, it can be.

The pressure that the container or electronic box must withstand must be considered. The box must be evacuated before filling, so it must withstand atmospheric pressure on the outside. Once filled, the box must withstand the internal vapor pressure of the boiling liquid. Because air has been removed from the box, the pressure is the vapor pressure of the liquid. By definition, a liquid boils when its vapor pressure equals that of its environment. Because its environment is its own vapor pressure, it is always at the boiling point. The maximum pressure the box must withstand is the liquid vapor pressure at the maximum temperature expected. Some vapor pressure curves of dielectric fluids are presented in Fig. 8.53.

To maintain the boiling process, condensation must occur to return the vapor to the fluid state and keep the liquid level covering the electronics. Basic equations for boiling and condensation are now presented, along with some information on surface and fluid parameters required to predict the heat transfer coefficients.

The equation for nucleate pool boiling was developed by Rohsenow [13].

$$\Delta T = \frac{C_{sf} h_{fg} \mathrm{Pr}_1^{1.7}}{Cp_1} \left[\frac{q/A}{\mu_1 h_{fg}} \sqrt{\frac{\sigma}{g(\rho_1 - \rho_v)}} \right]^{0.33} \tag{8.61}$$

where C_{sf} = constant determined from experiment for combinations of liquid and surface material
Cp_1 = specific heat of the liquid
h_{fg} = heat or enthalpy of vaporization of the liquid
Pr = Prandtl number of the saturated liquid
q/A = heat per unit area
μ_1 = viscosity
σ = surface tension of the liquid-vapor interface
ρ_1 = density of the saturated liquid
ρ_v = density of the saturated vapor
g = gravitational constant

Values for C_{sf} for water and various surfaces can be found in Ref. 2. A simplified formulation is also given in Ref. 14 with coefficients for Freons and fluorocarbons.

A critical heat flux occurs with all boiling liquids at which the surface becomes coated with vapor and any increase in flux will cause dramatic increases in temperature; essentially a thermal runaway situation. This critical flux should be avoided and is defined by Zuber [15] and repeated in many texts as

$$q/A = \frac{\Pi}{24} h_{fg} \rho_v \left[\frac{\sigma g^2 (\rho_1 - \rho_v)}{\rho_v^2} \right]^{.25} \left(1 + \frac{\rho_v}{\rho_1} \right)^{0.5} \tag{8.62}$$

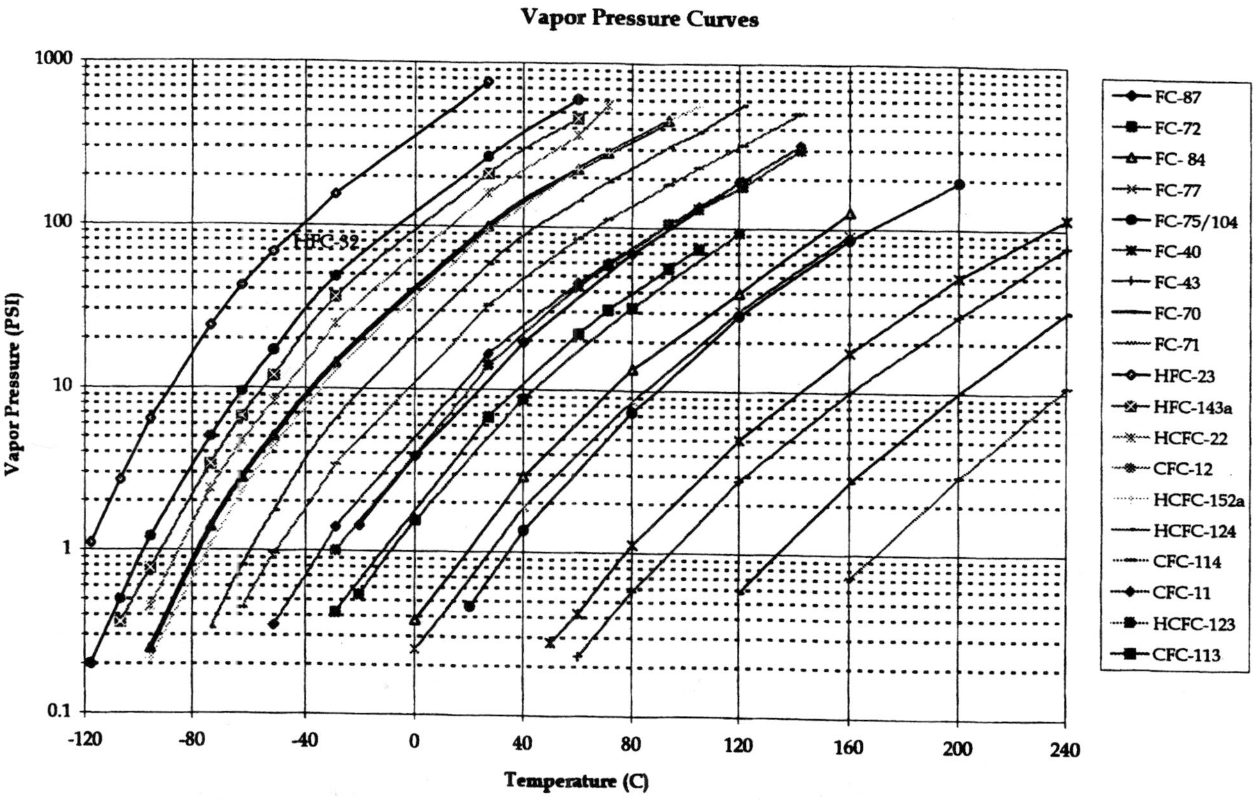

FIGURE 8.53 Vapor pressure curves of selected thermal fluids. *(Data provided by 3M and DuPont.)*

Other types of two-phase cooling, which utilize solid-to-liquid and solid-to-gas phase change, are used in electronics. These are not explored here.

Boiling heat transfer can be used in situations where the vapor is expendable, but the usual situation is one where condensation replenishes the liquid so that a close loop system is used to transfer the heat (via boiling and condensing) to a prime coolant, such as air.

In condensation flow, the transition to turbulent flow occurs at a Reynolds number of 1800. The Reynolds number is calculated as

$$\mathrm{Re}_f = \frac{4\dot{m}}{x\mu_f} \tag{8.63}$$

where x is either the perimeter (if the flow is in a tube) or the flow depth (if the flow is on a vertical wall).

For laminar flow on a vertical wall or tube, the following equation is suggested by McAdams [4], which includes a 20 percent increase in the heat transfer coefficient to account for common disturbances in the flow, which make it depart from true laminar flow.

$$h = 1.13\left[\frac{\rho_1(\rho_1 - \rho_v)g \sin\phi h_{fg}k^3}{L\mu(T_g - T_w)}\right]^{0.25} \tag{8.64}$$

where L is the length of the vertical wall or fin and ϕ is the angle between the wall and vertical, if it is inclined.

If the flow is turbulent, the following equation by Kirkbride [16] is suggested.

$$h = 0.0077\ \mathrm{Re}_f^{0.4}\left[\frac{\rho_1(\rho_1 - \rho_v)gk^3}{\mu^2}\right]^{1/3} \tag{8.65}$$

8.5 NUMERICAL METHODS IN THERMAL ANALYSIS

With the advent of digital computers, it is possible to solve many heat transfer problems by numerical procedures. Such methods permit thermal analysis to be carried out in situations that are too complex for exact, closed-form solution methods or hand calculation. The present section describes numerical methods for thermal analysis of the electronic packages. Many numerical approaches are available.

Many computer programs are available commercially. In general, a physical continuum is replaced by information at discrete spatial and temporal locations called *nodes*. The nodes can be arranged along patterned grid lines, as with finite difference schemes, or in almost-arbitrary arrangements, as with finite element schemes. Both finite difference and finite element approaches are commonly used for conduction problems, whereas the finite difference method is more widely used for convective heat transfer. Many references are available that provide background information in mathematical fundamentals, conduction, convection, and fluid mechanics. (See Refs. 2–4, 11, 13, 17, and 18.)

For those planning to use numerical methods, take several precautions at the outset. First, various means should be used, where possible, to check the quality and validity of a numerical solution. These include mesh refinements, comparison with experimental results, comparison with analytical solutions, and hand checking of results. Accurate prediction of chip-junction temperature is critical in the design process to optimize for cost, weight, and electrical performance, while meeting reliability and integrity requirements. Direct Numerical Simulation (DNS) software provides multichip module thermal analyses with automatic mesh generation, high accuracy, high feature resolution, reduced analyst and cycle times, large node capability (6 million nodes for VAX 2100), and numerous options for the visualization of results. In this section, Direct Numeral Simulation is introduced, along with a detailed thermal trade study of material stack-ups for a high-power GaAs FET used in a multichip module design.

Material choices for substrates, solders, epoxies, and thermal spreaders or tabs are considered in each multichip module design and an accurate thermal analysis tool is required to predict the impact of each material selected. For example, it would be less expensive to bond a substrate to a thermal spreader with conductive epoxy than soldering with a gold-tin preform, but the epoxy can increase junction temperatures significantly. The use of thermal vias in low-temperature or high-temperature ceramic substrates will improve thermal performance, but will increase cost, weight, and possibly lower the yields. A study by T. Poulin et al. [3] empirically determined the conductance of Green Tape via arrays for a specific material stack-up. However, it is shown that the conductance of the entire material stack-up with via arrays depends highly on the interface materials used.

Thermal analysts typically utilize numerical modeling, which is tedious and error prone, when trying to resolve to length scales of chip junctions less than 1.0 micrometer. The methods of modeling complex systems are usually inadequate, are subject to assumptions based on user experience, and can only be validated by comparison with experiment. In contrast, DNS requires no assumptions; thus, the results are more accurate than measurements because of the inaccuracies associated with controlling test parameters, instrumentation, and equipment. Temperature profiles of electronic devices are typically measured with infrared-scanning or liquid-crystal techniques [19], where infrared scanning is simple, but spatial resolution is limited and liquid-crystal phase-transition visualization is somewhat subjective. DNS software has been compared to problems that have exact mathematical solutions and has been proven to be accurate and mathematically valid.

8.5.1 Solution Method

The wide spread of the powerful computer has made DNS extremely valuable for solving problems that are not susceptible to the analytic methods. DNS basically solves the three-dimensional unsteady heat equation, where the equation in tensor notation is

$$\rho Cp \frac{DT}{Dt} = \frac{\partial}{\partial x_j}\left(k \frac{\partial T}{\partial x_j}\right) + Q \tag{8.66}$$

Or where repeated index means summation, Q is the heat generation per unit volume, ρ, Cp, and k are the density, specific heat, and thermal conductivity, respectively. If the equation

$$\rho Cp\left(\frac{\partial T}{\partial t} + U_j \frac{\partial T}{\partial x_j}\right) = \frac{\partial}{\partial x_j}\left(k \frac{\partial T}{\partial x_j}\right) + Q \tag{8.67}$$

is rewritten in cartesian coordinates:

$$\rho Cp\left(\frac{\partial T}{\partial t} + U\frac{\partial T}{\partial x} + V\frac{\partial T}{\partial y} + W\frac{\partial T}{\partial z}\right) = \frac{\partial}{\partial x}\left(k\frac{\partial T}{\partial x}\right) + \frac{\partial}{\partial y}\left(k\frac{\partial T}{\partial y}\right) + \frac{\partial}{\partial z}\left(k\frac{\partial T}{\partial z}\right) + Q \tag{8.68}$$

where U, V, and W are velocities in x, y, and z directions, respectively. Most analyses are performed for steady state operation. The result is the following equation, after integrating with respect to volume:

$$\iint_{\bar{V}}\int \nabla \times k\nabla T\, d\bar{V} + Q \times \text{Vol} = 0 \tag{8.69}$$

where ∇ represents the Laplacian operator. Using Green's divergence theorem equation (Eq. 8.69), it can be simplified to

$$\iint_S k\nabla T \times \vec{n} dS + Q \times \text{Vol} = 0 \tag{8.70}$$

where $\vec{n}$ is a normal vector of the surface. The integral on the left represents the net flow of heat into any arbitrary fixed region. Generally speaking, a closed-form solution for this heat equation can be obtained via the separation of variable technique for the simple boundary condition. The final form of the solution, however, is expressed as double infinite series of the eigen functions, which can only be evaluated numerically. Therefore, instead of seeking an expansive exact solution, we only wish to establish the temperatures at any of the nodal points within the body. Applying Fourier's law of conduction to evaluate the inflow of heat through each of the four boundaries of the control volume about point (i,j,k) gives

$$-k_x\Delta y\Delta z \frac{\partial T}{\partial x}\rangle_{i^{-1/2},j,k} + k_x\Delta y\Delta z \frac{\partial T}{\partial x}\rangle_{i^{+1/2},j,k} - k_y\Delta x\Delta z \frac{\partial T}{\partial y}\rangle_{i,j^{-1/2},k} + k_y\Delta x\Delta z \frac{\partial T}{\partial y}\rangle_{i,j^{-1/2},k}$$
$$- k_z\Delta x\Delta y \frac{\partial T}{\partial z}\rangle_{i,j,k^{-1/2}} + k_z\Delta x\Delta y \frac{\partial T}{\partial z}\rangle_{i,j,k^{+1/2}} + Q_{i,j,k}\Delta x\Delta y\Delta z = 0 \tag{8.71}$$

The ½ in the subscripts refers to evaluation at the boundaries of the control volume which are halfway between mesh points. Approximating the derivatives by central difference:

$$-k_x\Delta y\Delta z \frac{(t_{i-1,j,k} - T_{i,j,k})}{\Delta x} + k_x\Delta y\Delta z \frac{(T_{i+1,j,k} - T_{i,j,k})}{\Delta x} + k_y\Delta x\Delta z \frac{(T_{i,j-1,k} - T_{i,j,k})}{\Delta y}$$
$$+ k_y\Delta x\Delta z \frac{(T_{i,j+1,ik} - T_{i,j,k})}{\Delta y} + k_z\Delta x\Delta y \frac{(T_{i,j,k-1} - T_{i,j,k})}{\Delta z} + k_z\Delta x\Delta y \frac{(T_{i,j,k+1} - T_{i,j,k})}{\Delta z}$$
$$+ Q_{i,j,k}\Delta x\Delta y\Delta z = 0 \tag{8.72}$$

Dividing through by the finite volume,

$$\Delta x\Delta y\Delta z$$

a descretized equation can be obtained for the heat-conduction equation:

$$k_x \frac{(T_{i-1,j,k} - 2T_{i,j,k} + T_{i+1,j,k}}{(\Delta x)^2} + k_y \frac{(T_{i,j-1,k} - 2T_{i,j,k} + T_{i,j+1,k}}{(\Delta y)^2}$$
$$+ k_z \frac{(T_{i,j,k-1} - 2T_{i,j,k} + T_{i,j,k+1}}{(\Delta z)^2} + Q_{i,j,k} = 0 \tag{8.73}$$

This descretized equation can be solved numerically. A large number of textbooks have been written about the numerical method. For a typical module analysis, i and j can run up to 528 and k can go up to 24 to resolve the minimum-length scale (order of micrometer) involved in the module design.

Electronic circuit geometry and material information is entered into a Direct Numerical Simulation data file that is user-friendly and simple to modify. Computer-generated layouts from electrical design or drafting can be transferred directly to DNS data file format to significantly reduce analysis cycle time. Trade studies are generated by simply changing one line in the data file.

8.5.2 Component and Bare Chip Assemblies

In the previous section on the analysis of CCAs, the internal temperatures of the electronic components was characterized by θ_{jc}. Often, the θ_{jc} characterization is overly simplistic. This is especially true when dealing with GaAs MMICs, which respond very quickly to power pulses. In these, the transient nature of the internal temperature rise of the chip must be considered. Also, the characterization of the chip temperature rise by using θ_{jc} implies that the chip is isothermal. Although we know it is not, there is often no information to define it better. This lack of information leads to the treating of the chip as isothermal, which maximizes the heat transfer area used for the chip to CCA or the case to CCA. This results in optimistic temperature predictions.

The need for chips to do more per square inch has resulted in annual increases in the power density that the thermal designer is faced with having to cool. The increased use of GaAs in microwave applications has pushed the watt density (i.e., heat flux) to extreme levels because of the reduced size of the chip circuitry (often submicron dimensions) and increasing absolute-power dissipations. The actual point of maximum power dissipation in these chips is a few microns below the surface of the chip and often reaches a flux greater than 6000 watts/inch [19]. The dimensions are so small that measurement by physical means is nearly impossible. Liquid-crystal measurements have been able to define temperatures in the area around the source/drain sites, but at best give an average. The best measurement is the calibration and measurement of the base-to-emitter voltage V_{be}, which is sensitive to temperature. These measurements have proven reasonably accurate, although some averaging does occur. Analysis using the DNS method, has given excellent results when node sizes are small enough to resolve the minute geometry of the chip circuitry.

In contrast to the high-power GaAs chips, new high-temperature superconducting circuits challenge the thermal designer to keep chips with small power dissipations at or near 75 K (−198°C).

8.5.3 Chip Analysis and Characterization

Accurate representation of bare chips requires a chip thermal analysis, which requires detailed knowledge of the chip construction and layout.

Many new chip technologies are emerging using new chip materials. This is especially true for very high power and very high frequency devices. Some of the new materials being used in chip development are listed here, along with the device technologies of current interest or those showing the most promise. Some of these materials have very high thermal conductivities, which looks nice for the thermal designer, except that they are being made with the expectation of being able to dissipate many times more power than present devices. This puts the cooling problem right back where it started, with the additional problem of having to absorb much more power once it gets off the chip.

GaAs	MESFET, HBT, PHEMT, CHBT, IMPATT
InP	HEMT, HBT, PIN
SiC	SIT, MESFET
InSb	HEMT
SiGe	HBT
GaN	IGFET
Diamond	MESFET, MEMS, VME
YBCO	High-temperature, superconducting Josephson junction

The thermal analysis of a chip of any sort is complicated. Silicon chips, in relation to GaAs MMIC chips, have been easier because of their relatively high thermal conductivity and large heat-dissipating areas. GaAs chips can have extremely small heat-generating volumes with high heat densities. This makes their time constants very short and the evaluation of the response to peak powers during power pulses must be examined. In contrast, silicon chips have time constants long enough that a steady state analysis using average power dissipation is usually adequate.

The complexity of chip structures necessitates analysis using either symmetry or some substructure that is far enough from other structures that no interaction takes place. This type of chip characterization is shown by the analysis of two cells of a GaAs MMIC having the structure shown in Fig. 8.54. This shows the structure like an X-ray picture through the 0.004-inch thickness of the chip around the two cells that each have two sources. The circular structures are vias that extend through the chip, just like plated-through holes in a CCA. The entire structure shown here is 200 × 230 μm. Figures 8.55 through 8.66 show alternately deeper layers into the chip and the temperature profiles of each layer. The actual heat dissipation is several layers down inside the chip and this can be seen in the temperature cross-section view of Fig. 8.67, which shows the hottest location well inside the chip. Notice that this DNS analysis is steady state, but a transient analysis must also be done to define the temperature rise of this hot spot during a peak power pulse. Once the rise is known, a θ_{jc} can be defined as

$$\theta_{jc} = \frac{\Delta T_{\text{peak}}}{\bar{P}} \tag{8.74}$$

This resistance can be used for the next higher level analysis (i.e., chip package, CCA, or multichip package).

8.5.4 Multichip Packages/Chip-on-Board CCAs

This section applies to assemblies referred to by a variety of names (such as MCP, MCM, COB, or IMA), which are all multiple chips on a substrate. The substrate can be Duroid, high-temperature ceramic (HTC and HTCC), low-temperature cofired ceramic (LTCC), or various types of circuit materials used for different applications.

The MCM shown in Fig. 8.68 is constructed using low-temperature cofired ceramic (LTCC) as the substrate and has a variety of passive and active devices, both

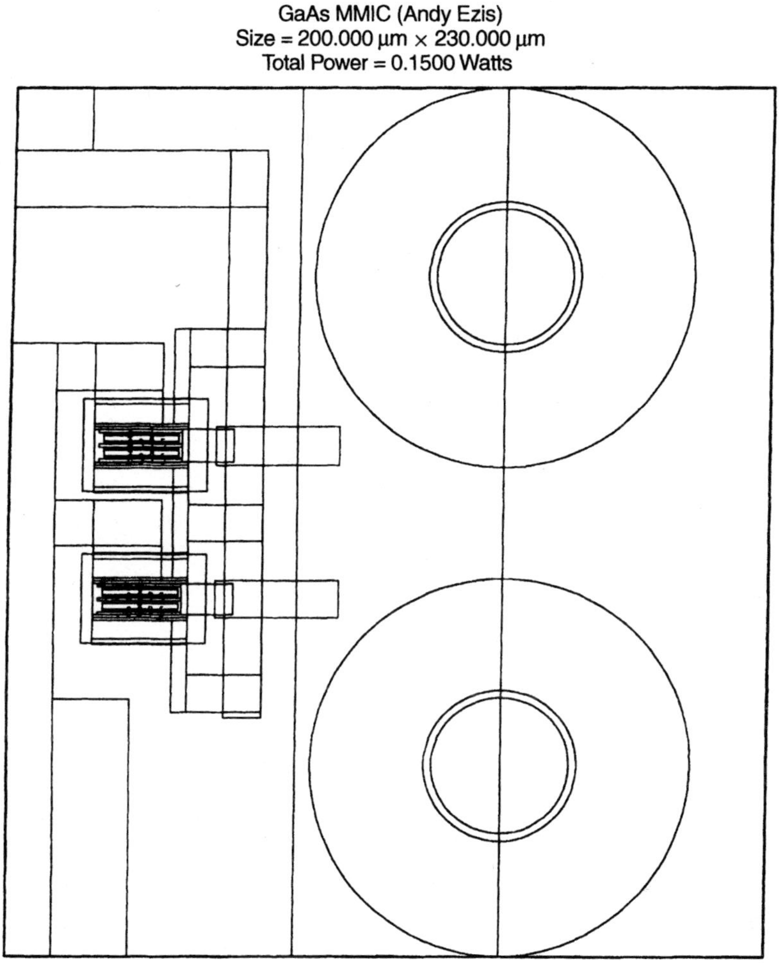

FIGURE 8.54 GaAs chip structure.

digital and microwave. There are also both silicon and GaAs active devices. Figure 8.69 shows how this chip was modeled using DNS. This view shows how vias under high power-dissipating devices are placed to enhance heat transfer through the relatively low conductivity substrate. The effect of the low conductivity of the substrate combined with the comparative short distance for heat to travel downward is illustrated graphically in Figs. 8.70 to 8.72. These show very little interaction between chips, except those that are very close together.

Finally, Fig. 8.73 shows a comparison between the physical thickness of each layer used in the construction and the *thermal thickness*, or temperature rise caused by thicknesses of the materials under component U6 (refer to Fig. 8.69 for the location of U6).

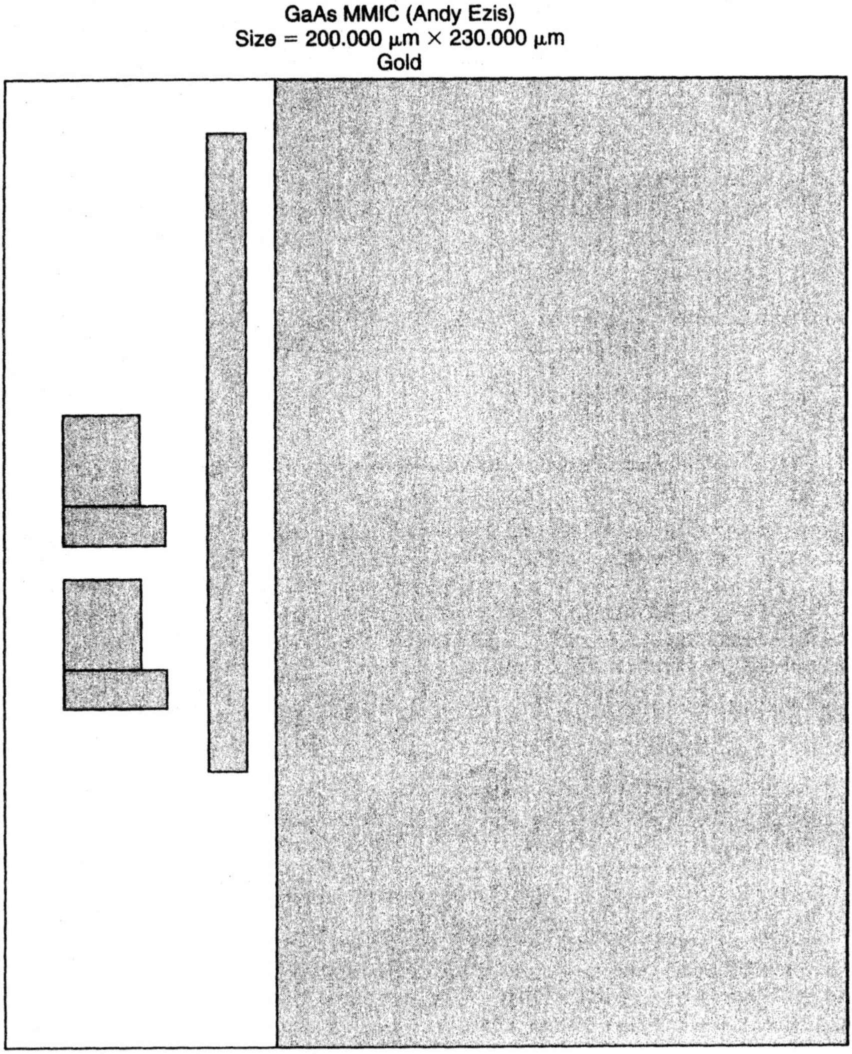

FIGURE 8.55 Top layer of the chip.

8.6 AIR HANDLING

Air motion provides greatly improved heat transfer, compared to natural convection. Used in electronic packaging, the impetus for air motion is provided by either the vehicle or facility in which the electronics is installed or, most usually, by fans or blowers.

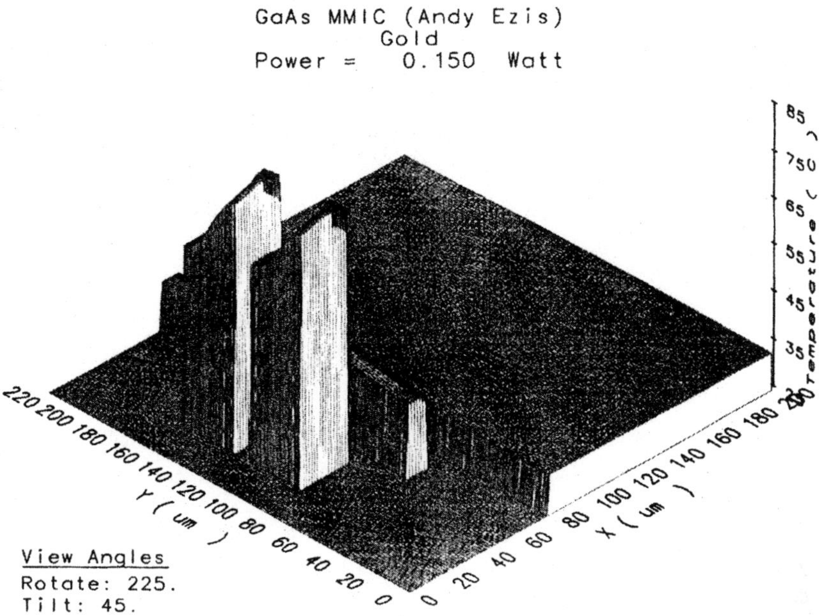

FIGURE 8.56 Temperature of the top of the chip.

8.6.1 Fans and Blowers

Fans and blowers are governed by a simple set of laws, known as the fan laws, which characterize the performance of a blower, with respect to speed, flow, pressure, density, and power. These are as follows:

$$W = K_0 \rho N \tag{8.75}$$

$$\text{CFM} = K_1 N \tag{8.76}$$

$$\Delta P = K_2 \rho N^2 \tag{8.77}$$

$$\text{HP} = K_3 \rho N^3 \tag{8.78}$$

where W = mass flow rate in lb_m/min
ρ = air density in lbm/ft.^{3}
N = fan speed in rpm
CFM = volumetric flow rate in cubic feet per minute
ΔP = pressure developed by the fan in inches of H_2O
HP = horsepower required by the fan motor
K_0 to K_3 = proportionality constants that depend on the fan size and design

The fan laws can be used to find the properties of a desired fan based on known fan properties when the fan type is unchanged. For example:

$$\text{CFM}_2 = \text{CFM}_1 \left(\frac{\text{Size}_2}{\text{Size}_1}\right)^3 \left(\frac{\text{RPM}_2}{\text{RPM}_1}\right) \tag{8.79}$$

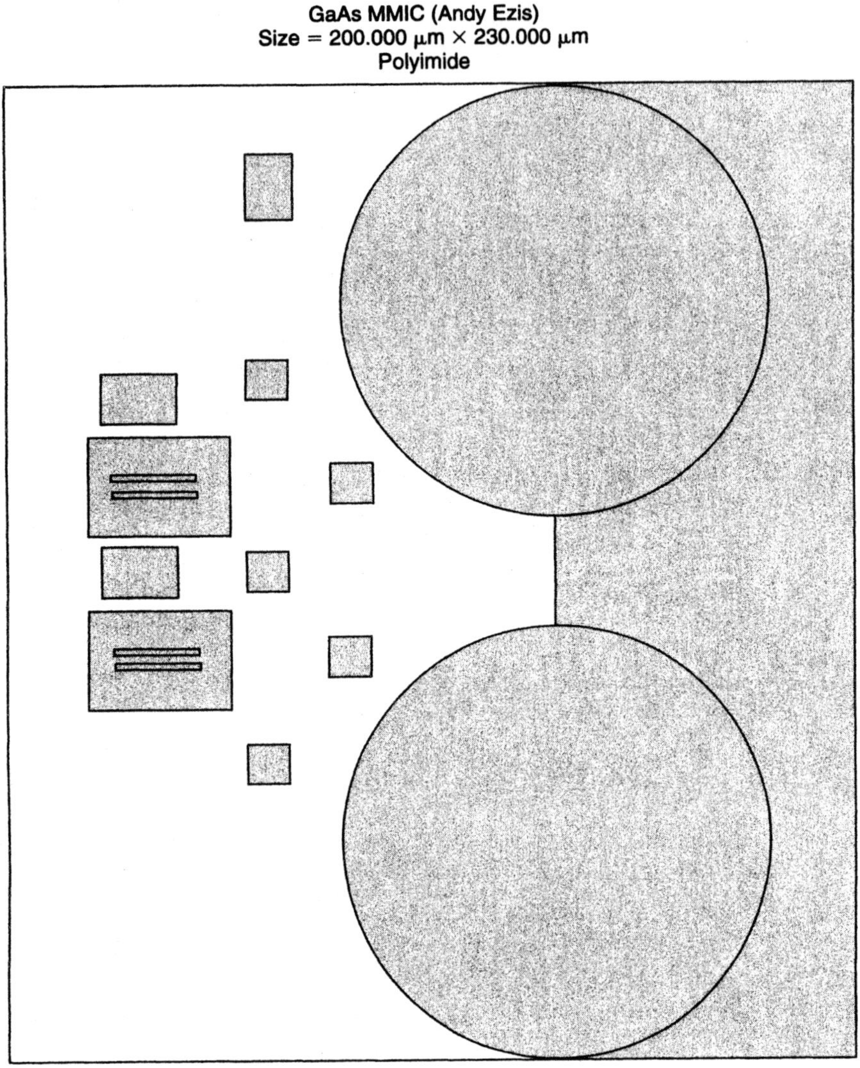

FIGURE 8.57 Chip structure at the polyimide layer.

$$\Delta P_2 = \Delta P_1 \left(\frac{\text{Size}_2}{\text{Size}_1}\right)^2 \left(\frac{\text{RPM}_2}{\text{RPM}_1}\right)^2 \frac{\rho_2}{\rho_1} \tag{8.80}$$

$$\text{HP}_2 = \text{HP}_1 \left(\frac{\text{Size}_2}{\text{Size}_1}\right)^5 \left(\frac{\text{RPM}_2}{\text{RPM}_1}\right)^3 \frac{\rho_2}{\rho_1} \tag{8.81}$$

Equations courtesy of EG&G ROTRON [20].

From Eqs. 8.79 through 8.81, it is easy to see that obtaining high flow rate has a high cost. It is less costly to increase the speed than fan size. Also notice that the

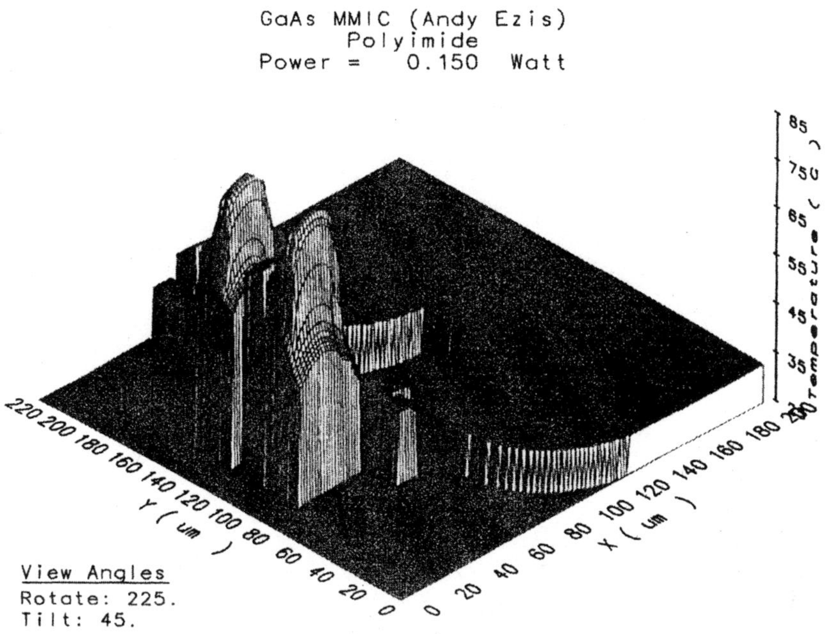

FIGURE 8.58 Temperatures at the polyimide layer of the chip.

CFM is independent of density. When dealing with fan performance at altitude, it is important to remember that the fan is a constant-CFM machine.

When choosing a fan, the flow rate and pressure capability of the fan must be matched to the pressure drop characteristics of the electronic unit or system being cooled. This is illustrated in Fig. 8.74, which shows the pressure of the fan curve dropping with increasing flow and the pressure resistance of the electronic unit increasing with increasing flow rate. Where the two lines cross is the operating point of the system. Knowing the desired flow rate, a fan must be chosen to cross the system operating curve as near the desired operating point as possible. One method of characterizing fans, which is an aid in choosing the most efficient fan for an application, is specific speed N_s. This is defined as follows:

$$N_s = \frac{N(\text{CFM})^{1/2}}{P^{3/4}} \tag{8.82}$$

where N = rpm
CFM = volume flow rate in ft^3/min
P = operating point static pressure in inches of H_2O

The maximum $N = 0.9 \times 120 \times f/P$, where f is the line frequency and P is the number of poles (always even).

The desired pressure and cfm allows the computation of a specific speed, which can be matched to fan characteristics (Fig. 8.75).

The fan curve in Fig. 8.74 shows a smooth fan curve with monotonically decreasing pressure capability as the flow rate increases. This is true for many types of fans

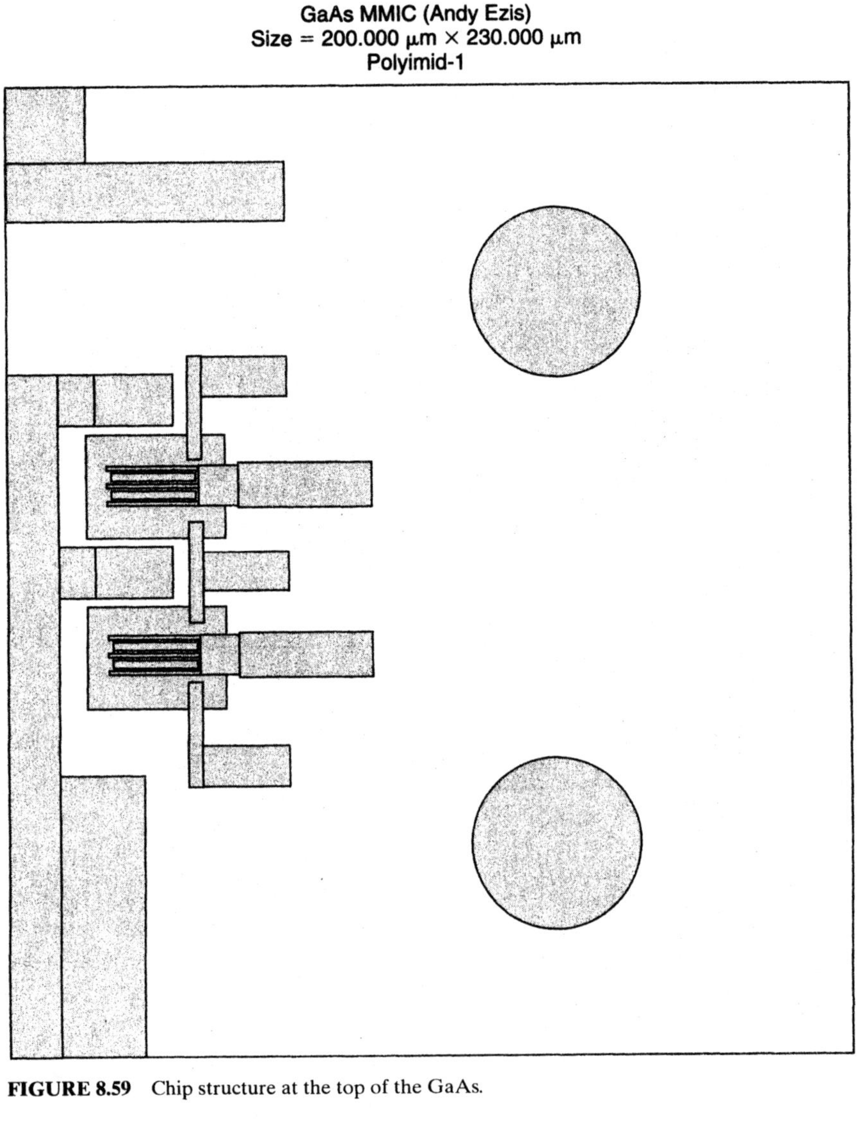

FIGURE 8.59 Chip structure at the top of the GaAs.

and is largely a function of the blade design. Figure 8.76 illustrates the relative shape of the performance curve for three types of impeller designs. Notice that the more forward curved the impeller design, the higher is the free delivery-pressure capability. When a fan has a dip in the performance curve, as is characteristic of forward-curved centrifugal and vaneaxial blowers, it is important to be sure that the system pressure curve does not fall in that area. A lower cfm and or higher pressure blower must be chosen to move the system pressure curve to the right (i.e., into the negative slope area of the blower performance curve).

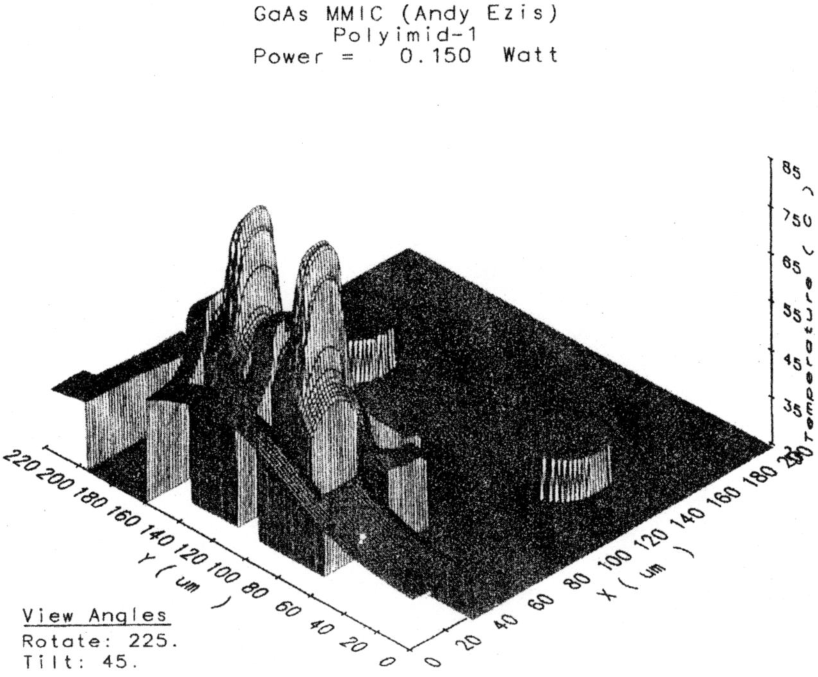

FIGURE 8.60 Temperatures at the top of the GaAs chip.

8.6.2 Altitude Considerations

In the previous section, it was stated that the fan or blower is a constant volume device. Blower motors turn at a constant speed that is determined by the number of commutators and the power line frequency. This means that they always deliver a specified volume flow rate. Because cooling is determined by mass flow rate, the amount of cooling thus varies with density; at altitude, the mass flow rate is, therefore, greatly reduced. The system pressure resistance must be calculated at the lower density and the blower chosen to provide the required cfm, defined by that condition. However, the blower will deliver the same volume flow at sea level (usually, much more than is needed) and the motor, blower size, and power requirements can be very large. As an example, consider the following situation.

Suppose that a 500-watt unit must be cooled at 30,000 feet. It is desired that the air temperature rise be limited to 30°C and a fan chosen to accomplish this.

$$\dot{m} = \frac{0.1316 \times 500}{30} = 2.193 \text{ lb/min}$$

At 30,000 feet, the air density is 0.0261 lb/ft^3 and the unit pressure drop is calculated to be 2 in. of H_2O at this density and flow rate.

At sea level, the density is 0.0706 lb/ft^3. Because both the unit-pressure drop and the fan-pressure capability are directly proportional to density, the fan must supply 84 ft^3/min × 0.0706 lb/ft^3 = 5.93 lb/min at sea level and will require 2 in. of H_2O ×

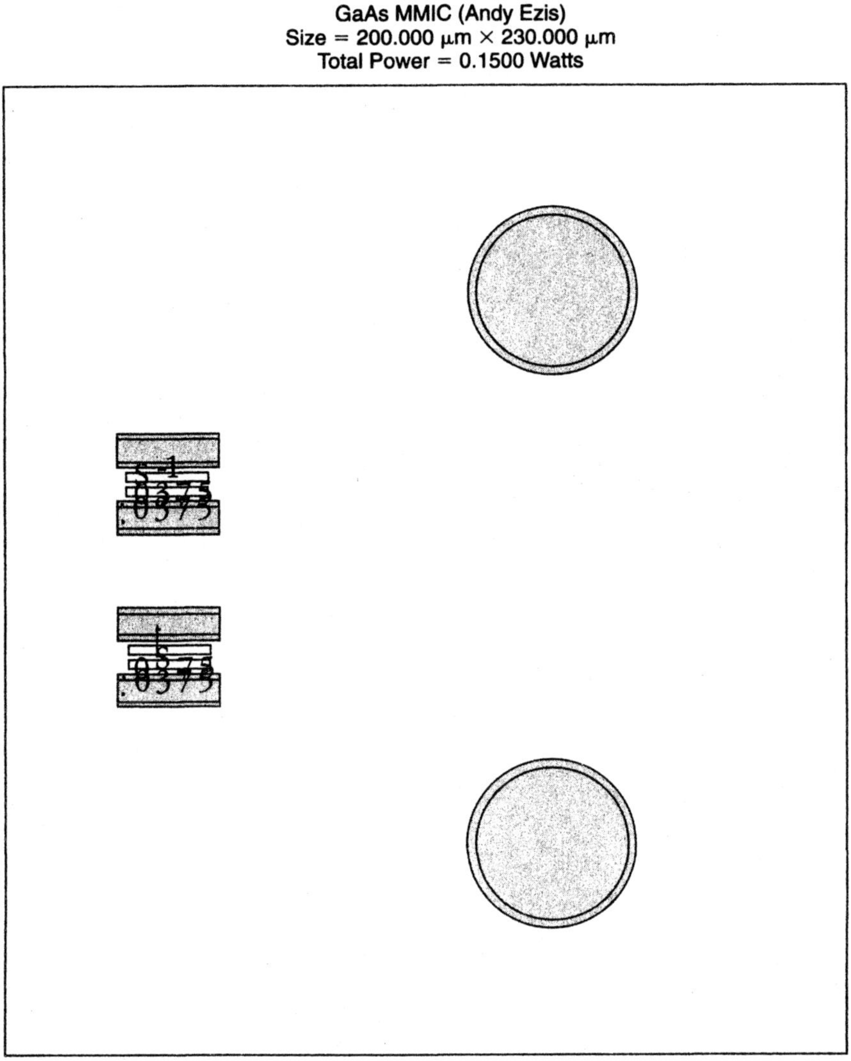

FIGURE 8.61 Structure 0.4 microns inside the GaAs where power is dissipated.

0.0706/0.0261 = 5.4 in. of H_2O. The unit still dissipates 500 watts and the air-temperature rise is only 11°C (i.e., overcooled). Ignoring the motor and impeller efficiency, the power required is

$$HP = \frac{84 \times 5.4 \times 144}{27.67} \text{ ft-lb/min} \times 3.03 \times 10^{-5} = 0.072 \text{ hp}$$

whereas only 0.0266 hp is actually needed to cool the unit (from Eq. 8.81).

To use a fan with the size and power requirements better matched to the cooling needs of the system, two schemes are usually used. The most straightforward solu-

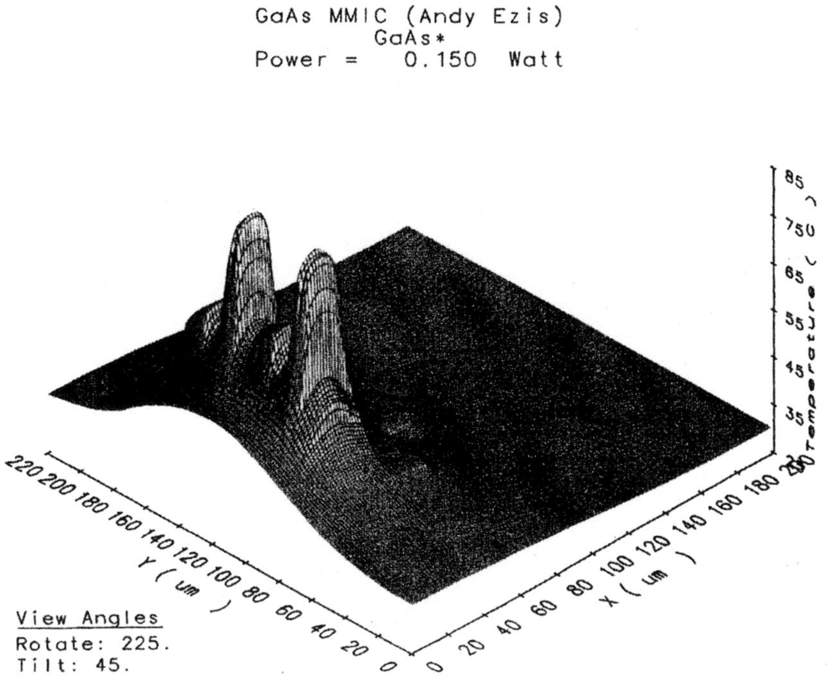

FIGURE 8.62 Temperature at 0.4 microns inside the chip.

tion is to use a two-speed blower. The blower can run at low speed, delivering less cfm at sea level, then, at higher altitudes, can increase its speed to provide the needed cooling. From the fan equations, power requirements increase as rpm cubed. Therefore, the fan should be designed or chosen for the proper size, speed, and altitude to provide low-altitude cooling.

Another method is to use a *limited-slip* blower. This type of construction allows the impeller shaft to turn slower than the motor when there is a high load as a result of high-density air on the impeller. This allows a smaller, lower-power motor because it does not have to do as much work as a nonslip blower at sea level. The amount of slip is determined by the manufacturer.

8.7 COLD-PLATE DESIGN

The cold plate is and has been an important device in cooling electronics. Although its use has been mostly in military electronics, it is now being used more in high-powered areas of commercial electronics, such as power supplies and transmitters. The cold plate is a forced convection–cooled plate that absorbs heat from one or both surfaces (Fig. 8.77). These two surfaces are used to mount electronics. It can be considered to be a simplified heat exchanger and the same methods of design are used.

In practice, the heat sources are circuit card assemblies (CCAs) or subassemblies of electronics, such as occur in power supplies (i.e., transformers, diode assemblies,

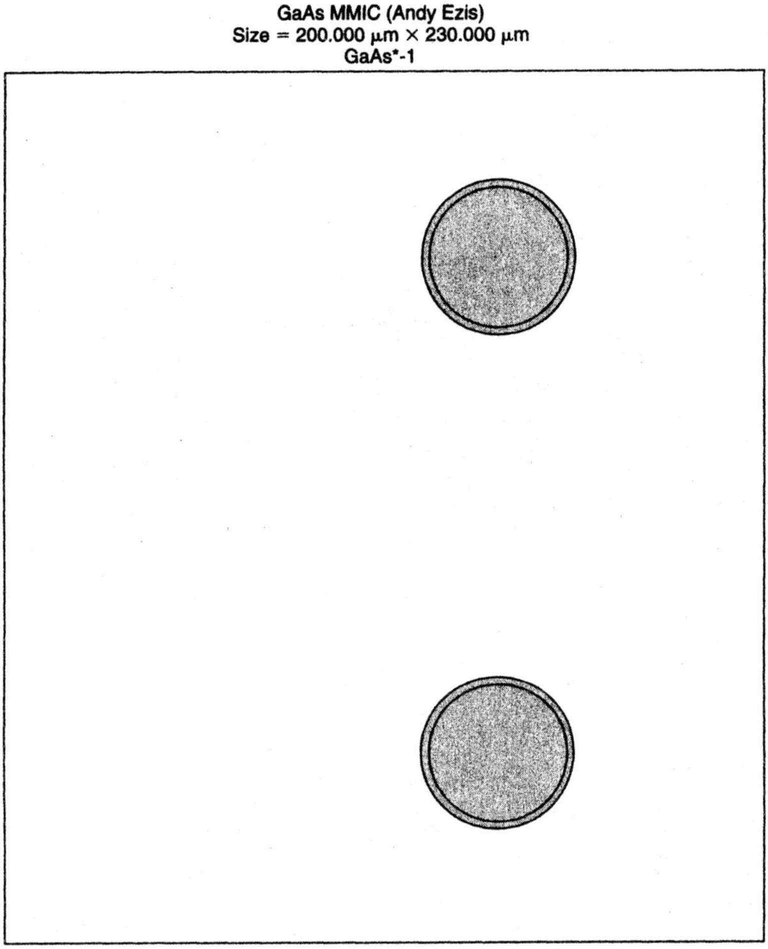

FIGURE 8.63 Chip structure at 1.1 microns.

and power transistors). In any configuration, the heat can be applied fairly uniformly or highly nonuniformly and can be applied on one side or both sides. In addition, there are two standard methods of the design of the cold plate: the most common is called the *NTU method,* which stands for the *number of thermal units* (also called the *effectiveness method*), and the other is commonly called the *rating method.* In either case, two equations generally define the final configuration of the cold plate: the rate equation:

$$Q = hS\,\Delta T_m \tag{8.83}$$

where h = heat transfer coefficient in watts/ft^2-°C
S = heat transfer surface area (wetted area) in ft^2
T_m = log mean temperature difference in °C

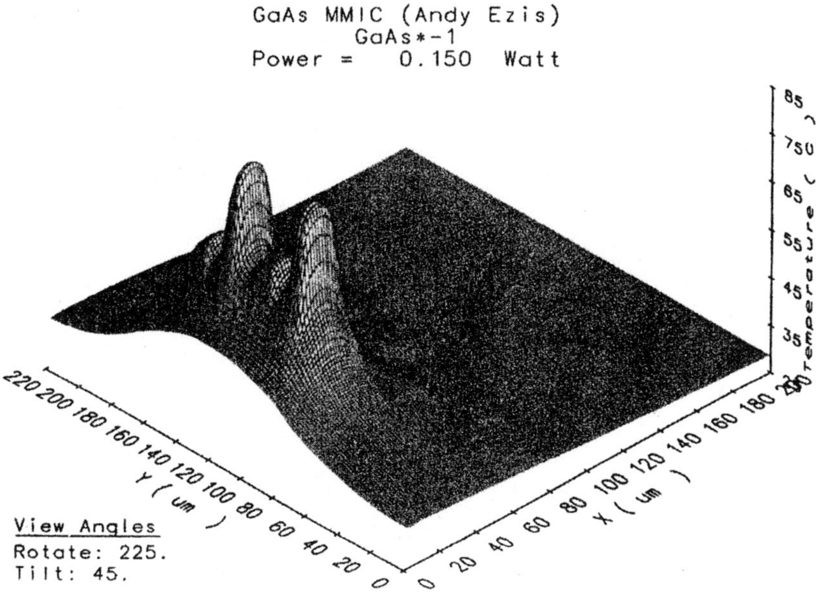

FIGURE 8.64 Chip temperature at 1.1 micron.

and the energy equation:

$$Q = \dot{m}\, C_p(T_{\text{out}} - T_{\text{in}}) \tag{8.84}$$

where m = mass flow of fluid in lb_m/min
C_p = specific heat of the fluid in watt-min/lb_m-°C
T_{out} = outlet temperature of the fluid in °C
T_{in} = inlet temperature of the fluid in °C

Obviously, these two equations must balance when the log mean temperature is defined. The definition depends on the heat distribution or the temperature distribution, as applied to the cold-plate surface. The most useful and straightforward definition is based on an isothermal surface (Fig. 8.78). For this configuration, the log mean temperature difference is defined as

$$\Delta T_m = \frac{(T_2 - T_1)}{\ln(T_0 - T_1/T_0 - T_2)} \tag{8.85}$$

where the temperatures are defined in Fig. 8.78.

By combining Eqs. 9.83, 9.84, and 9.85:

$$Q = \frac{hS(T_2 - T_1)}{\ln(T_0 - T_1/T_0 - T_2)} = \dot{m}C_p(T_0 - T_1) \tag{8.86}$$

from which the definition of NTU is derived:

$$\text{NTU} = \frac{hS}{\dot{m}C_p} = \ln\frac{T_0 - T_1}{T_0 - T_2} \tag{8.87}$$

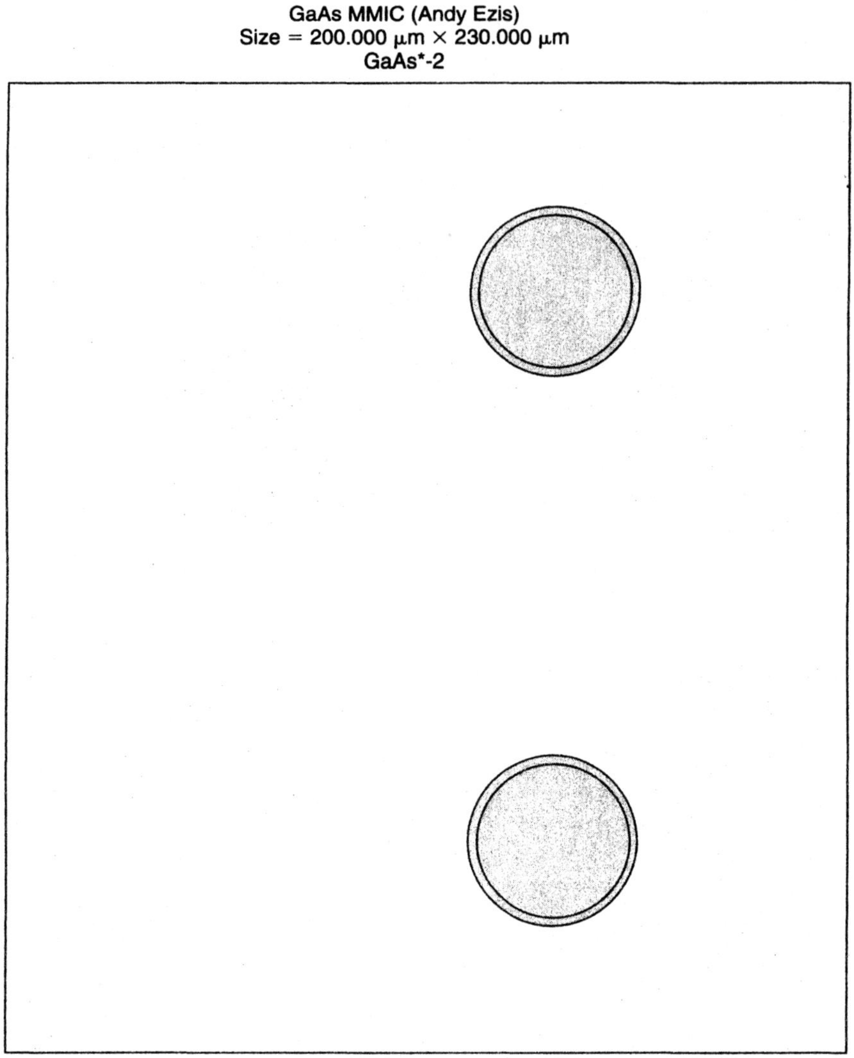

FIGURE 8.65 Chip structure at 2.1 microns.

Although this equation expresses all the properties of the cold plate, the actual design is a trial-and-error process, defined by the available space, fluid temperatures, and flow rates available; amount of pressure available; and, of course, the desired temperature of the cold plate T_0. Often, a trial configuration is defined, based on the space available to define an approximate hS, the flow rate is defined by the pressure drop (as covered in Sec. 8.2.3) and T_1 and T_2 (as defined by the energy equation, 8.84). This allows solution for T_0.

Equation 8.87 is often used in a different form by taking the antilog to get

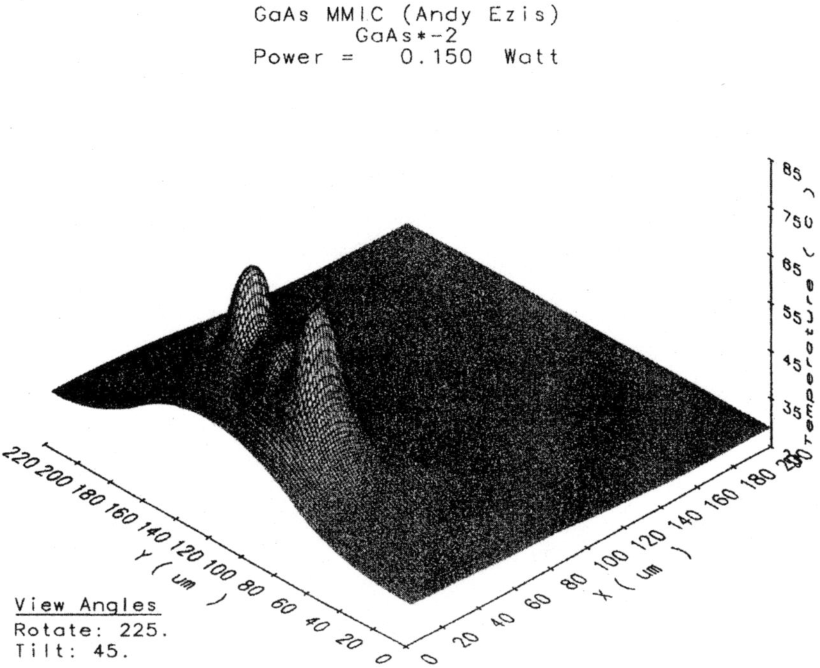

FIGURE 8.66 Chip temperature at 2.1 microns.

$$e^{\mathrm{NTU}} = \frac{T_0 - T_1}{T_0 - T_2} \tag{8.88}$$

A method used to get an initial trial sizing is to start with the definition of effectiveness for a cold plate ε:

$$\varepsilon = (1 - e^{-\mathrm{NTU}}) \tag{8.89}$$

and then choose a reasonable effectiveness (between 0.5 and 0.85) to solve for NTU and then use Eq. 8.86 or 8.87 to solve for other parameters. If the design is started with a calculation of the required effectiveness, it is called the *NTU* or *effectiveness method.* If the design is started from an assumed configuration and all heat transfer parameters are used, then refined to converge on a final configuration, the method is called the *rating method* (because it is based on the rate equation).

The previous text is based on an isothermal surface and is a good approximation of actual surface temperatures in most cases. The only way to obtain a truly isothermal cold plate is to have a very high thermal-conductivity material or to simulate it with a heat pipe. A condenser heat exchanger also provides an isothermal surface. Actual surface temperatures can vary and depend on the uniformity of the heat dissipation and the thickness and conductivity of the cold-plate surface material (faceplate thickness). For thin faceplates, the temperature profile might look more like that shown in Fig. 8.79, which reduces the actual log mean temperature difference and requires a slightly larger value of NTU to yield the same cold-plate surface tem-

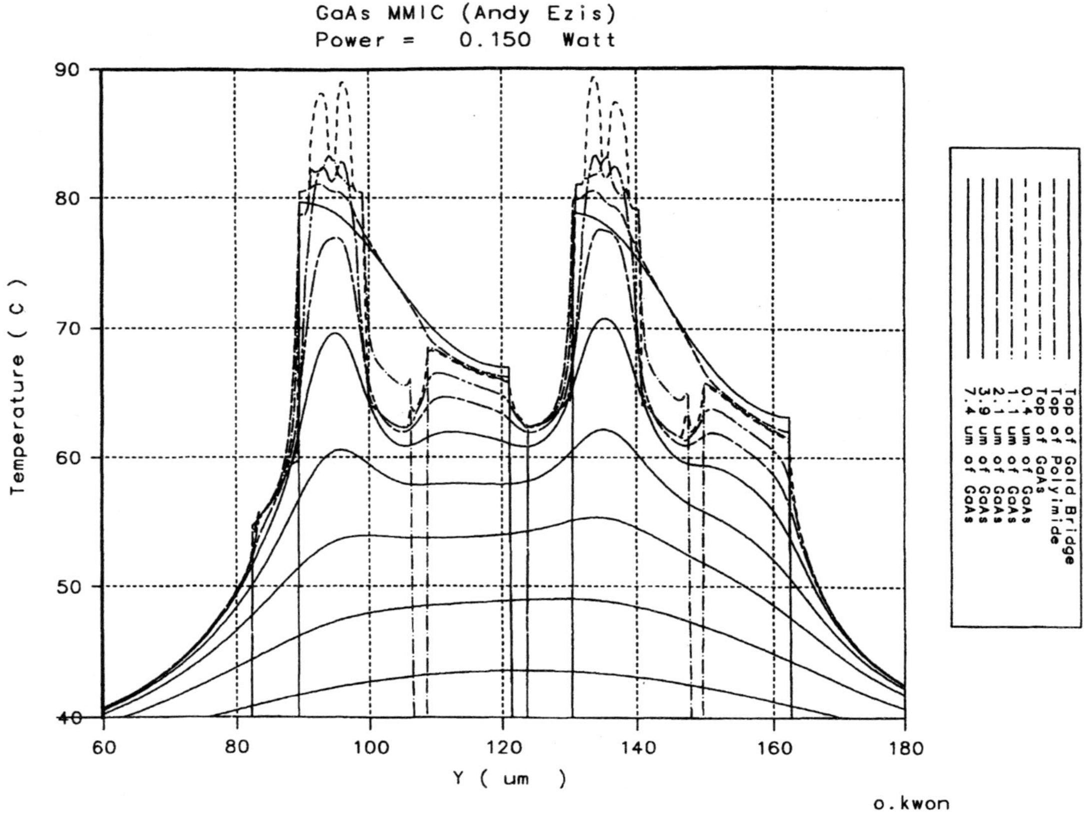

FIGURE 8.67 Chip-temperature profiles at all analysis levels showing the hot spots to be well inside the chip surface.

FIGURE 8.68 Multichip module (MCM) using a low-temperature cofired ceramic (LTCC) substrate.

perature. In this case, the more general definition of the log mean temperature difference used in heat-exchanger design should be used.

$$\Delta T_m = \frac{(T_{01} - T_1) - (T_{02} - T_2)}{\ln (T_{01} - T_1/T_{02} - T_2)} \tag{8.90}$$

The choice of the fins for the core of the cold plate is the heart of the design effort. The core material and the flow rate determine the hS and ΔP from which all the temperatures are determined. The fins can be pin, straight, wavy, strip (lanced-and-offset), or some combination. A comprehensive treatment of test data for many types of fins is presented in *Compact Heat Exchangers* by Kays and London [21].

Data for a strip-fin core is presented in Fig. 8.80 from Kays and London [21]. This presents the fanning friction factor on the upper curve and the heat transfer expressed as the product of the Stanton number and the Prandtl number to the two-thirds power (i.e., the Colburn Modulus). A lot of analysis of test data has been performed to define correlation equations usable for design, based on extensive test data available for air. The most common type of cold-plate core is either the strip fin or the straight fin. For straight fins having an aspect ratio α (defined as w/h in Fig. 8.81), of 3.0 to 4.0, Altoz in Ref. 14 gives the following expressions:

$\text{Re} \leq 2000$ and $L/D_h = 100$

$$f = \frac{18}{\text{Re}} \tag{8.91}$$

$$j = \frac{5.0}{\text{Re}} \tag{8.92}$$

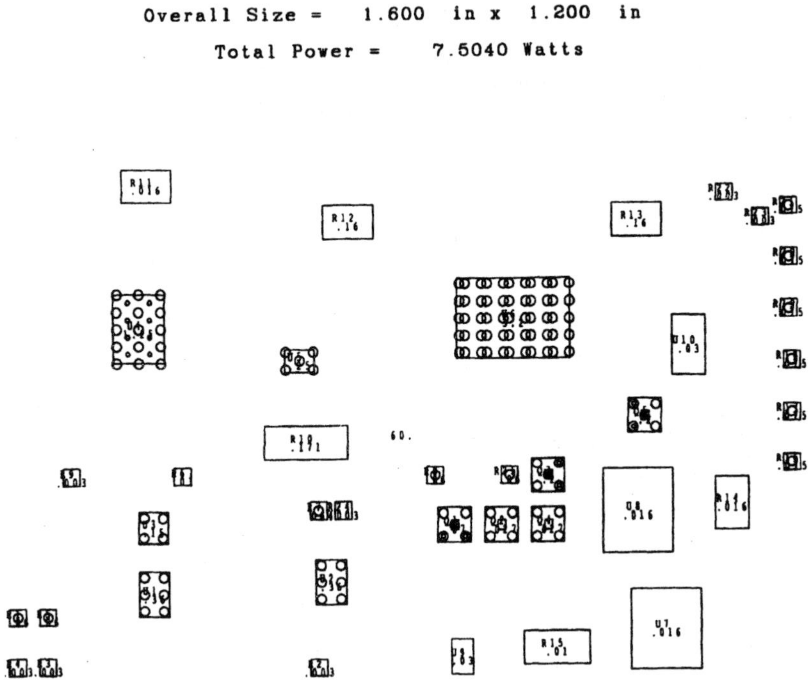

FIGURE 8.69 Thermal model of the MCM shown in Fig. 8.68 used for DNS thermal analysis.

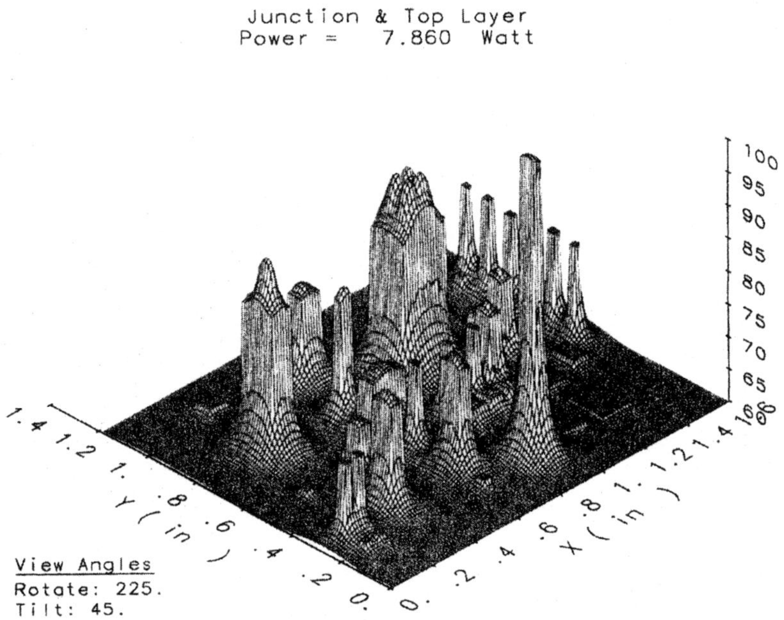

FIGURE 8.70 Junction and hot-spot temperatures of all the components mounted on the LTCC substrate.

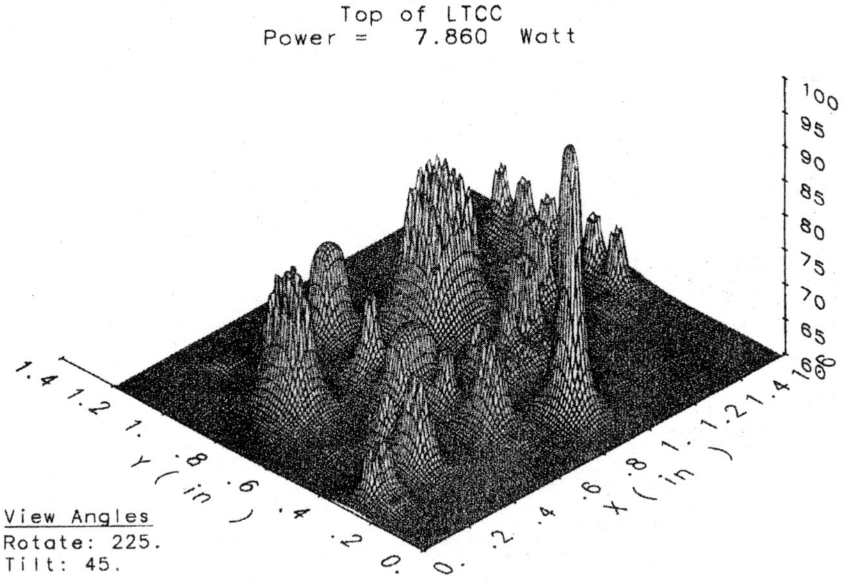

FIGURE 8.71 Top of the LTCC showing the effect of vias on depressing temperatures under chips.

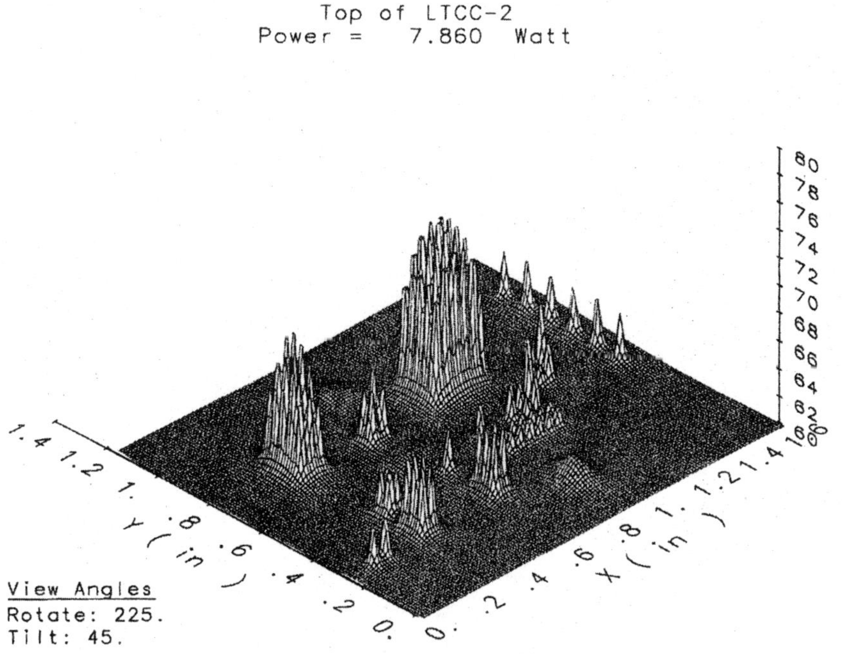

FIGURE 8.72 Bottom of the LTCC showing the increased temperature at via locations because of localized heat transfer supplied by the vias.

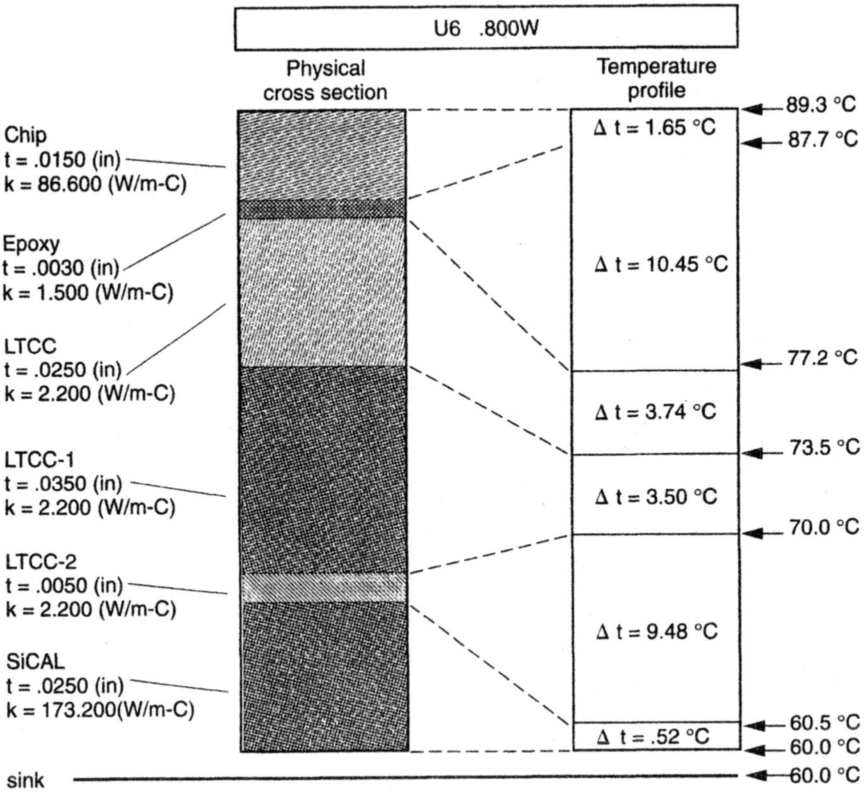

FIGURE 8.73 Physical and thermal stack-up diagram illustrates and aids in visualization of the sources of high resistance.

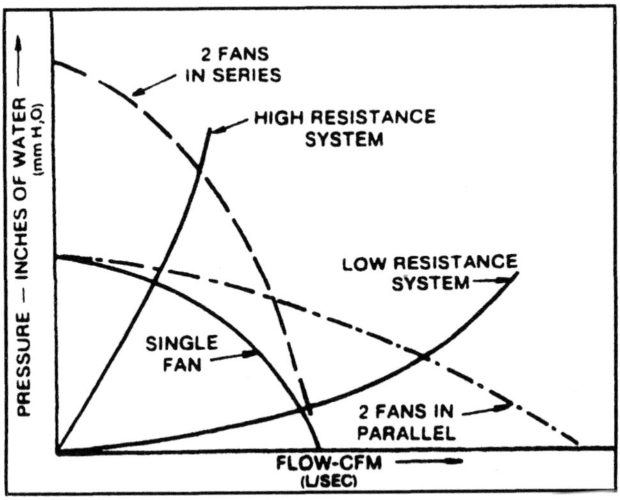

FIGURE 8.74 Fan pressure and system resistance must be matched to determine actual airflow. *(Courtesy of EG&G Rotron.)*

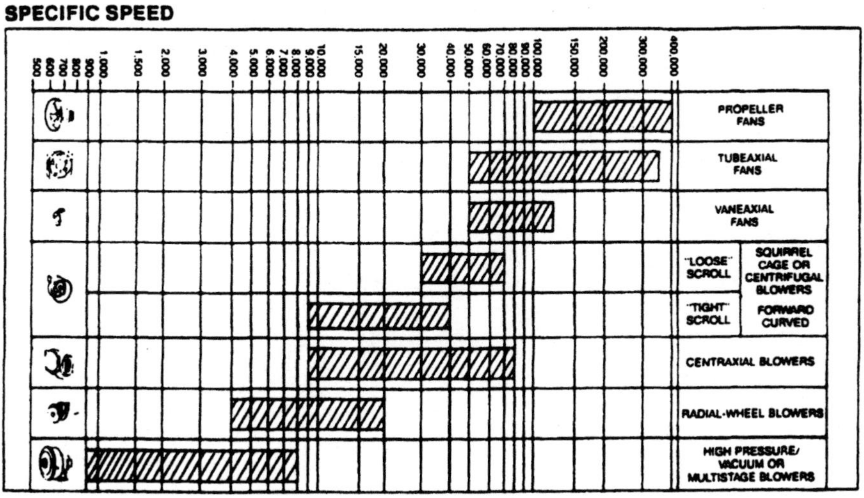

FIGURE 8.75 Specific speed characteristics of the fan or blower is an initial indication of the type of blower needed for the system. *(Courtesy of EG&G Rotron.)*

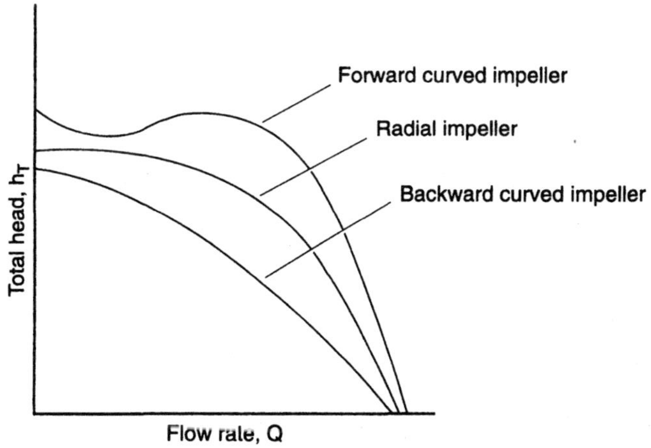

FIGURE 8.76 Characteristic shapes of fan curves relative to blade design.

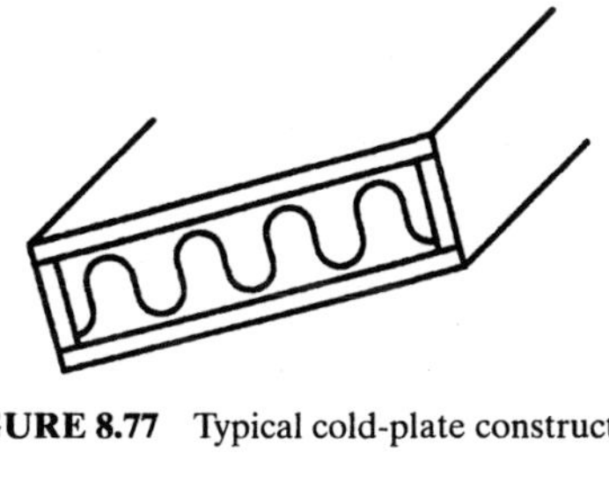

FIGURE 8.77 Typical cold-plate construction.

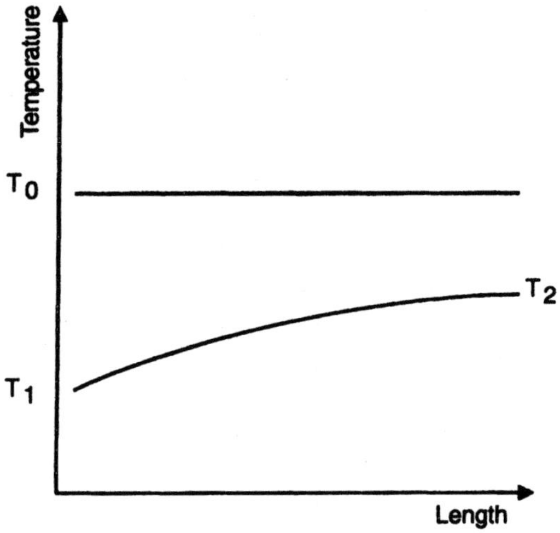

FIGURE 8.78 Typical fluid and cold-plate temperature profile.

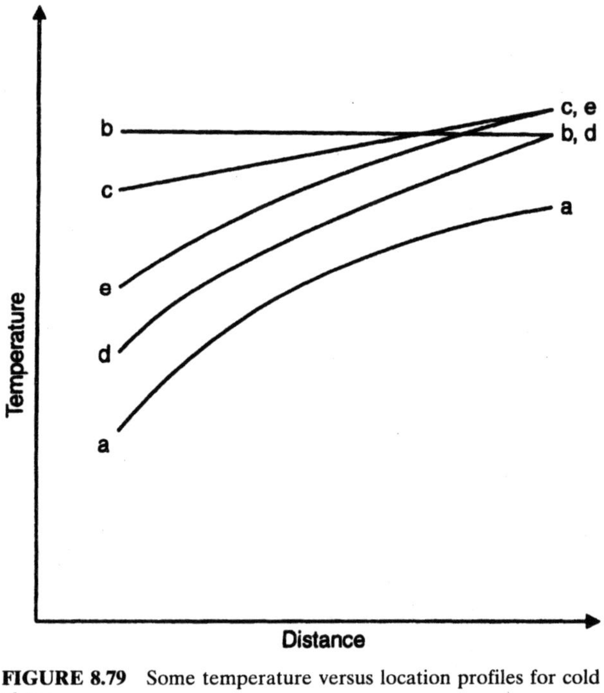

FIGURE 8.79 Some temperature versus location profiles for cold plates.

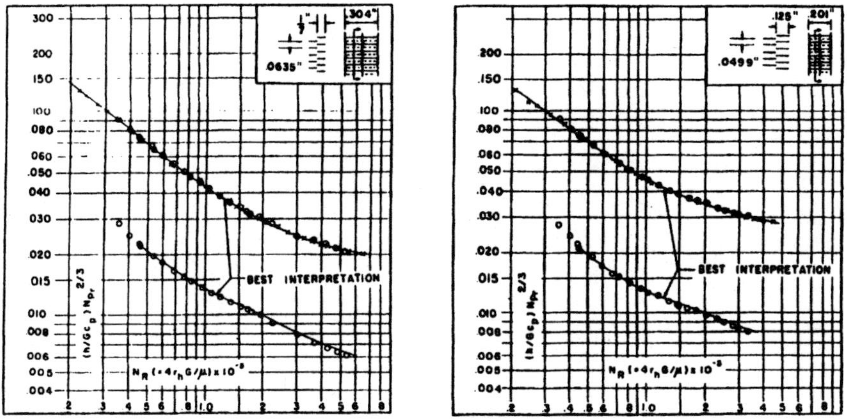

Fin pitch = 15.75 per in.
Plate spacing, $b = 0.304$ in.
Splitter symmetrically located
Fin length flow direction = 1/7 in.
Flow passage hydraulic diameter,
$4r_h = 0.006790$ ft
Fin metal thickness = 0.004 in., aluminum
Splitter thickness = 0.006 in.
Total heat transfer area/volume between plates,
$\beta = 526$ ft^2/ft^3
Fin area (including splitter)/total area = 0.859

Fin pitch = 20.06 per in.
Plate spacing, $b = 0.201$ in.
Splitter symmetrically located
Fin length flow direction = 0.125 in.
Flow passage hydraulic diameter,
$4r_h = 0.004892$ ft .0587 in.
Fin metal thickness = 0.004 in., aluminum
Splitter metal thickness = 0.006 in.
Total heat transfer area/volume between plates
$\beta = 698$ ft^2/ft^3
Fin area (including splitter)/total area=0.843

FIGURE 8.80 Strip-fin friction factors and Colburn Moduli. *(From Kays and London, Ref. 21.)*

For Re > 8000

$$f = \frac{0.048}{\mathrm{Re}^{0.2}} \tag{8.93}$$

$$j = \frac{0.025}{\mathrm{Re}^{0.22}} \tag{8.94}$$

For lanced and offset or strip fins, Weiting [22] gives the following relationships:
For Re ≤ 1000

$$f = 7.661 \left(\frac{x}{D}\right)^{-0.384} \alpha^{-0.092} \mathrm{Re}^{-0.712} \tag{8.95}$$

$$j = 0.483 \left(\frac{x}{D}\right)^{-0.162} \alpha^{-0.184} \mathrm{Re}^{-0.536} \tag{8.96}$$

For Re ≥ 2000:

$$f = 1.136 \left(\frac{x}{D}\right)^{-0.781} \left(\frac{t}{D}\right)^{0.534} \mathrm{Re}^{-0.198} \tag{8.97}$$

$$j = 0.242 \left(\frac{x}{D}\right)^{-0.322} \left(\frac{t}{D}\right)^{0.089} \mathrm{Re}^{-0.368} \tag{8.98}$$

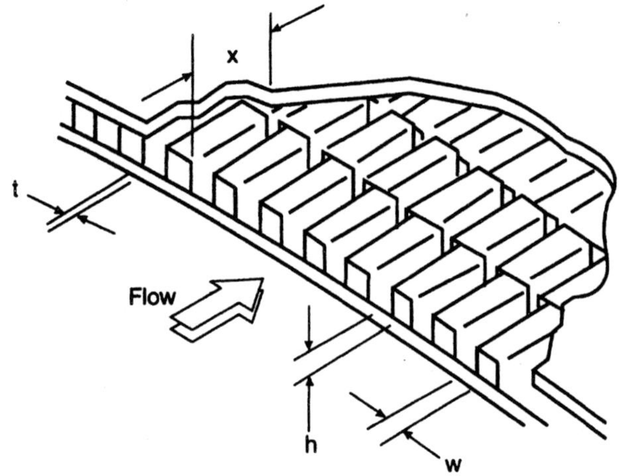

FIGURE 8.81 Typical rectangular strip fin.

All of the parameters of these equations are defined in Fig. 8.81, except α, which was defined as w/h, and D, which is the hydraulic diameter of a single opening between fins.

The pressure loss through the cold plate is the sum of the pressure losses caused by entrance and exit contractions and expansions and the viscous losses in the core. Entrance- and exit-loss coefficients K_c and K_e for a multiple square tube and triangular tube cores are shown in Figs. 8.82 and 8.83.

Many references discuss the entrance- and exit-pressure drops expressed as

$$\Delta P_c = \frac{\rho}{2g}(\overline{V}^2(1 - \sigma^2) + K_c\overline{V}^2) \tag{8.99}$$

$$\Delta P_e = \frac{\rho}{2g}(\overline{V}^2(1 - \sigma^2) + K_e\overline{V}^2) \tag{8.100}$$

where σ is the area ratio of open areas at the inlet and exit of the fins.

At the exit, pressure is recovered so that Eq. 8.100 represents an increase in pressure and must be subtracted from Eq. 8.99. However, the sign convention allows the recoverable terms to cancel and the irreversible losses to add. The friction loss of the core must be added to Eqs. 8.99 and 8.100 to define the total pressure loss of the cold plate.

8.7.1 Cabinet and Large-Enclosure Cooling

Large cabinet and electronic enclosures, such as are used in many ground-based systems, are usually cooled by natural or forced convection. The choice is determined by the heat density in the cabinet and the heat flux of the individual assemblies inside the cabinet. Natural convection has the advantage of being quiet, not requiring power, and never failing. Forced convection can handle much higher heat densities and can provide a greater flexibility in the internal arrangement and packaging

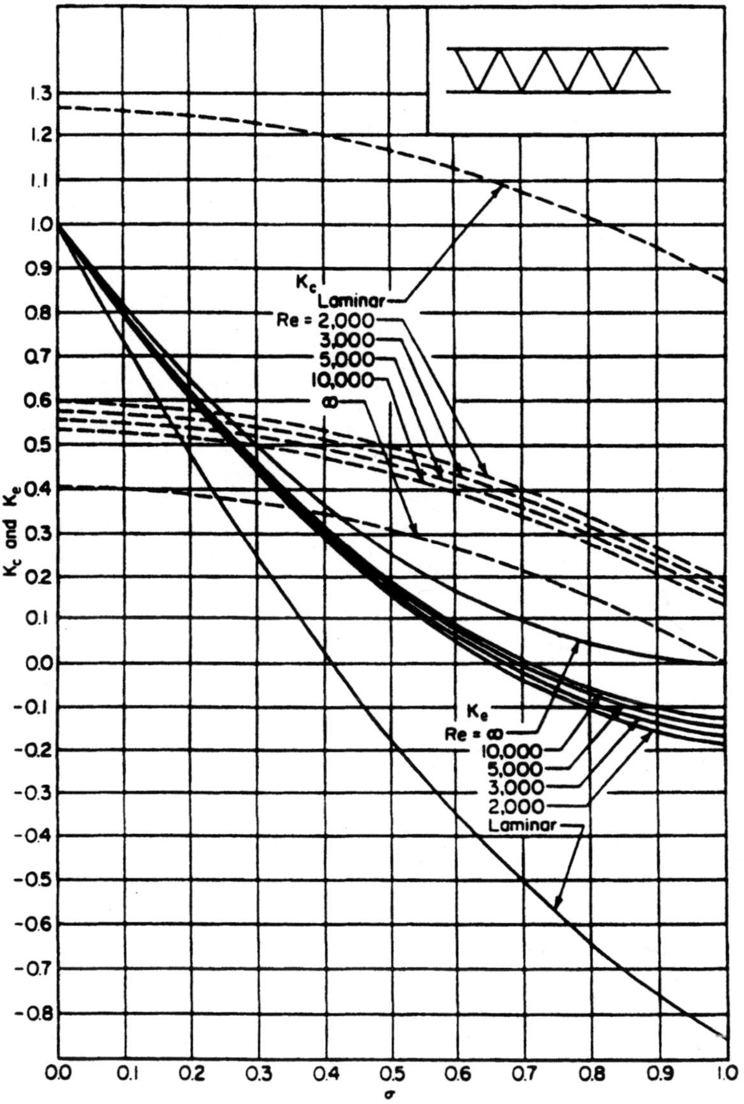

FIGURE 8.82 Entrance and exit pressure loss coefficient for a triangular tube with abrupt contraction entrance and abrupt expansion exit.

of the cabinet. The air inlet must be at the bottom of the cabinet and the exit at the top. Both should be as large as possible to allow maximum airflow. A maximum value for natural convection cooling given by Ross, Ref. 23 is 5.5 W/m^2-°C. A more general value is given by Altoz in Ref. 14 as 100 W/ft^3.

As with any convection system, the air temperature rises as it picks up heat. For proper cooling, it is desirable for each electronic assembly to get cool air and not air

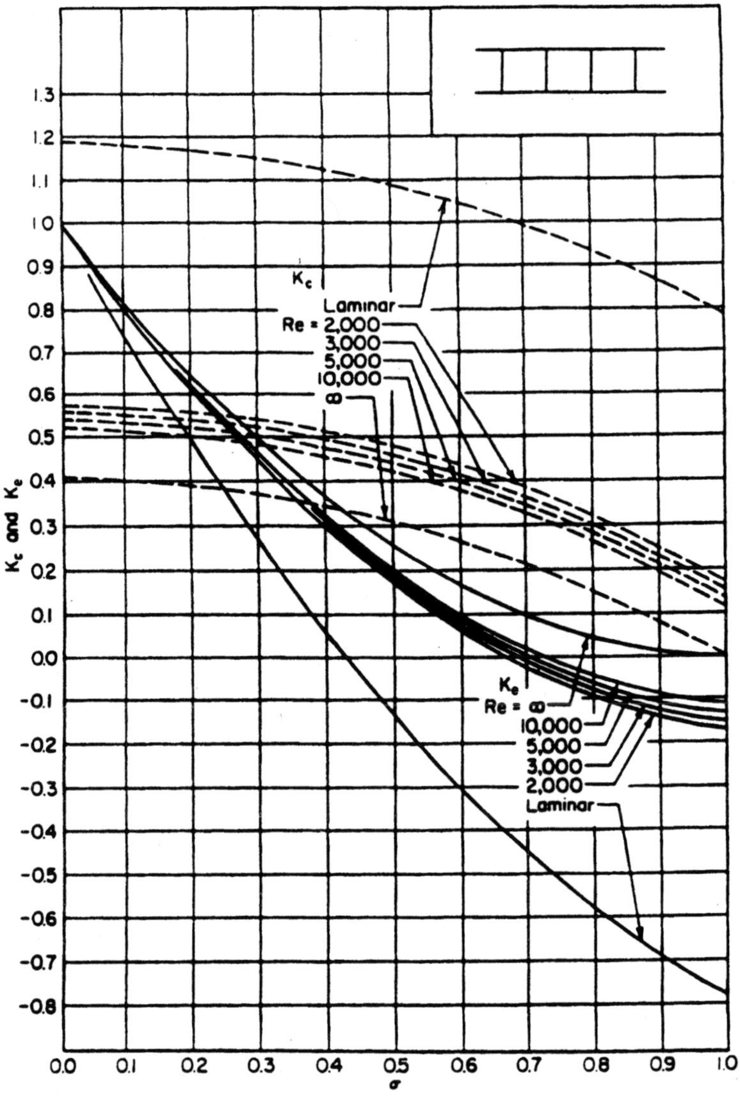

FIGURE 8.83 Entrance and exit pressure loss coefficient for a rectangular tube with abrupt contraction entrance and abrupt expansion exit.

that is exiting from another assembly. This can often be accomplished by proper arrangement and by including baffles to direct the air, such as in Fig. 8.77. Whether airflow is forced by a fan or buoyancy, the flow through the cabinet is a balance of the pressure loss through the system and the forces driving the air. The draft equation [14] gives the forcing pressure in inches of water.

$$\Delta P = 0.192\, H\rho \ln\left(\frac{T_2}{T_1}\right) \tag{8.101}$$

where ρ = average density of air in the cabinet in lb/ft^3
T_2 = exit absolute temperature of the air temperature in °R or K
T_1 = entrance or ambient absolute air temperature in °R or K
H = height of the column of air from the centroid of heat to the top of the cabinet in ft

Notice that putting the heat low in the cabinet generates a higher draft pressure. This could be a consideration in planning the layout of the cabinet. A computer program to generate both the draft pressure and pressure loss curves versus flow rate is available [14]. Also, be aware that this draft equation, Eq. 8.83, presumes that the entire air volume is heated in a series of heat absorptions. When the heat sources are arranged, as in Fig. 8.77, the equation should be applied for each rack, not the entire cabinet (Fig. 8.84).

Improper placement of large components within the cabinet can generate dead-air areas or local circulations of air that are warmer than desired. A detailed examination of internal flows can be performed using computational fluid-dynamics computer programs, such as Flowtherm [22] or IcePack [12]. These programs combine computational fluid dynamics with heat transfer to predict fluid velocity, pressures, and temperatures. Figure 8.85 is an analysis done using IcePack to define airflow and provide a visual representation so that dead or blocked areas can be easily located.

When fans are used to cool large enclosures, there is a choice of forcing air through the cabinet or drawing air through the cabinet. Each has advantages. A

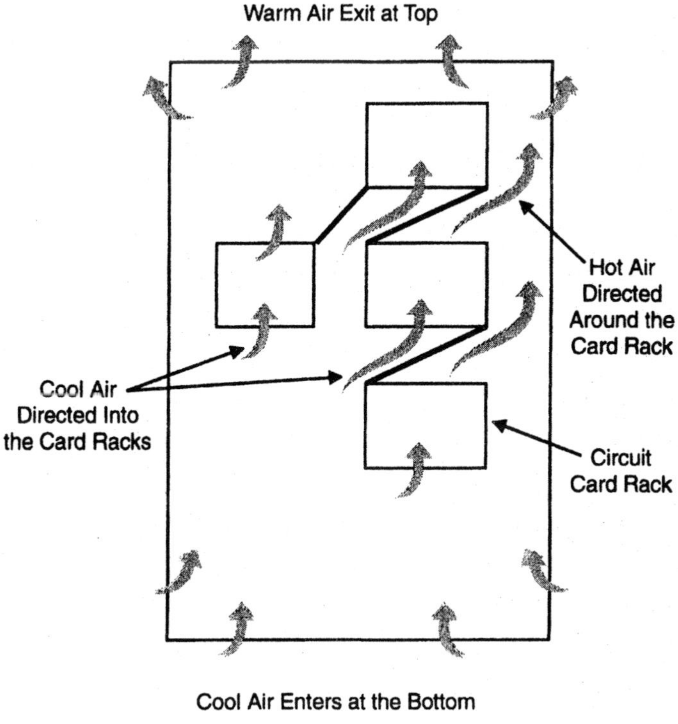

FIGURE 8.84 Circuit card racks arranged to induct cool air into each level.

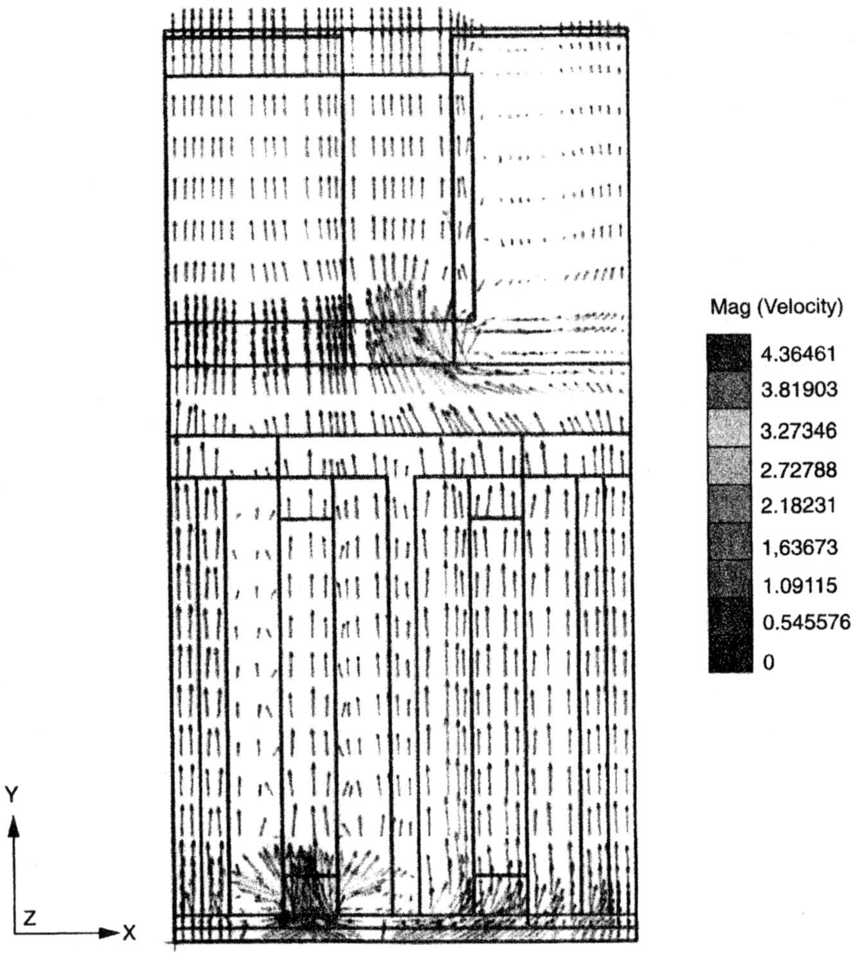

FIGURE 8.85 Computational fluid dynamics (CFD) used to define flow fields and temperature. *(Courtesy of Fluid Dynamics, Inc.)*

draw-through system allows entry points to be placed where wanted to bring cool air into the cabinet at desired locations. However, air entering the cabinet must be filtered; thus, multiple filters are required. Also leaks in the cabinet might bring unwanted air and dust into the enclosure, which can reduce the airflow in other areas (because the fan or fans will only handle a certain amount). Pressurizing the cabinet is a more usual scheme and has the advantage of not ingesting dust from leaks and not being as sensitive to leakage, but filters must still be used to keep the air entering the cabinet dustfree. Fans at the entrance of a pressurized cabinet are handling cooler air, so they are more reliable and smaller fans are needed because the air is denser. The fans heat the air as it passes through the fans, so the air in the cabinet is slightly warmer. As explained in Sec. 8.6, it is important to match the fan to the system pressure drop at the desired flow rate. Entrance and exit losses at the inlet and outlet are often a large part of the pressure loss in large cabinets. Be sure that they are both large enough that the desired flow rates are not overly restricted.

Fans for enclosures can be located at the bottom of the cabinet and supply air for the entire cabinet or individual trays of fans can be used for each card rack. Most of these are designed to slide out for easy removal and replacement of both fans and filters (Fig. 8.86).

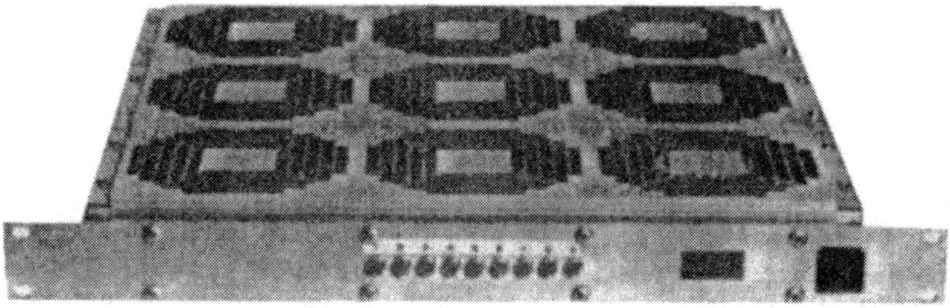

FIGURE 8.86 Fan tray for large cabinets.

It is not necessary that an enclosure be either natural or force convection cooled. It can be both and this is often the optimum solution—especially where local hot spots require higher flow rates. This is the case with most desktop personal computers, which are generally natural convection cooled, but have a small fan to cool the power supply and an even smaller fan to cool the processor chip. These fans don't make the air in the enclosure any cooler because they don't increase the flow through the enclosure. But they do reduce the temperature of individual pieces of electronics inside the enclosure.

8.8 SPECIAL THERMAL MANAGEMENT DEVICES

8.8.1 Thermoelectric Devices

Thermoelectric devices operate on the Peltier principle, which states that a current forced through the junction of two dissimilar materials will produce a cooling or heating effect at the junction, depending on the direction of the current flow. This is expressed mathematically as

$$Q = \alpha T I \tag{8.102}$$

where Q = heat generated or removed
α = Seebeck coefficient in volts/K
T = absolute temperature at the junction in K
I = current in amperes

The device is the opposite of a thermocouple for which an EMF is induced by temperature differences at the junctions. The operation of the thermoelectric device is a balance of opposing heating and cooling terms: the Peltier effect, the Joule effect (I^{2R} heating in the conductors), and the Thompson effect:

$$Q = \tau I \,\Delta T \tag{8.103}$$

where τ = Thompson coefficient in volts/K
I = current
T = temperature difference between the hot and cold junctions

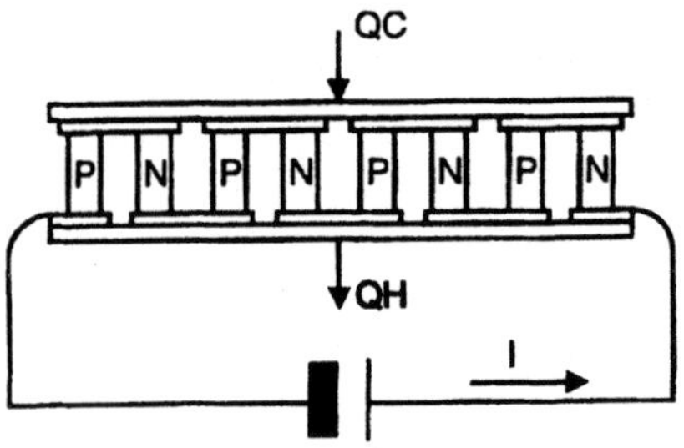

FIGURE 8.87 Construction and polarity of a refrigerating thermoelectric device (TED).

The sign of the Thompson coefficient changes with current direction so that at one junction, it cools, and at the other, it heats.

Modern devices are made of semiconductor material, usually some form of telluride (such as bismuth, antimony, or lead) oriented so that the current flows through n-p junctions on one side and p-n junctions on the other. This forms the *cold* side and the *hot* side junctions and looks much as shown in Fig. 8.87.

The amount of heat that can be absorbed on the cold side Q_c is expressed as

$$Q_c = n[I(\alpha_c T_c + \tfrac{1}{2}\tau\,\Delta T) - \tfrac{1}{2}I^2R_m - G_m\,\Delta T] \tag{8.104}$$

where R_m = average electrical resistance of the couple
α_c = Seebeck coefficient at the cold junction temperature
τ = Thompson coefficient at the average temperature of the couples $(T_c + T_h)/2$
G_m = thermal conductance (kA/x) of the device couples and k is evaluated at the average temperature
ΔT = temperature difference between the hot and cold sides
n = number of couples

Obviously, the Peltier and Thompson effects are providing the cooling while the Joule effect and the first law of thermodynamics are opposing energy sources.

In practice, Eq. 8.104 can be simplified because the Thompson coefficient can be expressed in terms of the Seebeck coefficient. This allows the average value of the Seebeck coefficient α_m to be used to eliminate a term in the equation and yield the following:

$$Q_c = n(I\alpha_m T_c - \tfrac{1}{2}I^2R_m - G_m\,\Delta T) \tag{8.105}$$

The Thompson and Peltier effects are reversible so that at the hot side junction, the heat flow is

$$Q_h = n(I\alpha_m T_h + \tfrac{1}{2}I^2R_m - G_m\,\Delta T) \tag{8.106}$$

where the Seebeck coefficient is defined at the average couple temperature. For practical purposes, the heat flow at the hot junction is more simply defined as the sum of Eq. 8.104 plus the input power, $V \times I$ and the voltage is determined from

$$V = n(\alpha_m\,\Delta T + IR_m) \tag{8.107}$$

where n is the number of couples connected in series.

Caution must be used when using these devices. The most important consideration is the ability of the heat sink on the hot side of the device to absorb heat without changing temperature. The heat rejected at the hot side is the sum of the heat absorbed and the power input to the device. If this heat causes the heat sink temperature to rise, any gains in the temperature suppression of the cold side will be offset and possibly negated by the sink rise. Once this hurdle is passed, a secondary, but equally important consideration, is the interfaces. These should be very flat and smooth with a very thinly applied thermal-grease interstitial filler.

The operational efficiency is expressed as with most refrigeration systems as the coefficient of performance in refrigeration (COPR). Because the device can be used as a heater or heat pipe by reversing the current, a coefficient of performance in heating (COPH) is also used. These are simply the ratio of the amount of cooling or heating obtained to the input power.

$$\text{COPR} = \frac{Q_c}{nI(\alpha_m\,\Delta T + IR_m)} \tag{8.108}$$

$$\text{COPH} = \frac{Q_h}{nI(\alpha_m\,\Delta T + IR_m)} \tag{8.109}$$

Devices can be cascaded or stacked in smaller and smaller sizes to increase the ΔT capability. Each device must be able to absorb the heat from the device above it and still maintain a desired temperature depression so that the amount of heat that can be absorbed by the final stage is much smaller.

When designing with thermoelectric devices, it can be useful to know certain behavioral characteristics. Because the material properties of the couples vary with temperature, the performance of thermoelectric devices also varies with temperature (Fig. 8.88). Notice that the performance of the device improves with temperature up to approximately 100°C, then drops off. This is useful in the electronics cooling industry because electronics are often operating in the 70 to 110°C range.

Another useful characteristic is illustrated in Fig. 8.89, which shows the variation of COPR and Q_c with ΔT. The COPR drops below 1 when $\Delta T > 20$°C. It is useful to know that a 20°C temperature reduction can be obtained for the same power input as the power being removed.

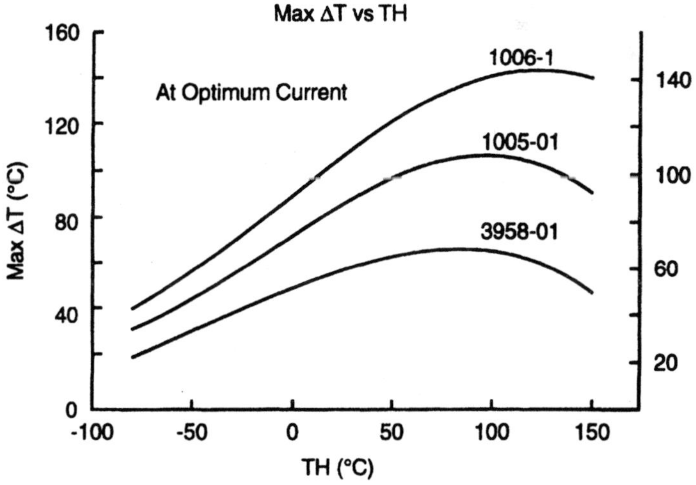

FIGURE 8.88 Performance characteristics vary with temperature.

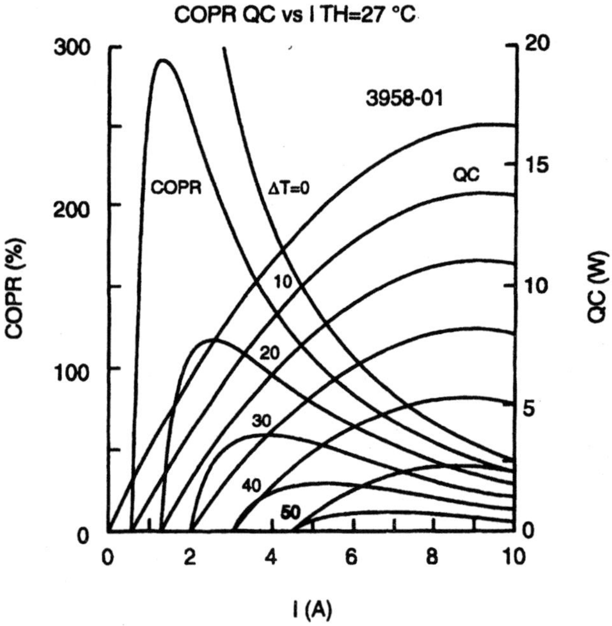

FIGURE 8.89 Large T_s or large Q_c require high currents and produce low efficiencies.

The coefficients depend on the material, and precise values are somewhat guarded by suppliers. However, the following values are useful for design calculations. Precise performance tables and curves can be supplied by manufacturers [24].

$$\alpha_m = 1 \times 10^{-9}(22{,}224 + 930.6T - 0.9905T^2) \tag{8.110}$$

$$R_m = 1 \times 10^{-8}(5112 + 163.4T - 0.6297T^2) \tag{8.111}$$

$$G_m = 1 \times 10^{-6}(62605 - 227.7T + 0.4131T^2) \tag{8.112}$$

The following computer program can be used to make design estimates. It uses the coefficient values presented in the previous paragraph.

```
5 REM ON ERROR GOTO 670
10 CLS:COLOR 20,0:PRINT" Use Ctrl-PrtSc if you want a hard copy of
output"
20 COLOR 7,0:PRINT:KEY OFF
30 READ S0,S1,S2,P0,P1,P2,K0,K1,K2
40 DATA 22224.,930.6,-0.9905,5112.,163.4,.6279,62605.,-277.7,.4131
41 KEY 15,CHR$(&H4)+CHR$(114)
42 ON KEY(15) GOSUB 180
43 KEY(15) ON
50 PRINT" This program contains five analysis options.":PRINT
60 PRINT"1. Generate performance properties similar to the curves in
the catalog."
```

```
70 PRINT"2. For a given voltage, determine refrigeration capacity and
current for a range of delta T s."
80 PRINT"3. For a given voltage and required refrigeration capacity,
determine the current and delta T."
85 PRINT"4. Determine a near optimum configuration for your voltage
and refrigeration requirements."
86 PRINT"5. Determine the proper device to use as a generator.":PRINT
90 PRINT" YOUR CHOICE":PRINT
100 LOCATE 15,15:COLOR 1,7:PRINT" 1 ":COLOR 7,1:LOCATE 15,19:PRINT"
for option 1"
105 LOCATE 15,40:COLOR 1,7:PRINT" 5 ":COLOR 7,1:LOCATE 15,44:PRINT"
for option 5"
110 LOCATE 17,15:COLOR 1,7:PRINT" 2 ":COLOR 7,1:LOCATE 17,19:PRINT"
for option 2
120 LOCATE 19,15:COLOR 1,7:PRINT" 3 ":COLOR 7,1:LOCATE 19,19:PRINT"
for option 3
125 LOCATE 21,15:COLOR 1,7:PRINT" 4 ":COLOR 7,1:LOCATE 21,19:PRINT"
for option 4"
127 LOCATE 23,15:COLOR 1,7:PRINT" E ":COLOR 7,1:LOCATE 23,19:PRINT" to
end"
130 A$=INPUT$(1):COLOR 7,0:IF A$="1" THEN CLS:GOTO 200
140 IF A$="2" THEN CLS:GOTO 460
150 IF A$="3" THEN CLS:GOTO 710
153 IF A$="4" THEN CLS:GOTO 2000
154 IF A$="5" THEN CLS:GOTO 1000
155 IF A$="E" OR A$="e" THEN CLS:END
160 PRINT" That option not available. Enter 1,2 or 3.":GOTO 130
180 LPRINT CHR$(27)"N"CHR$(6)
190 RETURN
200 PRINT"Program option 1-Generate performance curves"
210 INPUT"ENTER NUMBER OF COUPLES";N:NE=2*N
220 INPUT "ENTER LENGTH/AREA RATIO OF ELEMENT (1/cm.).
Default=[13.015]";LAMBDA
225 IF LAMBDA=0 THEN LAMBDA=13.015
230 INPUT "ENTER Th (Kelvin)";TH
240 INPUT"ENTER MAXIMUM CURRENT OF DEVICE-(AMPS) ";CURMAX
250 PRINT:PRINT"TABLES WILL BE PRINTED FOR CURRENT STEPS SPECIFIED UP
TO THE MAXIMUM CURRENT. YOU MUST ENTER THE CURRENT STEP YOU DESIRE."
260 INPUT "ENTER CURRENT STEP-";ST1
270 INPUT"ENTER STARTING CURRENT-";STC
280 FOR J=STC TO CURMAX STEP ST1:I=J:PRINT:PRINT "--FOR
CURRENT=";J:PRINT
290 FOR M=0 TO 4:TC=TH-10*M:DT=TH-TC
300 T=(TC+TH)/2:PRINT "DELTA T (KELVIN)-";DT
310 SM=(S0+S1*T+S2*T^2)*1E-09
320 RHOM=(P0+P1*T+P2*T^2)*1E-08
330 KM=(K0+K1*T+K2*T^2)*.000001
```

```
340 Z=SM^2/(RHOM*KM)
350 S=SM*NE:DT=TH-TC
360 R=RHOM*LAMBDA*NE:K=KM*NE/LAMBDA
370 'note that voltage equation has a fudge factor to match curves
380 V=(I*LAMBDA*RHOM+SM*DT)*NE*1.1
390 QC=S*I*TC-I^2*R/2-K*DT
400 PRINT "QC=";QC,"VOLTAGE=";V
410 NEXT M:NEXT J
411 PRINT:COLOR 2,0:PRINT"Do you want to repeat option 1? 'Y' or 'N'"
412 Y$=INPUT$(1):IF Y$="Y" OR Y$="y" THEN COLOR 7,0:CLS:GOTO 200
413 PRINT"Do you want terminate this program?"
415 Y$=INPUT$(1):IF Y$="y" OR Y$="Y" THEN END ELSE COLOR 7,0:CLS:GOTO
50
460 PRINT"Program option 2-Performance at fixed voltage"
470 INPUT"ENTER NUMBER OF COUPLES";N:NE=2*N
480 INPUT "ENTER LENGTH/AREA RATIO OF ELEMENT (1/cm.).
Default=13.015";LAMBDA
485 IF LAMBDA=0 THEN LAMBDA=13.015
490 INPUT "ENTER Th (Kelvin)";TH
500 INPUT"ENTER OPERATING VOLTAGE OF DEVICE-";V
510 PRINT:PRINT"TABLE WILL BE PRINTED FOR CURRENT, HEAT ABSORBED AND
DELTA T." 520 PRINT:PRINT" FOR TED VOLTAGE OF ";V;" VOLTS"
530 PRINT" QC CURRENT DELTA T (KELVIN)"
540 FOR M=0 TO 40:TC=TH-M:DT=TH-TC
550 T=(TC+TH)/2
560 SM=(S0+S1*T+S2*T^2)*1E-09
570 RHOM=(P0+P1*T+P2*T^2)*1E-08
580 KM=(K0+K1*T+K2*T^2)*.000001
590 Z=SM^2/(RHOM*KM)
600 S=SM*NE:DT=TH-TC
610 R=RHOM*LAMBDA*NE:K=KM*NE/LAMBDA
620 'note that voltage equation has a fudge factor to match curves
630 I=(V/NE/1.1-SM*DT)/LAMBDA/RHOM
640 QC=S*I*TC-I^2*R/2-K*DT
650 PRINT QC,I,DT
655 NEXT M:PRINT:COLOR 2,0:PRINT"Do you want to repeat option 2? 'Y'
or 'N'"
660 Y$=INPUT$(1):IF Y$="Y" OR Y$="y" THEN COLOR 7,0:CLS:GOTO 460
665 PRINT"Do you want terminate this program?"
666 Y$=INPUT$(1):IF Y$="y" OR Y$="Y" THEN END ELSE COLOR 7,0:CLS:GOTO
50
670 A$="PLEASE TURN PRINTER ON"
680 IF ERR=24 OR ERR=25 OR ERR=27 THEN PRINT A$ ELSE 700
685 PRINT"YOU MAY HAVE TO RE-ENTER 'Ctrl-PrtSc' BECAUSE PRINTER WAS
OFF."
690 INPUT"PRESS <ENTER> WHEN PRINTER IS READY";Y$:RESUME
700 PRINT" ERROR NO.-";ERR:STOP
```

```
710 PRINT"Program option 3-Specific operating point"
720 INPUT"ENTER NUMBER OF COUPLES";N:NE=2*N
730 INPUT "ENTER LENGTH/AREA RATIO OF ELEMENT (1/cm.).
Default=13.015";LAMBDA
735 IF LAMBDA=0 THEN LAMBDA=13.015
740 INPUT "ENTER Th (Kelvin)";TH
750 INPUT"ENTER OPERATING VOLTAGE OF DEVICE-";V
760 INPUT"ENTER HEAT TO BE ABSORBED (QC)-";QC
770 PRINT" V Qc CURRENT Th Tc DELTA T "
780 PRINT:Z$="##.# ##.#### ##.#### ### ### ###"
790 T=TH:DT=0:TC=TH
800 SM=(S0+S1*T+S2*T^2)*1E-09
810 RHOM=(P0+P1*T+P2*T^2)*1E-08
820 KM=(K0+K1*T+K2*T^2)*.000001
830 Z=SM^2/(RHOM*KM)
840 S=SM*NE
850 R=RHOM*LAMBDA*NE:K=KM*NE/LAMBDA
860 'note that voltage equation has a fudge factor to match curves
870 I=(V/NE/1.1-SM*DT)/LAMBDA/RHOM
880 QC2=S*I*TC-I^2*R/2-K*DT
885 REM PRINT QC2,I,DT
890 IF QC2>QC+.2 THEN T=T-.2:TC=2*T-TH:DT=TH-TC:GOTO 800
910 PRINT USING Z$;V,QC2,I,TH,TC,DT
920 PRINT" INPUT POWER=";I*V
930 PRINT:COLOR 2,0:PRINT"Do you want try another voltage and Qc?-'Y'
or 'N'"
940 Y$=INPUT$(1):IF Y$="Y" OR Y$="y" THEN COLOR 7,0:CLS:GOTO 750
941 PRINT"Do you want to repeat option 3 (change the TED parameters)?"
942 Y$=INPUT$(1):IF Y$="y" OR Y$="Y" THEN COLOR 7,0:CLS:GOTO 710
950 PRINT"Do you want terminate this program?"
960 Y$=INPUT$(1):IF Y$="y" OR Y$="Y" THEN END ELSE COLOR 7,0:CLS:GOTO
50
1000 PRINT"Program option 5-Power generation"
1005 INPUT"ENTER REQUIRED CURRENT";I
1006 INPUT"ENTER REQUIRED OPERATING VOLTAGE-";V
1010 INPUT "ENTER Th (Kelvin)";TH
1015 INPUT"ENTER Tc (Kelvin)";TC
1020 PRINT" V CURRENT Th Tc DELTA T "
1025 Z$="##.# ##.#### ### ### ###"
1030 T=(TH+TC)/2:DT=TH-TC
1035 SM=(S0+S1*T+S2*T^2)*1E-09
1040 RHOM=(P0+P1*T+P2*T^2)*1E-08
1045 KM=(K0+K1*T+K2*T^2)*.000001
1050 Z=SM^2/(RHOM*KM)
1055 S=SM*NE
1060 PRINT USING Z$;V,I,TH,TC,DT:PRINT
1065 EOC=2*V
```

```
1070 R=V/I
1075 NE=EOC/SM/DT
1080 LAMBDA=R/RHOM/NE
1085 NC=NE/2
1090 PRINT"LOAD RESISTANCE=DEVICE RESISTANCE=";R
1100 PRINT"NUMBER OF COUPLES REQUIRED=";NC
1105 PRINT"LAMBDA REQUIRED=";LAMBDA
1110 QC=NE*KM*DT/LAMBDA:EFF=V*I*100/QC
1115 PRINT"HEAT TRANSPORTED THROUGH THE GENERATOR (WATTS) =";QC
1120 PRINT"GENERATOR EFFICIENCY=";EFF
1130 PRINT:COLOR 2,0:PRINT"Do you want try another voltage and cur-
rent?-'Y' or 'N'"
1140 Y$=INPUT$(1):IF Y$="Y" OR Y$="y" THEN COLOR 7,0:CLS:GOTO 1000
1150 PRINT"Do you want terminate this program?"
1160 Y$=INPUT$(1):IF Y$="y" OR Y$="Y" THEN END ELSE COLOR 7,0:CLS:GOTO
50
2000 PRINT"Program option 4 to design optimum configuration"
2010 N=71:NE=2*N:LAMBDA=13.015 'These values are meaningless
2040 INPUT"ENTER DESIRED DELTA T-";DT:NE=2*N
2050 INPUT "ENTER Th (Kelvin)";TH
2060 TC=TH-DT:T=(TC+TH)/2
2070 INPUT"ENTER SUPPLY VOLTAGE AVAILABLE-";V
2080 INPUT"ENTER TOTAL HEAT TO BE ABSORBED (QC)-";QC
2090 GOSUB 2200 'Series optimization
2100 GOSUB 3000 'Parallel optimization
2110 GOTO 3900
2200 'Subroutine for series optimization
2210 GOSUB 4000
2212 ILAM=KM*DT*(1+SQR(1+Z*T))/SM/T
2214 NE=V/(ILAM*RHOM+SM*DT):N=NE/2
2215 LAMBDA=.5:GOSUB 4000
2220 GOSUB 4050
2230 IF QC2<QC-.25 THEN 2420 ELSE LAMBDA=LAMBDA+.1:GOTO
2220
2420 PRINT" V Qc CURRENT Th Tc DELTA T COUPLES LAMBDA"
2430 PRINT:Z$="##.# ##.#### ##.#### ### ### ### ### ##.###"
2440 PRINT USING Z$;V,QC2,I,TH,TC,DT,N,LAMBDA
2460 CNTR=0:PRINT" INPUT POWER=";PWR
2470 RETURN
3000 'Subroutine for parallel optimization
3020 PRINT:PRINT" For two Parallel strings:"
3030 QC1=QC:QC=QC/2
3060 GOSUB 2220
3100 'subroutine for three parallel strings
3120 PRINT:PRINT" For three parallel strings:"
3130 QC=QC1/3
3160 GOSUB 2220
```

```
3170 RETURN
3900 CNTR=0:PRINT:COLOR 2,0:PRINT"Do you want to try another voltage
and Qc?"
3910 Y$=INPUT$(1):IF Y$="y" OR Y$="Y" THEN QC2=0:QC3=0:COLOR
7,0:CLS:GOTO 2070
3920 PRINT"Do you want to repeat option 4 for another design point?"
3930 Y$=INPUT$(1):IF Y$="Y" OR Y$="y" THEN COLOR 7,0:CLS:GOTO 2010
3940 PRINT"Do you want to terminate the program?":Y$=INPUT$(1)
3950 IF Y$="Y" OR Y$="y" THEN END ELSE COLOR 7,0:CLS:GOTO 50
4000 SM=(S0+S1*T+S2*T^2)*1E-09
4010 RHOM=(P0+P1*T+P2*T^2)*1E-08
4020 KM=(K0+K1*T+K2*T^2)*.000001
4030 Z=SM^2/(KM*RHOM)
4040 S=SM*NE
4050 R=RHOM*LAMBDA*NE:K=KM*NE/LAMBDA
4060 'note that voltage equation has a fudge factor to match curves
4070 I=(V/NE/1.1-SM*DT)/LAMBDA/RHOM
4080 QC2=S*I*TC-I^2*R/2-K*DT:PWR=V*I
4085 REM PRINT "QC2=";QC2,"I=";I
4086 CNTR=CNTR+1:IF CNTR>500 THEN PRINT:PRINT"**QC too high or V too
low**"
4087 IF CNTR>500 THEN CNTR=0:GOTO 3900
4090 RETURN
```

8.9 HEAT PIPES

The heat pipe is an extremely efficient method of transferring heat from one area to another, such as from an electronic component to the edge of a CCA. Heat pipes have been considered and designed for electronics for many years, but haven't yet been commonly used except in space applications, although they are used in extreme problem areas where cost permits and volume impose restrictions on cheaper methods. The principle of operation of the heat pipe is based on boiling and condensing of a working fluid inside the pipe (see Sec. 8.4.4). This process is illustrated in Fig. 8.90, wherein heat enters the evaporator end of the pipe via conduction through the

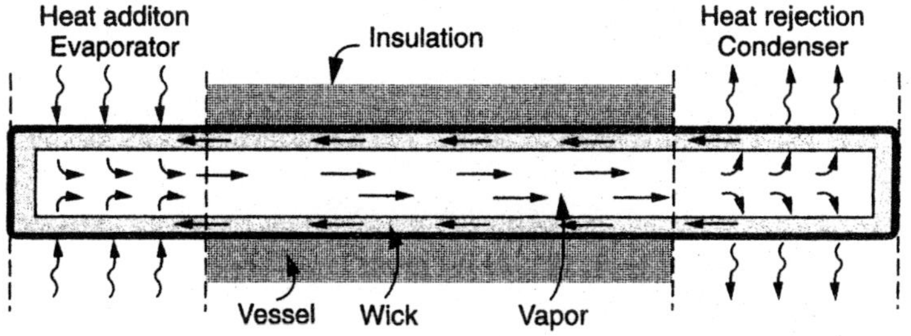

FIGURE 8.90 Typical heat pipe operation.

pipe material and conduction into a liquid-filled wick structure, which causes the liquid to change into a vapor phase. A small pressure differential exists between the evaporator area and the condenser area, caused by a small temperature difference. This pressure difference causes the vapor to flow to the condenser area, where it condenses within the wick structure and releases heat to be conducted out through the pipe walls to the ultimate sink. The liquid in the condenser wick travels back to the evaporator area via capillary action.

The wick can be made of many structures, such as wire screen, felt, sintered material, or just grooves in the pipe wall small enough to promote capillary pumping (Fig. 8.91). Of course, the cheapest heat pipe fabrication is to eliminate the wick by placing the condenser at the top and letting gravity return the liquid.

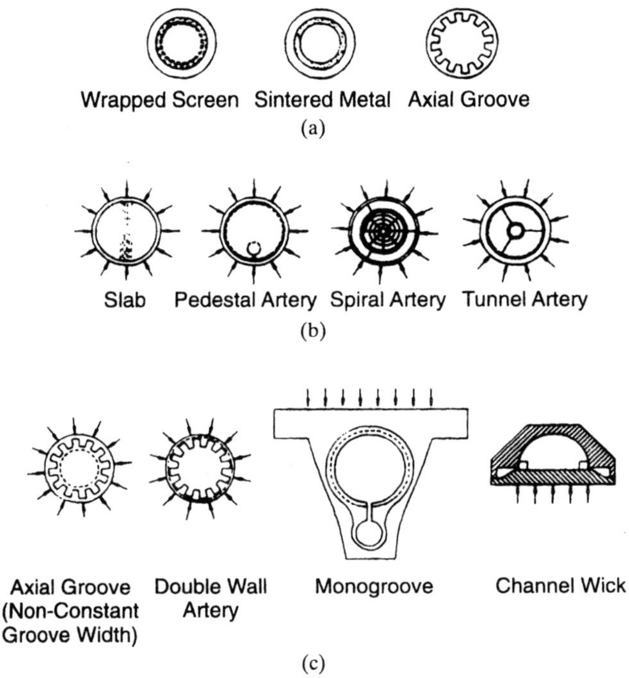

FIGURE 8.91 Heat pipe wick structures: (*a*) simple homogeneous; (*b*) current composite; and (*c*) advanced designs. *(From Ref. 11.)*

Several processes occur inside the heat pipe, which can limit its operation. The most prevalent of these is the limit on the capillary pumping action acting against gravity. Where the heat pipe is horizontal or used in space, this might not apply. However, too much heat can require more fluid to be returned to the evaporator area than a given wick structure can supply. This results in a drying out of the evaporator wick and total loss of heat transfer in the heat pipe; a catastrophic failure. The forces involved include the capillary pressure, axial hydrostatic pressure, radial hydrostatic pressure, liquid pressure drop, vapor pressure drop, and choked flow of the vapor. The capillary pressure must be greater than the fluid pressure drop and the axial hydrostatic pressure, in order for liquid to flow from the condenser area to the evaporator area. The axial hydrostatic pressure can help or impede flow, depending on the orientation of the evaporator and condenser to the gravity vector. If the condenser is higher than the evaporator, gravity assists operation by adding to the capillary pressure. Fig. 8.92 shows the impact of gravity on a typical heat pipe operation, with respect to orientation to a 1-g gravity field.

The capillary pressure, as expressed by Bar-Cohen and Kraus [11] is a function of the meniscus radii in evaporator and condenser wick structures (as shown in Fig. 8.93 and as expressed in Eq. 8.113).

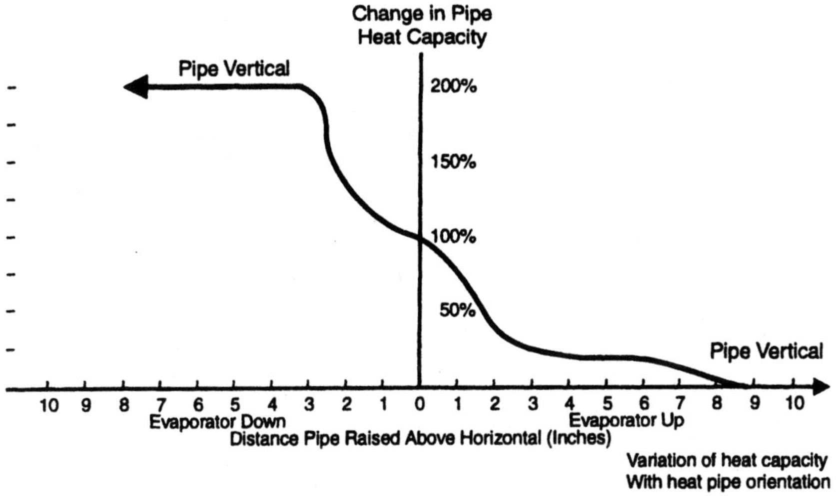

FIGURE 8.92 Pumping against gravity reduces the capacity of the heat pipe.

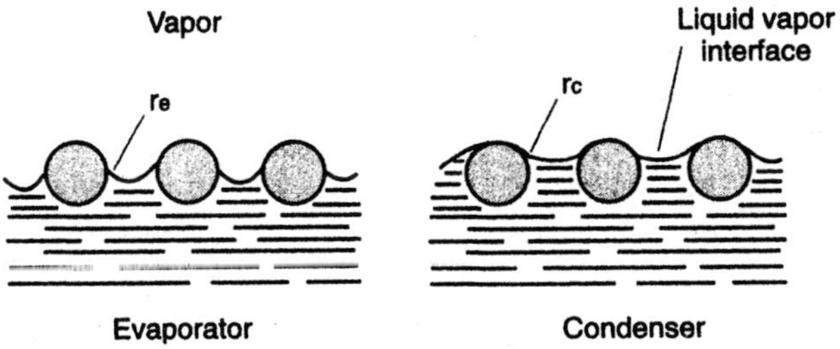

FIGURE 8.93 Radius of curvature of the meniscus in the evaporator and condenser vapor/liquid interface.

$$\Delta P_c = 2\sigma\left(\frac{1}{r_e} - \frac{1}{r_c}\right) \tag{8.113}$$

where σ = liquid surface tension
r_e = evaporator meniscus radius
r_c = condenser meniscus radius

In a properly operating heat pipe, the condenser is being flooded, so r_c is considered infinite and Eq. 8.113 reduces to

$$\Delta P_c = \frac{2\sigma}{r_e} \tag{8.114}$$

and effective r_e for a few-wick structure is given in Table 8.8.

The axial hydrostatic pressure is expressed by Bar-Cohen and Kraus [11] as

$$\Delta P_{ah} = \rho g L \sin\theta \tag{8.115}$$

TABLE 8.8 Expressions for Effective Capillary Radius*

Structure	r_e	Data
Circular cylinder Artery or tunnel wick	r	r = Radius of liquid passage
Rectangular grooves	w	w = Groove width
Triangular grooves	$w/\cos\beta$	w = Groove width β = Half included angle
Parallel wires	w	w = Wire spacing
Wire screens	$(w + d_w)/2 = \frac{1}{2}N$	d = Wire diameter N = Screen mesh number w = Wire spacing
Packed spheres	$0.41r_s$	r_s = Sphere radius

* From Chi, Ref. 25, and Bar-Cohen and Kraus, Ref. 11.

TABLE 8.9 Wick Permeability*

Structure	K	Data
Circular cylinder (artery or tunnel wick)	$r^2/8$	r = Radius of liquid flow passage
Open rectangular grooves	w/s	w = Groove width s = Groove pitch
Circular annular wick	$2r_h^2/fR_e$†	$r_h = r_1 - r^2$
Wrapped screen wick	$0.0082d^2\varepsilon^3/(1-\varepsilon)^2$	d = Wire diameter $\varepsilon = 1 - (1.05\,\Pi N d/4)$ N = Mesh number
Packed sphere	$0.0267r^2\varepsilon^3/(1-\varepsilon)^2$	r = Sphere radius ε = Porosity (dependent on packing mode)

* From Chi, Ref. 25, and Bar-Cohen and Kraus, Ref. 11.

† $fR_e = 4r_h\tau(\varepsilon A\rho\lambda)^2/\mu q^2$

where τ = friction shear stress at the liquid-solid interface
ε = wick porosity
A = wick cross-section area
ρ = liquid density
h_{fg} = latent heat of vaporization
μ = liquid viscosity

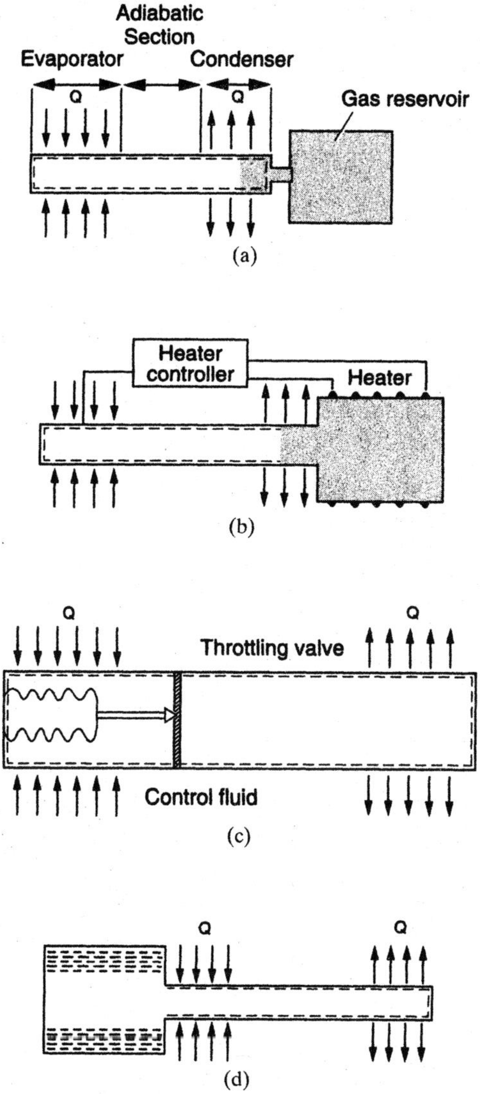

FIGURE 8.94 Variable conductance heat pipes: (*a*) passively controlled gas-loaded; (*b*) actively controlled gas-loaded; (*c*) vapor-modulated variable conductance; (*d*) liquid-modulated.

where e = liquid density
g = acceleration of gravity
L = length of the heat pipe
θ = angle between the heat pipe axis and horizontal

TABLE 8.10 Acceptable Combinations of Heat Pipe Material and Fluids

Fluid	Copper	Aluminum	Stainless Steel	Nickel	Refrasil
Water	X		X		X
Acetone	X	X			X
Ammonia		X	X	X	X
Methonol	X			X	X
Dow-A	X	X	X	X	X
Dow-E	X	X	X	X	
Freon-11	X	X	X		X
Freon-113	X	X	X		

The pressure drop of the liquid flowing through the wick is expressed by Bar-Cohen and Kraus [11] as

$$\Delta P_f = \left(\frac{\mu}{K A_w h_{fg} \rho}\right) L_{\text{eff}}\, q \tag{8.116}$$

where μ = liquid viscosity
ρ = liquid density
h_{fg} = latent heat of vaporization
$L_{\text{eff}} = 0.5L_e + L_a + 0.5L_c$ and L_e = evaporator length
L_a = length between the evaporator and condenser areas
L_c = condenser length
q = heat flow rate
K = wick permeability defined in Table 8.9

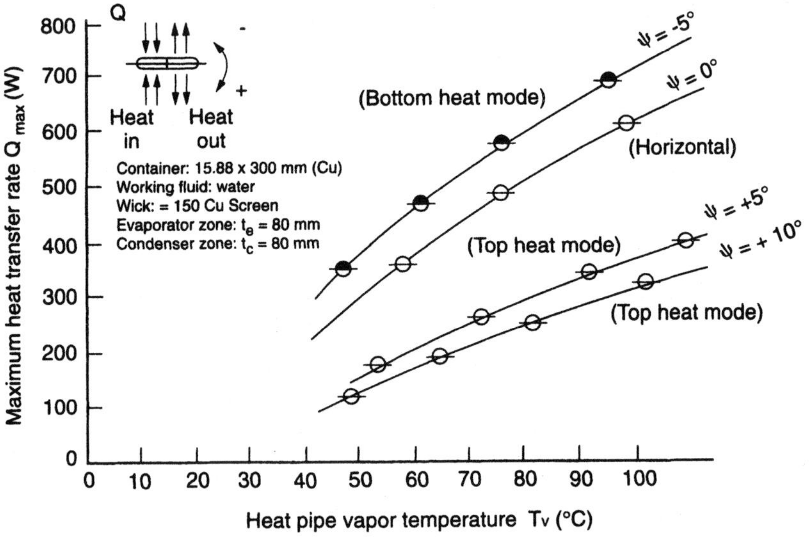

FIGURE 8.95 Maximum heat transfer of a copper-water heat pipe with a screen wick.

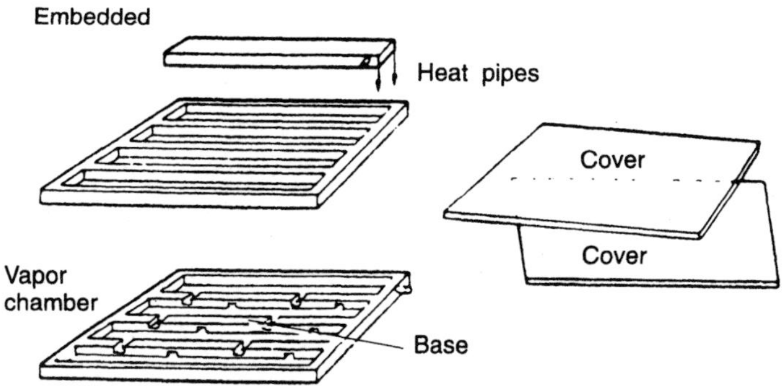

FIGURE 8.96 Basic embedded and vapor chamber concepts for cooling CCAs.

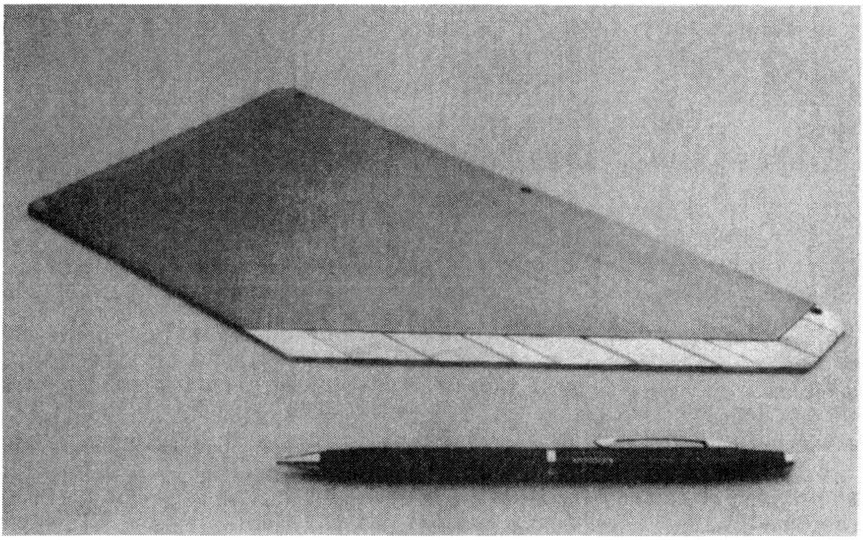

FIGURE 8.97 Embedded heat pipe in a CCA-pair heat sink.

The heat flow rate determines the mass flow rate and, in steady state operation, the mass flow rate of the liquid and vapor must be equal.

Expressions for the effective evaporator radius for some wick structures are given in Table 8.8 and values of K in Eq. 8.116 are given in Table 8.9.

Other forces and phenomena taking place in the heat pipe are not considered here, but a much more detailed treatise is given in Refs. 12, 22, and 26 through 30. Variations of the basic heat pipe structure have been successfully used to provide temperature control and one-directional (diode) heat transfer. Figure 8.94 shows some of these applications in which either excess gas or excess liquid is used to limit heat transfer. In Fig. 8.94*a*, a noncondensible gas is added to the working fluid. Because the vapor pressure curve is much steeper than the perfect gas law, the working fluid can easily compress the noncondensible gas into the reservoir

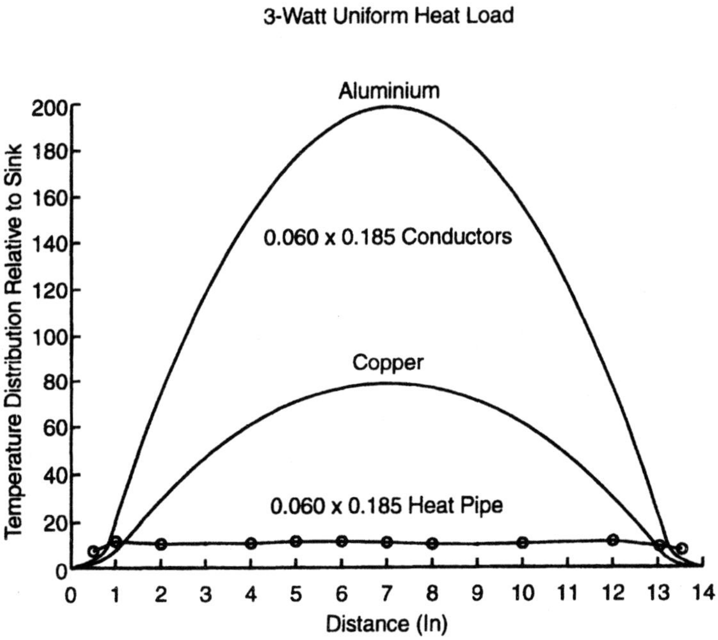

FIGURE 8.98 Comparison between the CCA heat pipe and metal conductor rails with a 3-watt uniform load.

and use the entire evaporator area. When the temperature of the evaporator starts to drop, the gas expands into the condenser area to reduce the heat transfer and maintain some temperature control. Where this passive scheme is not accurate enough, the method illustrated in Fig. 8.94*b* adds a heater to the reservoir to enhance the expansion of the gas. Excess liquid is used in 8.94*d* to make a diode heat pipe, whereby if the condenser end ever becomes hotter than the evaporator end, condensed liquid is collected in the reservoir and depletes the working liquid, which renders the heat pipe inoperative.

A major concern in the construction of heat pipes is material compatibility. Working fluids can be chosen for the temperature range and heat transport characteristics and the materials used in the wick and pipe structure must be compatible to avoid corrosion. Table 8.10 lists a few compatible material combinations.

The best heat transfer liquid, in most cases, is water and, when used with copper, provides excellent conduction properties to get heat into and out of the heat pipe. Figure 8.95 shows the operation of a copper-water heat pipe with a screen wick at different orientations, with respect to gravity.

Heat pipes are used on CCAs as individual heat pipes on rail-cooled boards (see conduction-cooled CCAs, Sec. 8.3) or are embedded in a frame structure (Figs. 8.96 and 8.97). The effectiveness of heat pipes in reducing the temperature to which components are subjected is illustrated in Fig. 8.97. In this figure, a heat pipe rail is constructed (as in Figs. 8.98 and 8.99) and is compared to aluminum and copper rails of equal size. The main temperature rise for the heat pipe rail occurs at the ends, where the heat flux is high, conducting out of the pipe and into the CCA heat sink.

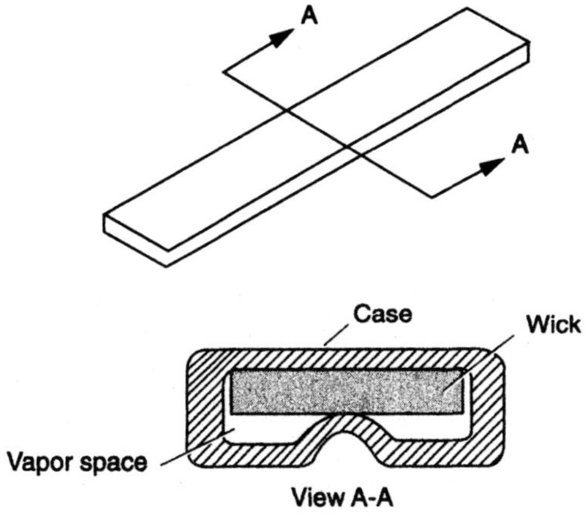

PHYSICAL DATA

- **Working fluid - methanol**
- **Case material - type 304 stainless steel**
- **Wick material - sintered type 304 stainless steel wire rovings**
- **Length - 6 in. to 14 in.**
- **Cross-sectional dimensions**
  - **— 0.076 in. x 0.220 in.**
  - **— 0.060 in. x 0.185 in.**
  - **— 0.040 in. x 0.160 in.**

FIGURE 8.99 Flat heat pipe rail design.

REFERENCES

1. J. R. Welty, *Engineering Heat Transfer,* John Wiley & Sons, New York, 1974.
2. J. P. Holman, *Heat Transfer, 4e,* McGraw-Hill, New York, 1976.
3. W. M. Rohsenow and J. P. Hartnett, *Handbook of Heat Transfer,* McGraw-Hill, New York, 1973.
4. W. H. McAdams, *Heat Transmission,* 3d edition, McGraw-Hill, New York, 1954.
5. Barnes Engineering Company, *Handbook of Infrared Radiation Measurement,* 1983.
6. NACA Technical Note 2836, "Radiation Interchange Configuration Factors," D. C. Hamilton and W. R. Morgan.
7. A. D. Kraus, *Cooling Electronic Equipment,* Prentice Hall, 1968.
8. W. M. Kays and A. L. London, *Compact Heat Exchangers,* 2d ed., McGraw-Hill, New York, 1964.
9. R. J. Philips, *Forced-Convection, Liquid-Cooled, Microchannel Heatsinks,* masters thesis, MIT, Cambridge, Massachusetts, 1987.
10. D. B. Tuckerman, *Heat-Transfer Microstructures for Integrated Circuits,* Ph.D. thesis, Stanford University, Stanford, California, 1984.

11. A. Bar-Cohen and A. D. Kraus, *Advances in Thermal Modeling of Electronic Components and Systems,* vol. 1, Hemisphere Publishing Corporation, New York, 1988.
12. A. Basilius, and C. P. Minning, "Improved Reliability of Electronic Components Through the Use of Heat Pipes," *37th National Aerospace and Electronics Conference,* Dayton, Ohio, May 20–24, 1985.
13. W. M. Rohsenow, "A Method of Correlating Heat Transfer Data for Surface Boiling Liquids," *Trans. ASME,* vol. 74, p. 969, 1952.
14. C. A. Harper, *Electronic Packaging and Interconnection Handbook,* McGraw-Hill, New York, 1991.
15. N. Zuber, "On the Stability of Boiling Heat Transfer," *Trans. ASME,* vol. 80, p. 711, 1958.
16. C. G. Kirkbride, "Heat Transfer by Condensing Vapors on Vertical Tubes," *Trans. AIChE,* vol. 30, p. 170, 1934.
17. H. Schlichting, *Boundary-Layer Theory,* McGraw-Hill, New York, 1968.
18. S. D. Conte, *Elementary Numerical Analysis,* McGraw-Hill, New York, 1965.
19. D. S. Steinberg, *Cooling Techniques for Electronic Equipment,* Wiley Interscience, New York, 1980.
20. EG&G ROTRON, Lancaster, Pennsylvania.
21. W. M. Kays and A. L. London, *Compact Heat Exchangers, 2d edition,* McGraw-Hill, New York, 1964.
22. A. R. Weiting, "Empirical Correlations for Heat Transfer and Flow Friction Characteristics of Rectangular Offset-Fin Plate-Fin Heat Exchangers," *ASME Journal of Heat Transfer,* pp. 488–490, August 1975.
23. K. H. Token, *A New Avionics Thermal Control Concept, ASME Intersociety Conference on Environmental Systems,* San Francisco, California, July 11–14, 1977.
24. *Thermoelectric Handbook,* 2d edition, Cambridge Thermionic Corporation, Cambridge, Massachusetts, 1972.
25. S. W. Chi, *Heat Pipe Theory and Practice,* McGraw-Hill, New York, 1976.
26. P. J. Brennan, and E. J. Kroliczek, *Heat Pipe Design Handbook,* prepared by B & K Engineering, Inc. for NASA, Goddard Space Flight Center, June 1979.
27. A. Basilius, H. Tanzer, and S. McCabe, "Heat Pipes for Cooling of High Density Printed Wiring Boards," *Proc. 6th International Heat Pipe Conf.,* Grenoble, France, pp. 531–536, 1987.
28. K. Token, "Trends in Aircraft Thermal Management," *Proc. Printed Wiring Board Heat Pipe Workshop,* Hughes Aircraft Company, Electron Dynamics Division, Torrance, California, 1986.
29. H. C. Hottel, "Radiant Heat Transmission," *Mechanical Engineering,* **52**, 1930.
30. T. Murose, K. Yoshida, J. Fujikake, T. Koizumi, and N. Ishida, "Heat Pipe Heat Sink HEAT KICKER for Cooling of Semi-conductors," *Furukawa Review* **2**:24–33, 1982.

INDEX

Please note that page numbers followed by an f *or* t *denote figures or tables.*

ABOUT THE AUTHORS

Dr. Jerry E. Sergent is the director of technology for TCA, Inc.

Al Krum is an engineering manager at Hughes Aircraft. He also teaches hybrid and MCM classes at UCLA and Berkeley, as well as for IMAPS.